Mechanisms and Effects of Pollutant-Transfer into Forests

Mechanisms and Effects of Pollutant-Transfer into Forests

Proceedings of the Meeting on
Mechanisms and Effects of Pollutant-Transfer into Forests,
held in Oberursel/Taunus, F.R.G., November 24–25, 1988

edited by

H.-W. GEORGII
Institute of Meteorology and Geophysics,
University of Frankfurt, F.R.G.

KLUWER ACADEMIC PUBLISHERS
DORDRECHT / BOSTON / LONDON

Library of Congress Cataloging in Publication Data

```
Meeting on Mechanisms and Effects of Pollutant-Transfer into Forests
  (1988 : Oberursel, Germany)
    Mechanisms and effects of pollutant-transfer into forests :
proceedings of the Meeting on Mechanisms and Effects of Pollutant
-Transfer into Forests, held in Oberursel/Taunus, F.R.G., November
24-25, 1988 / edited by H.-W. Georgii.
        p.    cm.
```

ISBN-13:978-94-010-6951-9   e-ISBN-13:978-94-009-1023-2
DOI: 10.1007/978-94-009-1023-2

```
    1. Forest flora--Effect of pollution on--Congresses.  2. Trees-
-Effect of pollution on--Congresses.  3. Forest ecology--Congresses.
I. Georgii, H. W.  II. Title.  III. Title: Pollutant-transfer into
forests.
QK750.M42  1988
581.5'2642--dc20                                      89-36642
```

ISBN-13:978-94-010-6951-9

Published by Kluwer Academic Publishers,
P.O. Box 17, 3300 AA Dordrecht, The Netherlands.

Kluwer Academic Publishers incorporates
the publishing programmes of
D. Reidel, Martinus Nijhoff, Dr W. Junk and MTP Press.

Sold and distributed in the U.S.A. and Canada
by Kluwer Academic Publishers,
101 Philip Drive, Norwell, MA 02061, U.S.A.

In all other countries, sold and distributed
by Kluwer Academic Publishers Group,
P.O. Box 322, 3300 AH Dordrecht, The Netherlands.

printed on acid free paper

TABLE OF CONTENTS

DEPOSITION OF ORGANIC COMPOUNDS

CASE STUDIES

INVESTIGATIONS ON FOG AND DEW

EFFECTS OF ATMOSPHERIC POLLUTANTS ON VEGETATION

PREFACE

In November 1988 the "Third Oberursel Symposium" devoted to the problems of input of pollutions into forest-ecosystems and their effects on plants or soil convened.

After several years of intensive research on the effects of pollutions on forest-ecosystems it is obvious that not a single specific pollutant can be made responsible but a mixture of several components act together or interact with each other.

The contributions of the workshop reflect to a large extend the results of research-projects which were started at the beginning of the eighties. They review our improved knowledge on the patterns of concentration, of the mechanism of wet and dry deposition and fog interception, modelling studies and the effect of the processes on plant-receptors and surfaces. Since the 1985 symposium the pathways of pollutants leading to biological damage have been examined and are more clearly recognised.

The book reflects the common interest and the continuous effort of scientists from many different disciplines to better understand the physical and chemical processes which finally lead to the observed damage of forest-trees.

Comparing the conclusions of the contributions of this book with the results of the first Oberursel symposium in 1981, our knowledge on the relevance of the different mechanisms leading to forest-decay has been considerably improved. The book indicates also in which directions future work should be concentrated.

Again, I have to thank the authors for their cooperation by submitting their recent research-results.

A major part in organising the meeting and in preparation of this book was played by Mrs. B. Schäfer and Mr. S. Grosch whom I would like to thank for their assistance.

Hans-Walter Georgii

DEPOSITION
INTO FOREST AREAS

DEPOSITION OF ATMOSPHERIC POLLUTANTS INTO A NORTH GERMAN FOREST ECOSYSTEM

W. Michaelis, M. Schönburg and R.-P. Stößel
Institute of Physics, GKSS Research Centre Geesthacht
P.O. Box 1160
D-2054 Geesthacht
Federal Republic of Germany

ABSTRACT. In a North German forest stand a measuring station has been set up with the aim of supplying information on the impact of atmospheric pollutants. The results suggest that the wet deposition of heavy metals does not significantly differ from that in stands with forest decline in Southern Germany. Due to the windward location the total input is predominantly determined by dry deposition. The sulphate and chloride inputs are within the variation intervals found at other sites while the total nitrogen deposition ranges towards the upper limit. The gas concentrations show a pronounced dependence on the meteorological conditions. The long-term average values were found to be higher than at other typical sites, but they are for most gases below the recommended limits. Ozone, however, requires particular attention. The long-term mean clearly exceeds the guidelines. In general, the impact of gaseous pollutants on the forest primarily occurs during short-time episodes with high peak concentrations. There are distinct relationships between the concentrations of SO_2 or O_3 on the one hand and that of CO_2 on the other. In each case the fluxes are of opposite sign.

1. INTRODUCTION

At the 'Postturm' site, Forest District Farchau/Ratzeburg, an interdisciplinary research programme on the relationships between environmental pollution and forest decline is being conducted by working groups of several institutions. Three field stations have been set up to measure inorganic and organic atmospheric pollutants, to perform plant-physiological experiments and to carry out soil-chemical analyses [1]. The site is located about 40 km north-east of the City of Hamburg. It is characterized by a soil with moderately podzolic braunerde and a mixed stand consisting predominantly of spruce with some pine and beech. Since 1982 a marked disease of the stand has been observed.

This contribution briefly describes the measuring station for deposition experiments [2, 3] and presents results obtained for trace elements, in particular for heavy metals, and for inorganic compounds.

3

H.-W. Georgii (ed.), Mechanisms and Effects of Pollutant-Transfer into Forests, 3–12.
© *1989 by Kluwer Academic Publishers.*

4

In previous investigations, a few years ago, rather high concentrations of heavy metals in the soil solution had been found [4]. The close neighbourhood of the Hamburg conurbation prompted the suggestion that gaseous pollutants might also contribute to the impact on the forest. Moreover, the stand is surrounded by areas with intensive agricultural utilization. Thus it was considered justified to supplement the biological and soil-chemical investigations by studies on the deposition of possibly harmful atmospheric constituents.

2. GENERAL OUTLINE OF THE MEASURING STATION

A schematic view of the station is shown in Figure 1. The sampling and sensor equipment is mounted on a 48 m high tower [2]. Rainwater samples are collected using an automatic device which is controlled by a moisture sensor. Undissolved constituents are strained off with a 0.45 μm filter. Samples of size-fractionated particulates are obtained by means of a high-volume sampler equipped with a 5-stage slotted cascade impactor. At a flow rate of 69 m³/h the aerodynamic particle diameter cutoffs at 50 % collection efficiency are about > 7.2, 3.0 to 7.2, 1.5 to 3.0, 1.0 to 1.5, 0.5 to 1.0 and < 0.5 μm (back-up filter). Cellulose paper filters are used as collection substrates. The flux onto the surface is derived with the aid of size-dependent deposition velocities which have been measured in field experiments with monodisperse test aerosols labelled with radioactive tracers [5]. It is assumed that the deposition velocity is proportional to the actual friction velocity.

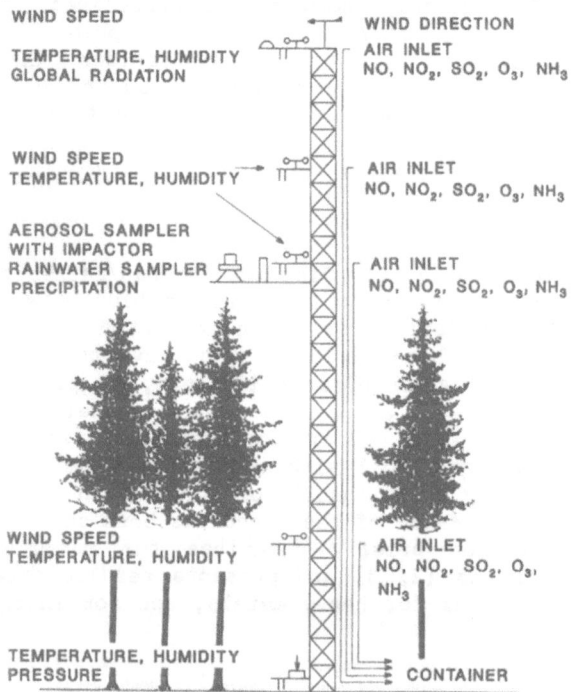

Figure 1.   Schematic view of the measuring station

Multielement trace analysis of both the aerosol and rainwater samples is performed by total-reflection X-ray fluorescence spectrometry [6]. The scope of information which is attainable by means of this method is illustrated in Figure 2 in which the particle-size separated concentrations of various elements are plotted. The ions SO_4^{-}, NO_3 and Cl^{-} in precipitation are determined by ion chromatography, NH_4^{+} is detected using flow-injection analysis in combination with a gas-sensitive electrode.

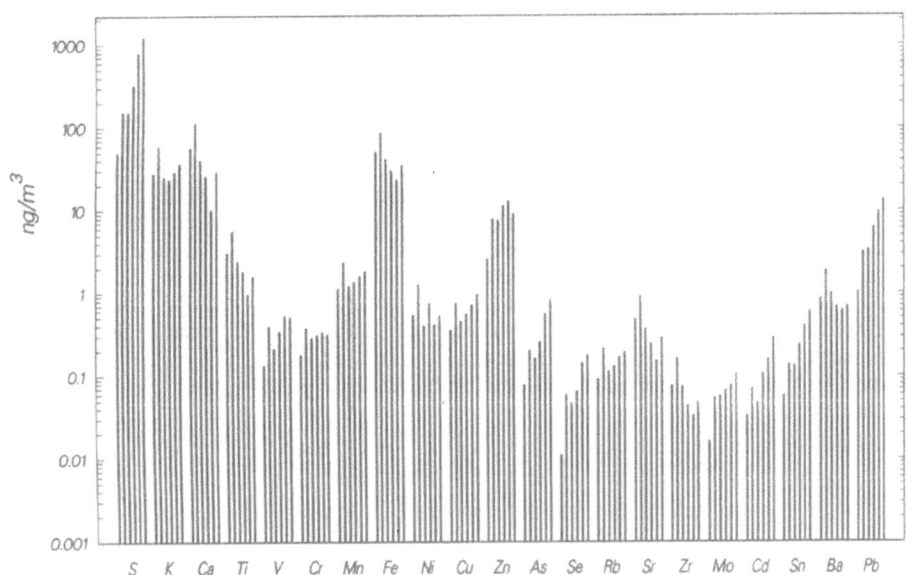

Figure 2. Differential element concentrations in air. Decreasing aerodynamic particle diameter from left (> 7.2 μm) to right (< 0.5 μm)

The gradient method is applied for studying the dry deposition of gases. Air sampling is undertaken at four heights: 48 m, 36 m, 28 m and 9 m (16 m at first). A four-lane piping system constructed of 2.5 m long teflon-covered steel modules with 80 mm inside dia. is mounted on the tower. The pipes are heated and thermally insulated in order to avoid condensation and to reduce wall effects. The flow velocity is about 11 m/s. Gas analyzers for SO_2, NO, NO_2, NH_3, O_3 and CO_2 are installed in an air-conditioned 20'-container. By means of a valve control system the air from the different levels is fed to one analyzer only for each gas in order to avoid difficult intercalibration problems. The gases are measured by UV-excited fluorescence, chemoluminescence, UV-photometry and IR-absorption, respectively. Before installation in the field the sampling system was subjected to careful tests in the laboratory [2,3]. Deviations of measured concentrations from the true values are always < 0.045 % per m pipe length. Appropriate corrections are applied to the analytical results obtained in the field. The turbulent diffusion coefficient is derived with the computer programme outlined in section 3.

Altogether 22 meteorological sensors are mounted on the tower. They measure horizontal wind speed, air temperature and atmospheric humidity at different heights, as well as wind direction, pressure, global radiation and precipitation intensity.

3.   TURBULENT FLUXES IN THE BOUNDARY LAYER

The evaluation of the analytical data requires the knowledge of several actual quantities: the friction velocity, the roughness height, the zero-plane displacement and the turbulent diffusion coefficient. A computer programme has been set up which derives these quantities from the 30 min-averages of the sensor data. The atmospheric stability is taken into account through the Monin-Obukhov length. The pertinent equations are solved numerically using an iterative procedure [3]. Assuming that the turbulent transport of gases occurs analogous to that of the sensible heat, the gas fluxes are calculated from the vertical concentration gradients and the heat diffusion coefficient. Details of the procedure, statistical results and specific aspects for a forest stand will be discussed elsewhere [7]. As an example, Figure 3 displays the azimuthal directional distribution of the zero-plane displacement height as obtained for the period February to December 1987 at the observation site. The plot is based on about 4400 complete profiles which meet several stringent criteria (wind speed > 1.5 m/s; Richardson number $|Ri| < 0.1$; correlation coefficient $\geq 1 - 10^{-6}$).

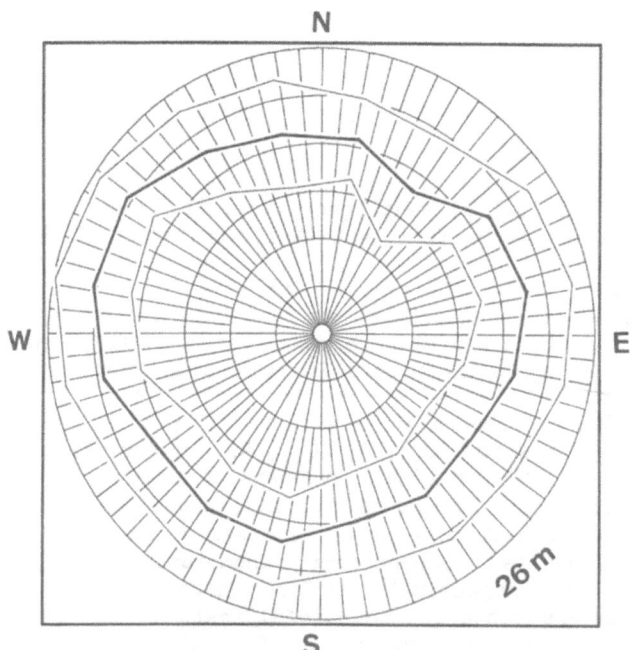

Figure 3. Directional distribution of the zero-plane displacement height in 1987

4. WET AND DRY DEPOSITION OF TRACE ELEMENTS

Results obtained for the wet deposition of 10 elements are summarized in Table 1. The data represent daily deposition rates averaged over the period May to December, 1987. Values for both the particulate phase in the rainwater samples and in the filtrate are specified. For comparison, data measured in 1984/1985 on Pellworm island in the German Bight [2, 3] and in 1980 in Hamburg [8] have been included in the table. The synopsis exhibits a rather good consistency. There seem to be no significant distinctions between these data and those of South German observation sites with remarkable forest decline. This is illustrated in Table 2 for the elements Cd and Pb [9, 10].

TABLE 1. Average daily wet deposition in $\mu g/m^2 d$

Element	'Postturm' filtrate May - December	'Postturm' part.	'Postturm' total 1987	Pellworm island total annual mean	Hamburg filtrate annual mean
V	1.7	0.5	2.2	2.5	–
Cr	0.5	0.6	1.1	0.43	–
Mn	10.8	1.9	12.7	6.8	–
Ni	2.6	0.4	3.0	1.8	–
Cu	3.9	0.9	4.8	2.3	19.3
Zn	73.6	1.8	75.4	25.1	80.9
As	1.7	0.1	1.8	0.74	–
Se	0.65	0.02	0.67	0.41	0.16(SeIV)
Cd	0.69	0.16	0.85	0.45	1.0
Pb	15.3	1.9	17.2	9.4	48

TABLE 2. Wet deposition of Cd and Pb: Comparison with results of other sampler stations [9, 10]. All data in $\mu g/m^2 d$

Observation site	Cd	Pb
'Postturm'	0.85	17.2
Rotenfels	0.60 – 0.82	12.6 – 32.9
Freudenstadt	0.58 – 0.91	16.2 – 21.6
Kälbelescheuer	–	32.6

A summary of the results on dry deposition is given in Table 3. The data shown represent averages over a period of 20 weeks (April to August, 1987). Ranges are listed instead of single values in order to point out the degree of uncertainty which is mainly influenced by uncertainties in the deposition velocity. Again, for comparison, long-term data observed on Pellworm island have been included in the table [2].

The results differ considerably though the concentrations at the reference height are of approximately the same order of magnitude. Here, the high interception deposition in a forest stand becomes apparent. Moreover, the windward location of the observation site strongly influences the results. Comparable rates for Cu, Zn, Cd and Pb have also been obtained elsewhere [11]. The total input is thus predominantly determined by dry deposition. In the presence of low pH values these high dry deposition rates might be responsible for the enhanced heavy metal concentration levels observed in the soil solution [4].

TABLE 3. Average daily dry deposition of trace elements in $\mu g/m^2 d$ for a period of 20 weeks

Element	'Postturm' Forest			Pellworm island Grass		
V	14	–	33	1.8	–	3.2
Cr	9	–	21	0.6	–	1.2
Mn	82	–	110	5	–	12
Ni	14	–	34	2.2	–	7.2
Cu	45	–	106	1.4	–	3.0
Zn	157	–	370	13.0	–	18.6
As	5	–	11	0.72	–	0.96
Se	1.1	–	2.7	0.21	–	0.28
Cd	4.7	–	10.5	0.25	–	0.77
Pb	68	–	160	8.9	–	11.8

5. ANIONS AND CATIONS IN RAINWATER

The results obtained for the period April 1987 to January 1988 are summarized in Table 4. Comparative data from sites in Southern Germany are also given [9, 12]. The forests in the Schönbuch region show few symptons of decline. In general, at the 'Postturm' site, the total nitrogen input seems to range in the upper half of the variation intervals. The rather high value of the NH_4^+ deposition is noteworthy.

TABLE 4. Average daily deposition of SO_4^{2-}, NO_3^-, Cl^-, F^- and NH_4^+ in $mg/m^2 d$ (April 1987 to January 1988)

	SO_4^{2-}	NO_3^-	Cl^-	F^-	NH_4^+
'Postturm'	10.2	8.0	2.1	0.041[a]	3.5
Schönbuch I [9]	6.5 – 10.8	4.6 – 6.8	1.8 – 5.3	–	1.8 – 2.1
Schönbuch II [12]	9.1 – 15.7	6.5 – 9.9	2.0 – 5.2	–	2.6 – 3.4
Rotenfels [9]	9.5 – 15.3	8.1 – 13.3	3.5 – 6.5	–	1.9 – 2.9
Freudenstadt [9]	7.8 – 18.1	6.5 – 10.3	3.1 – 6.9	–	1.8 – 2.2

[a] August 1987 to January 1988 only

6. GAS MEASUREMENTS

The measured gas concentrations show a pronounced dependence on the prevailing weather conditions. This is shown in Figure 4 by an episode in February 1987 which was characterized by a stable inversion layer. With the wind blowing from E.S.E. high peak concentrations of SO_2 were observed accompanied by rather moderate NO_x values. Here, the influence of high SO_2 emissions in the brown coal districts of the GDR becomes noticeable. When the wind shifted to S.S.W., the situation changed drastically. The dense motor-vehicle traffic in the FRG caused a remarkable increase in NO_x immission while the SO_2 concentration diminished considerably due to the stringent measures taken to reduce SO_2 emission. The general validity of this interpretation is strongly confirmed by a comprehensive statistical analysis of several tens of thousands of measured values [3].

The long-term averages (January to September 1988) were found to be as follows: SO_2 32.4 µg/m³; NO 3.4 µg/m³; NO_2 20.6 µg/m³ and O_3 90.5 µg/m³. For the first three gases the results are below the recommended maximum means. Therefore, with respect to long-term effects alone, no damage to the vegetation should be expected. The measured value for O_3, however, is clearly above the guidelines. The WHO recommendations, for instance, suggest a maximum of 60 µg/m³ during the vegetation period. The value observed for this period in 1987 was 84 µg/m³. Throughout, the long-term averages are higher than those at elevated sites in the Black Forest.

Nevertheless, the impact of gaseous pollutants primarily occurs during rather frequent short-time episodes with high peak concentrations (up to 300 µg O_3/m³ and 720 µg SO_2/m³). A good example of this is shown in Figure 5 on the left. It is of particular interest that enhanced SO_2 immissions are accompanied by a simultaneous rise in CO_2 concentration, which is often followed by a longer lasting increased level, especially

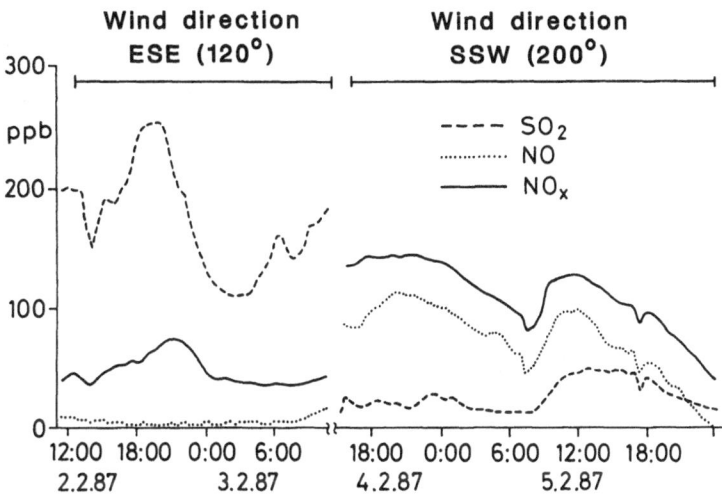

Figure 4. Gas concentrations under different weather conditions

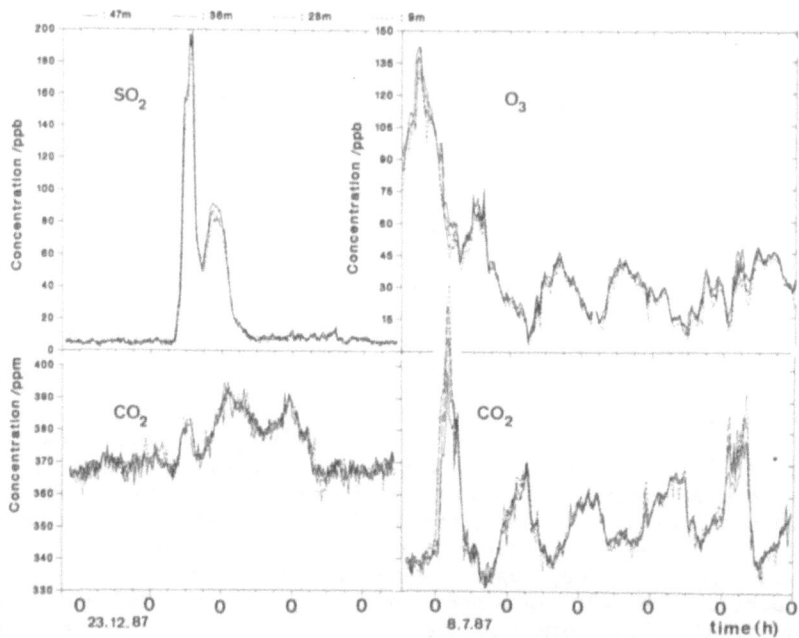

Figure 5. Concentrations of SO_2, O_3 and CO_2 during probable stress events

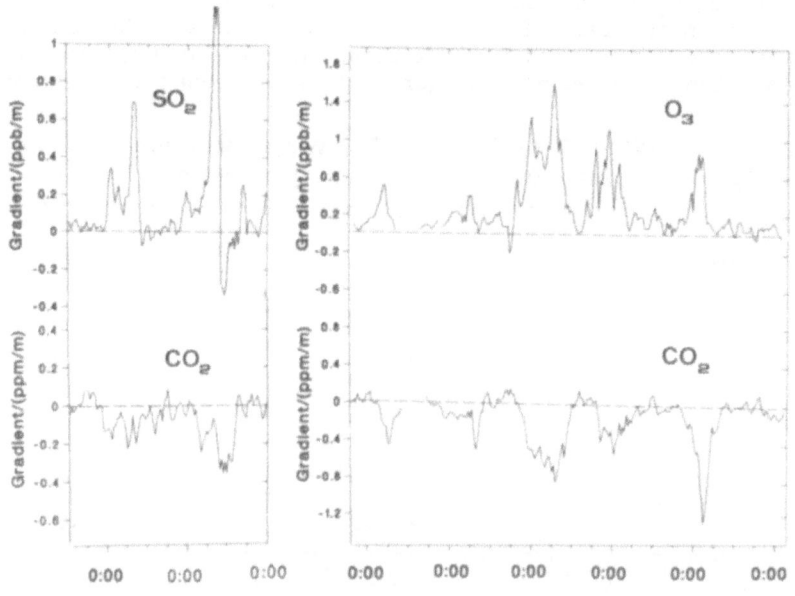

Figure 6. Interrelations between gradients of SO_2/O_3 and CO_2

at night, as is also shown in the figure. These phenomena happen at any hour, and also during the vegetation rest. The peaks are superimposed on the daily and seasonal variation, and the amplitudes often clearly exceed the normal increase in CO_2 concentration at night. One might speculate whether these relationships are connected with the integral respiration activity of the forest stand due to stress effects.

It is important to point out that such events are characterized by gradients of opposite sign for the two gases (Figure 6). Here, a positive sign denotes negative (downward) fluxes. The gradients specified hold for a reference height of 14 m above canopy. The fluxes observed for SO_2 in most cases range between 0 and - 5 µg/m²s. The release of CO_2 amounts to as much as 5 mg/m²s.

In the case of O_3 the findings are somewhat different. There seems to be a correlation between daytime O_3 concentration and the maximum value of CO_2 early in the morning of the next day (Figure 5). During the decrease of O_3 and the increase of CO_2 pronounced gradients occur for both gases, again with opposite sign (Figure 6). The fluxes of O_3 are in the order of 0 to -6 µg/m²s, those of CO_2 range between 0 and 10 mg/m²s. The data require further analysis. Interrelations between the gradients observed are summarized in Figure 7.

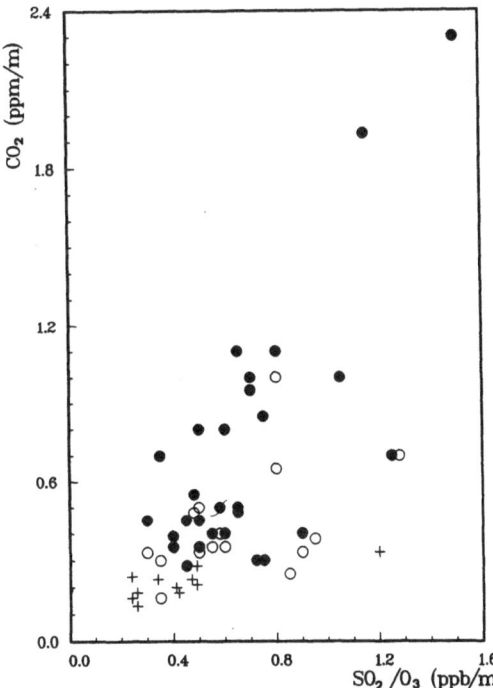

Figure 7.   Plot of CO_2 gradients vs. SO_2 (crosses) and O_3 (circles) gradients. Full circles during and open circles out of the vegetation period

7. OUTLOOK

So far, only a few percent of the collected data have been examined.
Besides extending and deepening the analysis, additional approaches are
envisaged. With regard to the heavy metal deposition, this comprises
comparisons with wet+dry samplers, the application of the gradient
method and the sampling of canopy drip. The further analysis of the gas
measurements will concentrate on the interrelationships between the
gases, the disclosure of their interaction with the forest ecosystem as
to which compartments (canopy, soil etc.) are effective, and the influence of meteorology and climate stress. Eddy-correlation experiments are
being prepared to complete the research programme.

REFERENCES

[1] Bauch, J. and Michaelis, W. (eds.) (1988) Das Forschungsprogramm
Waldschäden am Standort 'Postturm', Forstamt Farchau/Ratzeburg,
GKSS 88/E/55.

[2] Michaelis, W. (1988) 'Experimental studies on dry deposition of
heavy metals and gases', in H. van Dop (ed.), Air Pollution Modeling and its Application VI, Plenum Press, New York, pp. 61-74.

[3] Michaelis, W., Schönburg, M. and Stößel, R.-P. (1988) 'Trocken- und
Naßdeposition von Schwermetallen und Gasen', in [1], pp. 19-59.

[4] Rademacher, P., Bauch, J. and Michaelis, W. (1988) 'Einfluß der
Elementkonzentration der Bodenlösung auf den Elementgehalt in
gesunden und geschädigten Fichten des Standortes 'Postturm', in
[1], pp. 215-254.

[5] Jonas, R. (1984) Ablagerung und Bindung von Luftverunreinigungen an
Vegetation und anderen atmosphärischen Grenzflächen, Jül-1949.

[6] Michaelis, W. and Prange, A. (1988) 'Trace analysis of geological
and environmental samples by total-reflection X-ray fluorescence
spectrometry', Nucl. Geophys. Vol. 2, No. 4, 231-245.

[7] Schönburg, M. and Mengelkamp, H.-T. (to be published).

[8] Nürnberg, H.W., Valenta, P. and Nguyen, V.D. (1982) 'Wet deposition
of toxic metals from the atmosphere in the Federal Republic of
Germany', in: H.-W. Georgii, J. Pankrath (eds.), Deposition of
Atmospheric Pollutants, D. Reidel Publishing Company, Dordrecht,
pp. 143-157.

[9] Adam, K., Evers, F.H. and Littek, Th. (1987) Ergebnisse nieder-
schlagsanalytischer Untersuchungen in südwestdeutschen Wald-Ökosystemen 1981 - 1986, KfK-PEF 24.

[10] Mies, E. (1987) 'Elementeinträge in tannenreiche Mischbestände
des Südschwarzwaldes', Freiburger Bodenkundliche Abhandlungen,
Heft 18.

[11] Godt, J. (1985) 'Schwermetallbelastung des Teutoburger Waldes süd-
westlich der Stadt Detmold', Bielefelder Ökol. Beitr. 1, 7-16.

[12] Baumbach, G., Dröscher, F. and Mikisch, E. (1987) Staub- und Nie-
derschlagsuntersuchungen in Wäldern, Report No. 9, University of
Stuttgart.

DETERMINATION OF DRY DEPOSITION OF GASES OVER TREE TOPS BY MEASURED DATA AND A NUMERICAL MODEL

UWE HERRMANN and WOLFGANG JAESCHKE
Johann Wolfgang Goethe-Universität
Zentrum für Umweltforschung
Robert-Mayer-Str. 7-9
6000 Frankfurt/Main

ABSTRACT: In 1986 a project was started by the BMFT and the city of Frankfurt for research into the ongoing forest damage. Accordingly a measuring site with a meteorological tower of 51 m was constructed to record the micrometeorological parameters and thus to establish a diagnosis of the dry deposition of gaseous pollutants. Using a numerical model the evaluation of these micrometeorological data should provide mass-fluxes and the deposited masses onto the canopy of the Frankfurter Stadtwald.

1. Introduction

The ability of vegetation and soil to absorb atmospherical trace gases causes a dry deposition. In this way dry deposition is an effective self-cleansing of the atmosphere to discharge impurities from the air. Computational and experimental analysis of this dry deposition and the corresponding dry deposition velocity correlated with the structure of surface and canopy is a central problem in the "Waldschaden"-research because of its effects on vulnerable ecosystems.

The quantity of the deposition depends on the concentration of the pollutants in the vicinity of the surface or the canopy level and the overlaying airlayers. Another dominant factor for the deposition rate is the meteorological situation. The stratification in the surface layer and the resulting turbulent transport processes must be taken into account. Therefore a numerical model (Herbert und Kramm, 1985; Kramm 1986) of the surface-layer was used to describe these influences mathematically and to calculate the deposition by knowing the actual state of the boundary-layer near the surface.

The numerical model takes both the turbulent and the molecular effects near the surface into consideration. It is designed so that only the profile data of windspeed, dry- and wet-bulb temperature and the concentration of the pollutants in the turbulence-layer are required as input specifications. Due to the actual turbulent situation the model calculates the massflux of a chemical reactive trace gas and its dry deposition velocity.

2. Measurements

To obtain these vertical profiles of the concentration of the gases NO, NO_2, O_3 and SO_2, windspeed, dry- and wet-bulb temperature a 51 m high tower was erected in the Frankfurter Stadtwald. All these parameters are measured at 5 levels between 21 m and 50 m. The measure-platforms are mounted

13

H.-W. Georgii (ed.), Mechanisms and Effects of Pollutant-Transfer into Forests, 13–20.
© *1989 by Kluwer Academic Publishers.*

14

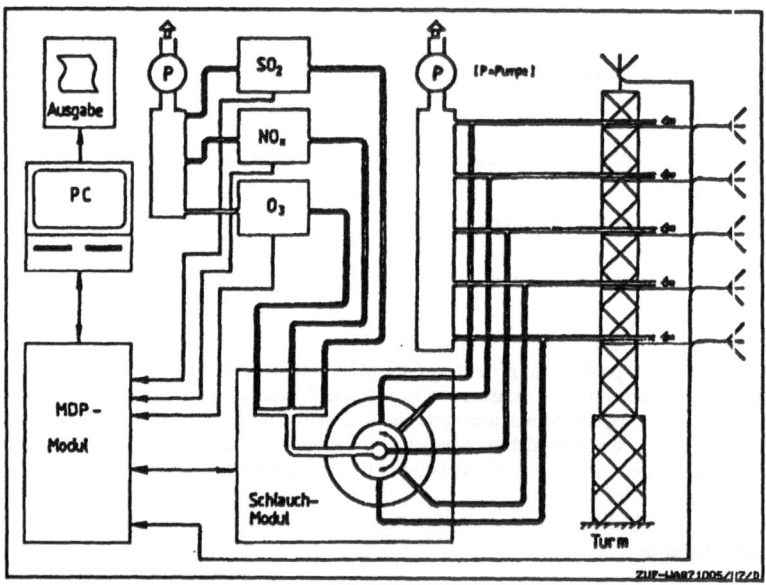

Fig.1   Data acquisition system

identically. The five booms face SW, the main wind direction. They carry a cup anemometer at the top and in 1/2 m distance from the tower, a psychrometer for dry- and wet-bulb temperatures and the opening of a 4 mm internal diameter gas sampling tube (Duyzer et al.,1983). With this teflon tube the sample of air is drawn to the chemical analyzers. The test-air is sucked continuously by a pump from all 5 measuring levels. To avoid pressure differences, all the tubes are of the same length. Thus the same sub-atmospheric pressure is present in all chemical analyzers. For a detailed description of the behavior of the analyzers in sub-atmospheric pressures, the condensation of water vapour and chemical reactions within the teflon tubes, see Sattler and Jaeschke (1987).

The test-air is drawn to the continuously working analyzers by a valvemodule in such a way, that air from only one level is measured for the atmospheric concentration of NO, NO_2, O_3 and SO_2 (Fig. 1). The measuring time of one minute is set on the t_{90}-time of the slowest working SO_2 analyzer. When the measuring time is over, a computer system switches the valvemodule to the next level. Within half an hour this cycle will be repeated on each of the five levels six times. At the end of this measuring-cycle of half an hour the mean values of the concentration of the pollutants and the meteorological parameters of all 5 levels are printed and stored on a floppy disc.

3.   Theory

The observed profile-data are the input for a diagnostical model of the surface layer by which mass-fluxes and deposition-velocities of the pollutants are determined. The premise of this model is that the vertical fluxes of momentum, sensible and latent heat and pollutants within the Prandtl- or surface-layer are nearly constant with height.

With this assumption of constant-flux-conditions for the first 50-100 m, vertical mass flux F_C can be expressed as

$$F_C = - (D + K_H) \, \delta C / \delta z \tag{1}$$

where D is the molecular diffusion coefficient of pollutant, K_H the eddy transfer coefficient for heat, C the concentration of pollutant and z the vertical distance from the ground. Integration from soil (z_s) to the upper border of the Prandtl-layer (z_p) leads to

$$F_C = - (C_p - C_s) / r_a = u_* C_* = const. \tag{2}$$

where u_* is the friction velocity, C_* the pollutant scale and r_a defines the following bulk-resistance

$$r_a = \int_{z_s}^{z_p} (D + K_H)^{-1} \, dz \tag{3}$$

which the surface-layer opposes to the transport of pollutants. To describe the vertical transports and the dry deposition of pollutants in the atmosphere, vertical mass flux F_C and the deposition velocity v_d are commonly used. This two quantities are connected by

$$v_d = - F_C / C(z) \tag{4}$$

4.   Results

The high burden of pollutants in the Rhein-Main-Area is illustrated in Fig. 2, 3 and 4 for the diurnal variation of NO, NO_2 and O_3. The concentration values are averaged over the month of May 1987. The NO-curve shows a clear gradient between the measuring levels with a maximum at 21 m and a minimum at 50 m during night and the early morning hours. The maximum of the NO-concentration is reached at the time of the morning rush-hour and then decreases to 10% of the maximum value. The vertical gradient decreases also and results in an almost constant concentration for the time interval from 9:00 to 18:00 (see Fig. 2). With the increase of radiation more and more Ozone can be generated (Fig. 3) which oxidizes the present NO to NO_2. A gradient can also be found, but here with a maximum at 50 m and a minimum at the level of the tree tops. The NO_2 -curve (Fig. 4) follows the expected reaction between the two responding pollutants.

With the help of the deposition model strength and direction of the pollutant fluxes were determined for some datasets during the vegetation period in 1987. The model inputs are marked with crosses in Fig. 5. The model calculates profiles for the individual components by a least squares fit. In addition to the calculated mass fluxes and deposition velocities of the individual pollutants potential temperature and specific humidity as well as specific boundary layer parameters like roughness length and displacement distance are calculated.

Fig. 6 represents a diurnal variation of the deposition velocity for NO on 23. of May 1987 at the height of 21 m. One can see a calculated diurnal evaluation of the amount of the deposition velocity of NO with minimal values during nighttime and 2 maxima during daytime. The first maximum

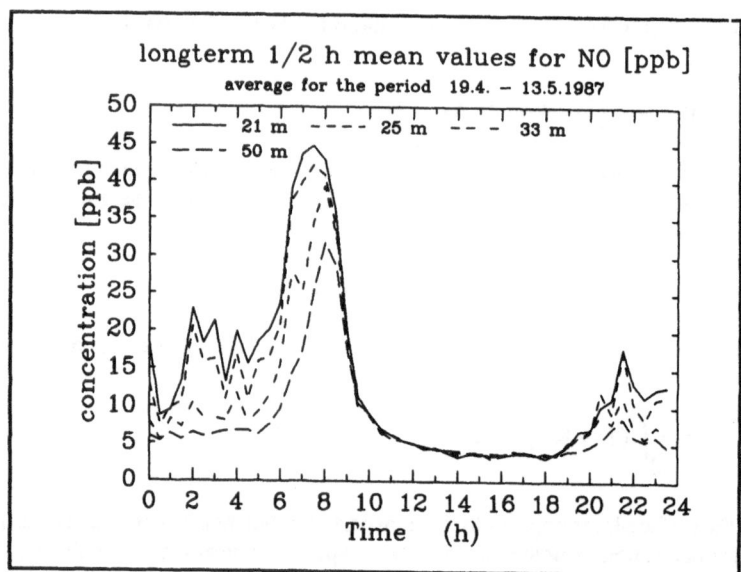

Fig.2   Longterm diurnal variation of NO

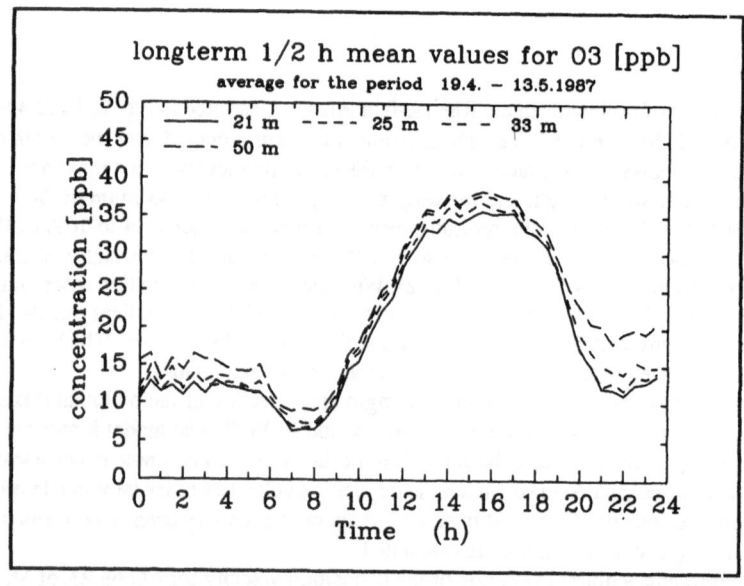

Fig.3   Longterm diurnal variation of O₃

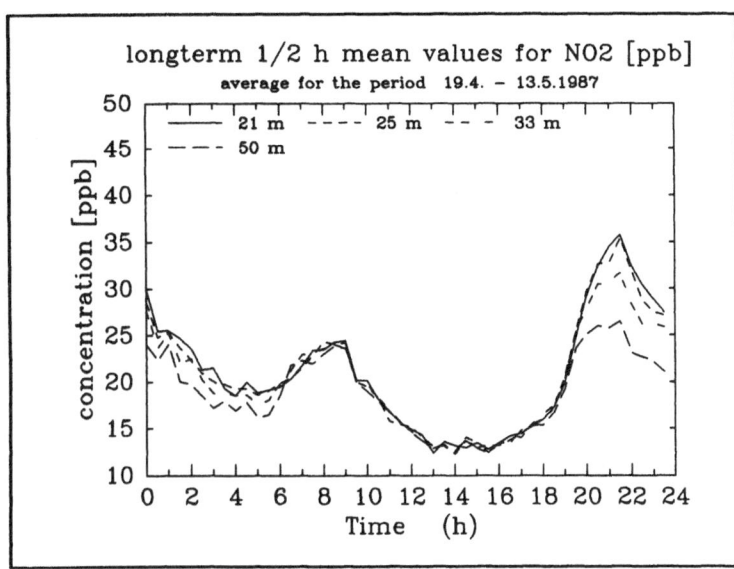

Fig.4   Longterm diurnal variation of NO₂

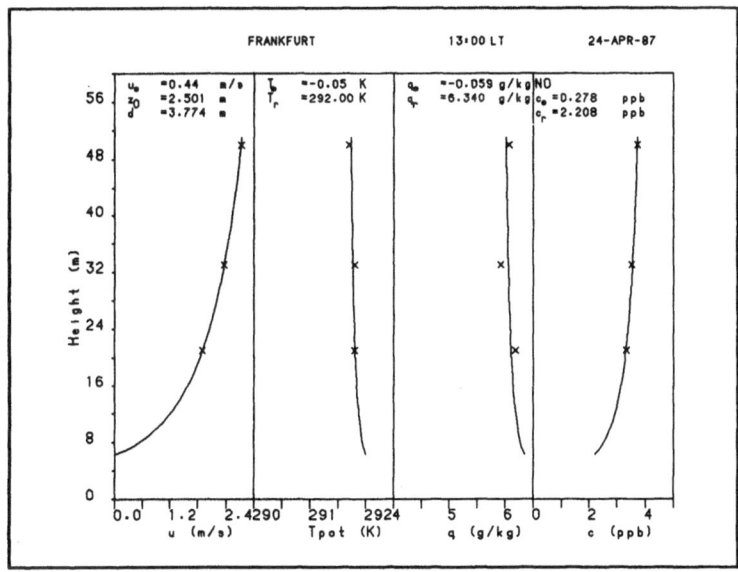

Fig.5   Calculated profiles of μ, T$_{Pot}$, q and C

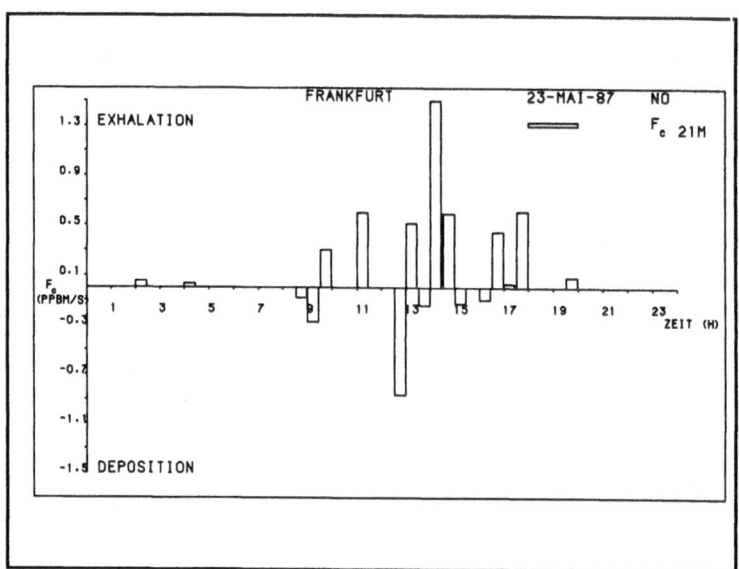

Fig.6   Diurnal variation of the deposition velocity of NO

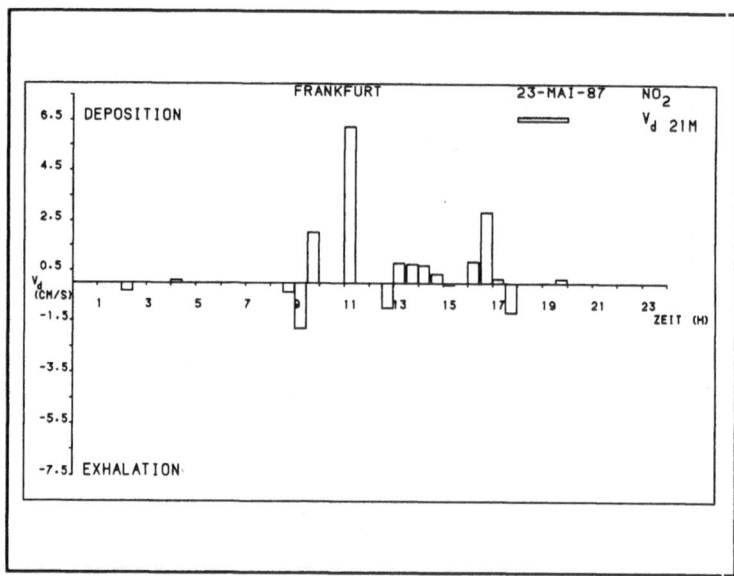

Fig.7   Diurnal variation of the deposition velocity of NO₂

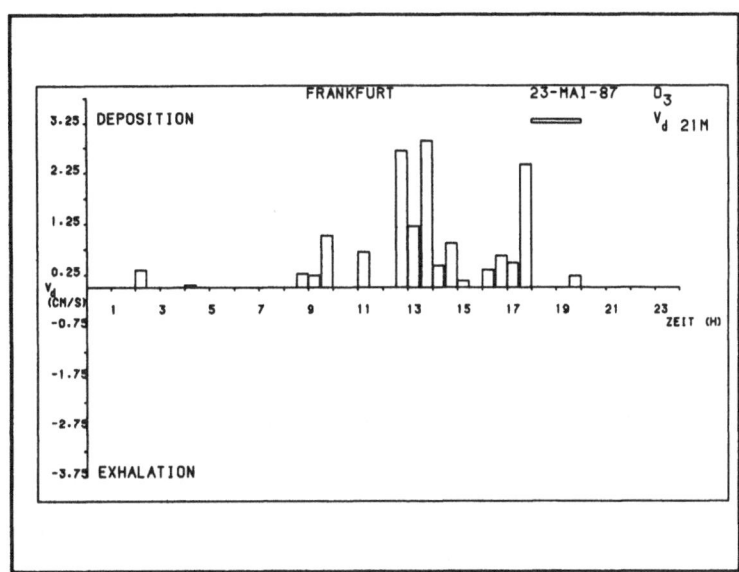

Fig.8   Diurnal variation of the deposition velocity of O₃

(7 cm/sec - corresponding to a flux of 0.6 ppb m/sec) is reached in the late morning. After reaching this, the amount of the deposition-velocities decreases and then reaches a second maximum of 12 cm/sec in the late afternoon. In the evening and night exhalations are more frequently computed, during daytime deposition and exhalation alternate.

A similar picture yields the amount of the deposition values of NO_2 at 21 m on the same day (Fig.7). It shows two maxima; the first of 6.5 cm/sec in the late morning and the second maximum of 2.5 cm/sec in the late afternoon. The computed deposition velocities for O₃ on this day reaches up to 3.2 cm/sec (Fig. 8). On occasions exhalations from the canopy were sporadically computed. A clear-cut explanation of this phenomenon has not yet been found.

5.   Conclusions

The strong fluctuation of the deposition values in amount and direction is mainly effected by the fluctuation of the observed gas-concentrations. One can often see the reversal of the concentration profile of NO. These exhalations can be explained by the NO-emissions coming from the neighbour-ing main-road and motorway traffic. Likewise a strengthened activity of micro-organisms in the ground can be seen (Höfken et al., 1986). In certain meteorological situations these emissions pene-trate the canopy into the overlying airlayers. These situations effect a reversal in the concentration gradient of the trace gases over the trees and result in the observed exhalation events. Ongoing eval-uations of the data-sets will show the relationship between exhalation and deposition on traffic, wind-direction and meteorological and climatic conditions.

6. Acknowledgements

This work is a result of a project "Untersuchung von Immissionsschäden im Frankfurter Stadtwald unter besonderer Berücksichtigung der Emission von Verbrennungsmotoren - Bestimmung der Deposition von Schadstoffen über Baumkronen mittels gemessener Daten und eines numerischen Modells." sponsored by the BMFT (037396 7).

7. References

Duyzer, J.H., Meyer, G.M., Van Aalst, R.M.(1983): Measurement of dry deposition velocities of NO, NO_2 and O_3 and the influence of chemical reactions. Atmos. Environment 17, 2117-2120.

Herbert,F., Kramm,G.(1985):Deposition reaktionsträger Substanzen,beschrieben mit einem diagnostischen Simulationsmodell der bodennahen Luftschicht. In: K.H. Becker, J. Löbl (Hrsg.): Atmosphärische Spurenstoffe und ihr phyikalisch-chemisches Verhalten. Springer Verlag Heidelberg, S. 190-209.

Hicks, B.B., Baldocchi, D.D., Meyers, T.P., Hosker, Jr., R.P., Matt, D.R.(1987): A preliminary multiple resistance routine for deriving dry deposition velocities from measured quantities. Water, Air and Soil Pollutant 36, 311-330.

Höfken, K.-D., Meixner, F., Müller,K.-P., Ehald, D.H. (1986): Untersuchungen zur trockenen Deposition und Emission von atmosphärischem NO, NO_2 und HNO_2 an natürlichen Oberflächen. Berichte der Kernforschungsanlage Jülich, Nr.2054.

Kramm, G.(1986): Modellrechnungen zur Bestimmung der trockenen Deposition atmosphärischer Spurenstoffe. IABG-Bericht B-TD 9003/02, München.

Sattler, T., Jaeschke, W. (1987): Automatisierte Bestimmung von Immissionsprofilen anorganischer Luftschadstoffe in verkehrsreichen Waldschneisen. Staub-Reinhaltung der Luft 47, 261-266.

PROFILES OF OZONE AND SURFACE LAYER PARAMETERS OVER A MATURE SPRUCE STAND

G. ENDERS[1], U. TEICHMANN[1], G. KRAMM[2]

[1]*Institute for Bioclimatology*
and Applied Meteorology
University of Munich
Amalienstraße 52
D-8000 München / W. Germany

[2]*Fraunhofer Institute for*
Atmospheric Environmental Research
Hindenburgstraße 43
D-8100 Garmisch-Partenkirchen
W. Germany

ABSTRACT. Measurements of ozone profiles above a mature spruce stand lead to the conclusion that either an ozone production takes place close to or within the canopy or a destruction well above it, because at certain times a decrease of ozone with height has been observed. Possible influences of terpenes to this phenomenon are discussed. Fluxes and deposition velocities of ozone are calculated on the basis of the gradient approach. In addition, surface layer parameters and aerodynamic stand characteristics derived from wind velocity, temperature, and humidity profiles are presented.

1.   Introduction

In spring 1986 the Institute for Bioclimatology and Applied Meteorology of the University of Munich began contineous profile and deposition measurements of air pollutants in the National Park "Bayer. Wald". There were mainly scientific reasons to choose this remote area 200 km northeast of Munich, even if running a permanently manned research station that far from the institute causes problems in infrastructure, transportation, service, and housing.

By the time, infrastructure and instrumentation of the station were improved step by step and became more and more attractive to other research groups. The cooperation, however, now must go beyond people working in the field, because after three years of measuring the preliminary results have to be analyzed and used by modellers, too.

2.   Site Description

The research field is located in the southern part of the "Große Ohe" watershed (Fig. 1), which serves as long-term reference catchment for water budget and water quality studies (Gietl and Rall 1986) and, therefore, is well equiped for physical and chemical analysis of fog, pure and bulk precipitation, throughfall, stemflow, surface and subsurface runoff. The size of the catchment is 19.1 km^2, the elevation reaches from 770 m to 1452 m, the percentage of forest cover is 98 % (70 % spruce, 28 % beech and others).

The specific site conditions at the spot "Schachtenau", where the gaseous deposition measurements take place, can be characterized as follows: Altitude 807 m; fairly level and uniform terrain to all direc-

21

H.-W. Georgii (ed.), Mechanisms and Effects of Pollutant-Transfer into Forests, 21–35.

Figure 1. Geographical location of the deposition station "Schachtenau" within the experimental watershed "Große Ohe" of the National Park "Bayer. Wald".

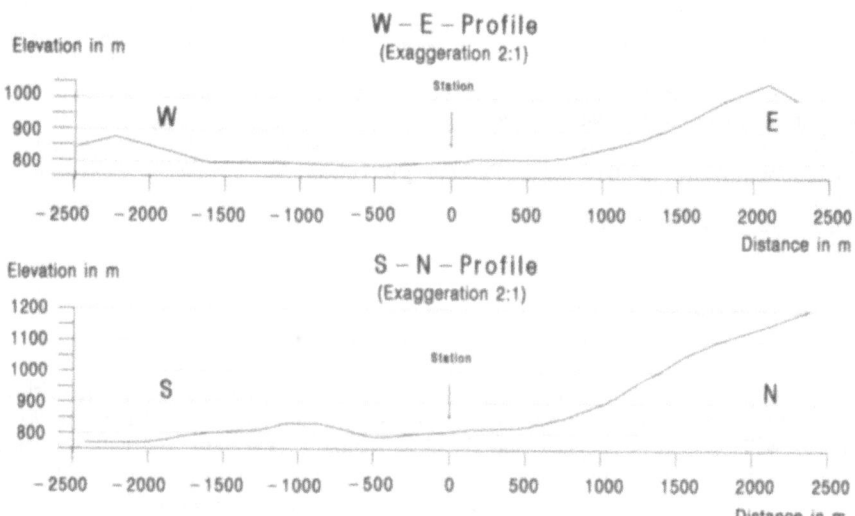

Figure 2. Horizon profiles at "Schachtenau", elevation exaggerated by factor 2.

tions up to a distance of 500 m (1500 m to the westerly main wind direction, see Fig. 2); completely forested with 86 % spruce (Picea abies (L.) Karst.) and 16 % beech. The spruces are 80 to 95 years old with an average tree height of 28 m and needles down to 16 m; the density of the stand is 712 trees/ha.

Because the National Park stretches over an area of 170 km^2, there are no local sources of SO_2 and NO_x in the vicinity of the station. If sometimes smaller slash burning is required for forest management reasons, the station people will be notified in advance so they can react by shutting down the system or marking questionable data.

3. Instrumentation and Principles of Measurement

The center piece of the deposition station is a 51 m tall tower, which serves as platform for most of the instruments used (Fig. 3). Horizontal wind speed, dry- and wet-bulb temperature (11 levels), short-wave and long-wave radiation (2 levels), wind direction, air pressure, precipitation, and soil heat flux are measured at a constant rate of 10 s. Up to now the chemical air analysis is mainly done by using a gradient system (since 1988 also an eddy correlation system for SO_2, NO_2, O_3, sensible and latent heat is available, which will be completed by NO in 1989).

Design and performance of the gradient system are reported in detail by Enders and Teichmann (1986) and Teichmann and Enders (1988), so that here a more general description will do: The air from different levels is pulled down the tower through separated and heated PTFE tubes to accumulation units, where SO_2 and NO_x are trapped specifically for further analysis (the "NO_x" trap in fact is specific only to NO_2, so NO is converted to NO_2 by oxydization before entering the trap). Each of those units consists of two traps, of which one is absorbing, while the other one is in the desorption mode or waiting for it. Absorption and desorption are forced by different thermal conditions. Three by three accumulation devices - absorbing air from different levels simultaneously, but desorbing one after the other - are connected to SO_2 and $NO/NO_2/NO_x$ analyzers (ML 8850/fluorescence, ML 8840/chemolumi-nescence). The latter one are switched to the NO_x channel, because the NO_2 partially may be reduced to NO again during desorption. In total six accumulation units for each gas species are available allowing simultaneous measurements of NO_x and SO_2 in six levels respectively. Typical absorption/desorption times under normal operation are 8 min/1 min for NO_x, and 20 min/5.5 min for SO_2.

For special purposes (e.g. a higher time resolution) the traps may be bypassed. This reduces the number of measuring heights, which simultaneously can be operated, to two, but allows the separate analysis of NO and NO_2 in each of the choosen levels.

Three lines of the tubing system split up to feed three ozone analyzers (ML 8810/UV absorption) operated at the same sampling rate as the meteorological sensors (10 s).

The high air stream velocities in the tubing system with delay times of only a few seconds and the small pipe diameter of 4.75 mm result in considerable pressure drops, which influence the conditions inside the measuring cell of each analyzer and lead, when not corrected, to errors in the results. Therefore, much time is spent again and again first for the basic calibration of the analyzers, then for measuring flow and pressure at the inlets to accumulation units and analyzers, and finally to calculate individual correction terms for each device. This procedure is checked from time to time by moving up one of the analyzers to the tower, connecting it to the air inlet of the gradient system by only a short PTFE manifold and comparing its measures to that analyzer, which remains at the bottom.

4. Results of Measurement

Again and again an "irregular" behavior of ozone above the canopy layer has been observed. This phenomenon needs further and extended investigation, because it may lead to a better understanding of ozone reaction and deposition processes in coniferous forests.

24

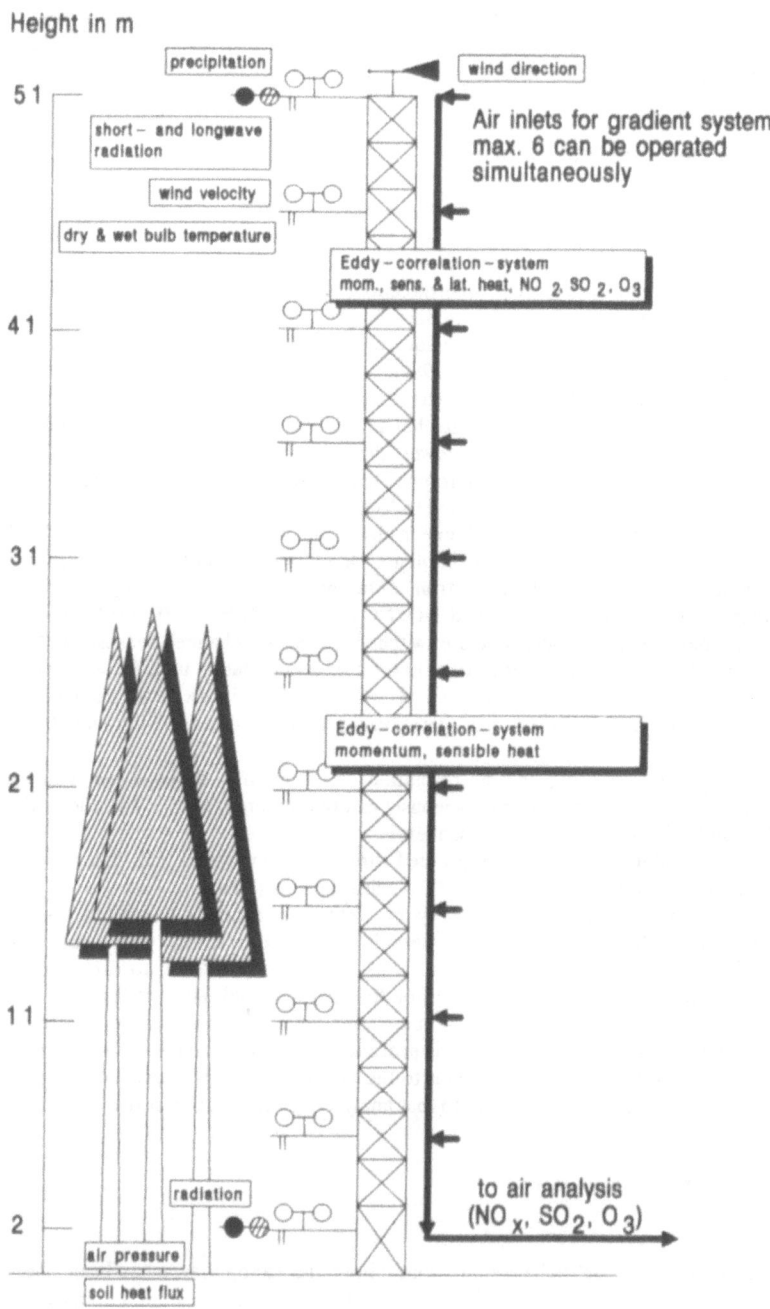

Figure 3. General instrumentation at the station "Schachtenau".

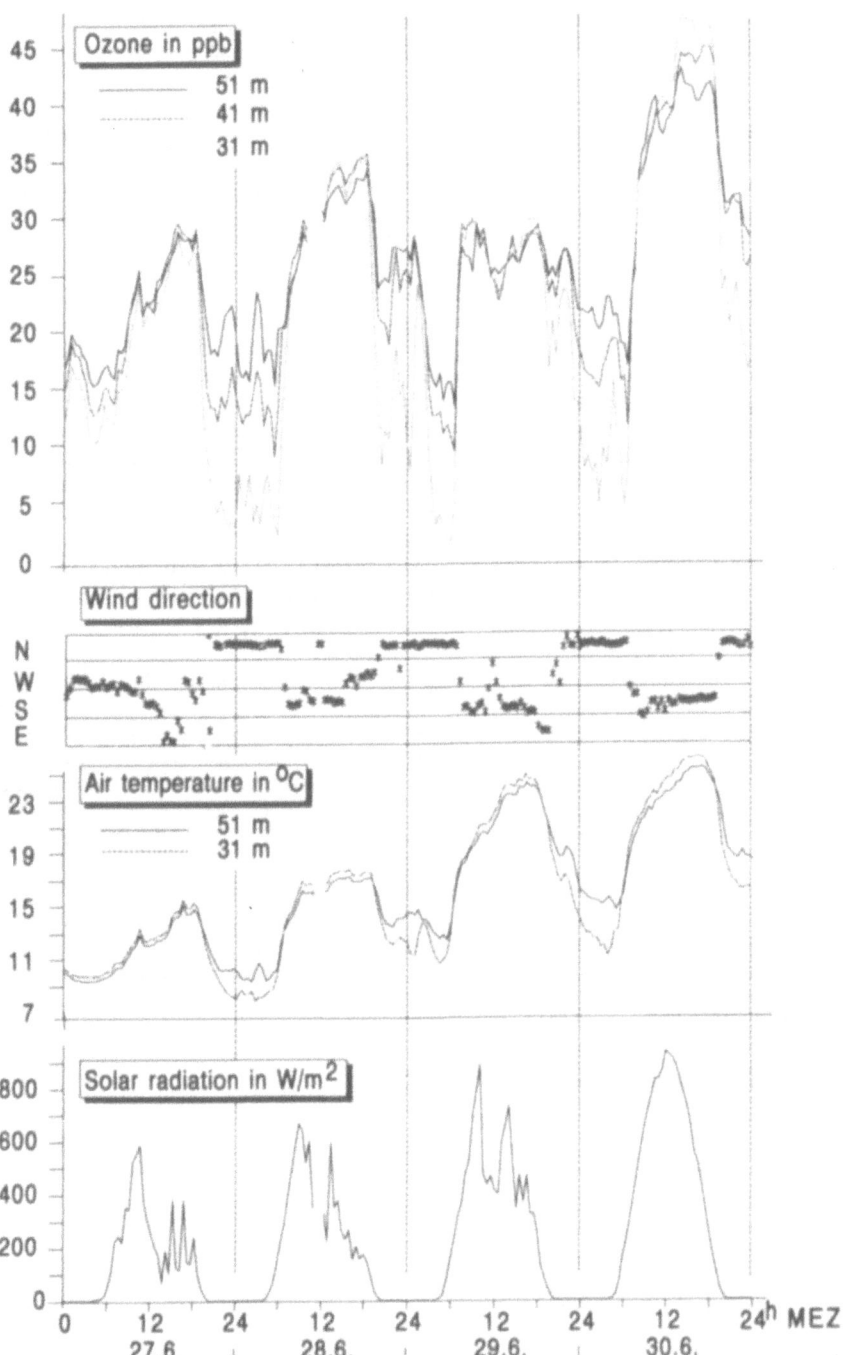

Figure 4. Daily cycles of ozone, air temperature, wind direction, and solar radiation over the canopy at "Schachtenau", June 27-30, 1987. All data given are 30-min averages.

As an example, the daily pattern of O_3 at three levels above the canopy (51 m, 41 m, 31 m) for the period June 27-30, 1987 are given in Fig. 4. The overall meteorological conditions are characterized by wind direction, air temperature and global radiation. All data are 30-min averages.

The ozone concentrations at all three levels show pronounced daily cycles with a late afternoon maximum, which is slightly delayed compared to the maximum of the global radiation, and a minimum during night. From time to time the minimum is broken up by a partly considerable increase in concentration. That may be caused by vertical transfer of ozone, when short-term reinforcement of turbulence happens - which can be initiated, for example, by vertical wind shearing - resulting in disturbances of the nocturnal boundary layer (Winkler 1980). As may be seen from Fig. 4, the station can be affected by nocturnal cold air drainage from the ridges to north (see also Fig. 3), especially during a high pressure situation, which was the case in the period described. This down-hill wind may originally be responsible for those processes. Generally during these four days the wind velocities (51 m level) were extremely low averaging to 1.6 m/s and reaching a maximum of only 2.6 m/s.

The main energy transfer level is the canopy space as clearly is shown in the comparison of the 31 m and 51 m air temperatures: A relative overheating of the canopy layer by absorption of solar energy during day takes place, which leads to instability of the air layer above, while long-wave emission causes cooling and stable conditions during night, when no short-wave radiation uptake exists.

Examining now the vertical gradient of ozone, one finds at all nights the expected decrease of concentration, the closer one gets to the canopy. After sunrise and a general increase of O_3 in all heights, however, the situation changes. The major slope in increase occurs at the 31 m level, the smallest at 51 m height, which different behavior may even lead to inverse vertical gradients. During the first two days of the period under examination this process ist almost negligible, but gets stronger with increasing solar radiation, until on June 30, the day with maximum insolation, this process results in differences of 5ppb/20 m (30-min averages).

This observation is not a unique case. To get a provisional idea of the frequency of this phenomenon, a close examination of all measurements from June to September 1987 has been made. Within that period inverse ozone gradients above the canopy has been observed on 80 % of all days. The duration of those events ranged from only a few minutes to more than 10 hours and their occurence was not limited to pronounced radiation weather, though they were stronger marked under those conditions than otherwise. In some few cases, inverse gradients have been observed even at night. Also the size of the gradients varied considerably and was not always significant, but reached 10 ppb/10 m in single cases.

It should be stressed, however, that days with high insolation not necessarely cause inverse gradients. We also found some days, when despite undisturbed solar radiation throughout the day only the "normal" increase of ozone concentration with height has been observed.

If deposition velocities and fluxes are calculated by a model like the one described next, anavoidably an exhalation of ozone must result from negative ozone gradients.

5. Theoretical Background

Dry deposition or exhalation of quasi-inert atmospheric trace gases can be described by (Kramm 1989a)

$$F_c = - \frac{c_h}{r_a + r_s} (1 - \frac{c_i}{c_h}) = \text{const.}, \tag{1}$$

wherein F_c is the vertical mass flux density, c_i the concentration within the vegetation or soil, c_h the concentration at the height z_h, the top of the atmospheric surface layer (often called the Prandtl layer),

r_a the total bulk resistance of the air against mass transfer and r_s the corresponding resistance of the vegetation or soil. The ratio $\mu = c_i/c_h$ determines the flux direction (Kramm 1989a), i.e.

$$F_c \begin{cases} < 0 & , \quad \text{if } \mu < 1 & \text{(deposition)} \\ = 0 & , \quad \text{if } \mu = 1 & \text{(compensation)} \\ > 0 & , \quad \text{if } \mu > 1 & \text{(exhalation)} \end{cases} \tag{2}$$

If $\mu << 1$, the deposition flux may be approximated by (Garland 1978)

$$F_c = - v_d \, c_h \quad , \tag{3}$$

where $v_d = (r_a + r_s)^{-1}$ is the deposition velocity at the height z_h, which not only depends on the transfer ability of the Prandtl layer, but also on the sorption properties of the receptor (vegetation, soil, etc.).

Since the transfer properties of the fully turbulent regime of the Prandtl layer differ considerably from the transfer properties of the underlying molecular-turbulent sublayer, an approach for the total bulk resistance in is specified by setting $r_a = r_t + r_s$, where

$$r_t = \frac{1}{u_* \, \varkappa} \left(\ln \frac{z_h - d}{\delta - d} - \Psi_h(\varsigma_h, \varsigma_\delta) \right) \tag{4}$$

denotes the purely turbulent resistance and

$$r_{mt} = \frac{1}{u_*} \left(\frac{U}{u_*} + B^{-1} \right) \tag{5}$$

the molecular-turbulent resistance over a rough surface. Herein, u_* is the friction velocity, $\varkappa = 0.4$ the von Kármán constant, d the zero-displacement, δ the lower boundary of the fully turbulent regime, $\Psi_h(\varsigma_h, \varsigma_\delta)$ the dimensionless integral stability function for heat and matter, U_δ the wind velocity at the height δ and B the sublayer Stanton number. The integral stability function is given by (Paulson 1970; Kramm and Herbert 1984)

$$\Psi_h(\varsigma_h, \varsigma_\delta) = \begin{cases} -\dfrac{5}{L} (z_h - \delta) & \text{for } L > 0 \quad \text{(stable)} \\ 0 & \text{for } L \to \infty \quad \text{(neutral)} \\ 2 \ln \dfrac{1 + y_h^2}{1 + y_\delta^2} & \text{for } L < 0 \quad \text{(unstable)} \end{cases} \quad , \tag{6}$$

with $\varsigma_\delta = (\delta - d)/L$ and $\varsigma_h = (z_h - d)/L$, the non-dimensional heights,

$$y_{\delta,h} = \Phi_m^{-1}(\varsigma_{\delta,h}) = (1 - 16 \, \varsigma_{\delta,h})^{1/4} \quad , \tag{7}$$

the reciprocal expressions of the dimensionless wind gradients in the unstable case at the two heights δ and z_h, and

$$L = \frac{u_*^2}{\varkappa \dfrac{g}{\Theta_m} (\Theta_* + 0.608 \Theta_m q_*)} = \text{const.} \qquad , \qquad (8)$$

the Monin-Obukhov stability length. Determination of the height-invariant scaling parameters u_*, Θ_*, q_* and c_* is based on the relations

$$u_* = \frac{\varkappa (U_h - U_\delta)}{\ln \dfrac{z_h - d}{\delta - d} - \Psi_m(\varsigma_h, \varsigma_\delta))} \qquad (9)$$

$$\left\{\begin{matrix} \Theta_* \\ q_* \\ c_* \end{matrix}\right\} = \frac{1}{u_* r_t} \left\{\begin{matrix} \Theta_h - \Theta_\delta \\ q_h - q_\delta \\ c_h - c_\delta \end{matrix}\right\} = \text{const.} \qquad , \qquad (10)$$

where the dimensionless integral stability function for momentum is given by (Paulson 1970; Kramm and Herbert 1984)

$$\Psi_m(\varsigma_h, \varsigma_\delta) = \begin{cases} \Psi_h(\varsigma_h, \varsigma_\delta) & \text{for } L > 0 \\ 0 & \text{for } L \to \infty \\ 2 \ln \dfrac{1 + Y_h}{1 + Y_\delta} + \ln \dfrac{1 + Y_h^2}{1 + Y_\delta^2} - 2 \arctan \dfrac{Y_h - Y_\delta}{1 + Y_h Y_\delta} & \\ & \text{for } L < 0 \end{cases} \qquad (11)$$

The sublayer Stanton number is a measure of the difference in the transfer rates of momentum and sensible heat as well as matter to and from a rough surface (Chamberlain 1968); B can be considered as a function of the special Reynolds number $Re_* = u_* z_r/\nu$ (ν is the kinematic viscosity) and the Prandtl number ν/α (α is the thermal diffusivity), respectively, of Re_* and the Schmidt number Sc/D (D is the molecular diffusion coefficient for trace gases or water vapour). Functions of the form

$$B = b^{-1} Re_*^{-m} \begin{cases} Pr^{-n} & \text{for heat} \\ Sc^{-n} & \text{for trace gases or water vapour} \end{cases} \qquad (12)$$

were derived by Owen and Thomson (1963) for heat transfer and by Chamberlain (1968) for the thorium-B and water vapour transfer from wind tunnel investigations. Chamberlain obtained a result for the mass transfer in the range $0.33 < Re_* < 33$, which is similar to that of Owen and Thomson for heat transfer.

Equations (4) to (11) described a close system for determining r_t and the fluxes for momentum ($\tau = \rho_a u_*^2$; ρ_a is the air density), sensible heat ($H = -c_{p,o} \rho_a u_* \Theta_*$; $c_{p,o}$ is the specific heat for dry air at constant pressure), water vapour ($Q = -\rho_a u_* q_*$) and trace gases

$$F_c = - u_* c_*$$ (13)

as functions of wind velocity (U), potential temperature (Θ), specific humidity (q) and trace gas concentrations (c) at the to heights δ and z_h.

With respect to δ, it is useful to take the first observation level above the characteristic height of the roughness elements z_r. Such a quantity is Nikuradse's (1933) equivalent sand roughness, which is approximately equal to 30 z_o (z_o is the roughness length). In contrast, Kramm and Herbert (1984) have been suggested the relation $z_r = z_o + d$. This suggestion is based on the requirement that the vertical profile of the mean horizontal wind velocity extrapolates to zero at this height.

The sorption resistance may be described as a function of the physical and biological processes taking place within the receptor. However, a description of the sorption resistance proves to be particularly difficult because most of these processes cannot be mathematically formulated in an acceptable manner. Since F_c can be estimated only from the data observed in the fully turbulent part of the Prandtl layer, the sorption resistance

$$r_s = - \left(\frac{c_h}{F_c} + r_a \right)$$ (14)

may be determined indirectly, but the functional relationships mentioned above are not revealed.

6. Discussion

According to Ripperton et al. (1967), the attitude of ozone in the atmospheric boundary layer can be described as a result of four physical and chemical processes: production, destruction, advection, and turbulent transport, which either single or combined produce the daily and saisonal variations in concentration as well as the vertical differences.

During day a photochemical production of O_3 takes place, mainly by photolysis of NO_2 (15) followed by oxidization of the highly ground state triplet-P oxygen atom (16). Simultaneously, the NO molecule resulting from that photolysis of NO_2 allows a destruction of ozone (17) and, therefore, leads to a stationary equilibrium of ozone production and destruction (Leighton relationship), through which only low O_3 concentrations can exist. In the presence of radicals like HO_2 and RO_2 this equilibrium may be forced to shift towards an increased ozone production by the oxidization of NO forming NO_2 and an OH radical (18). As will be shown later, those radicals, which also can result from reactions like Eq. (19), are very important in the O_3 cycle.

$$NO_2 + h\nu \rightarrow NO + O(^3P)$$ (15)

$$O(^3P) + O_2 + M \rightarrow O_3 + M \qquad \text{(M usually } O_2 \text{ or } N_2\text{)}$$ (16)

$$O_3 + NO \rightarrow NO_2 + O_2$$ (17)

$$HO_2 + NO \rightarrow NO_2 + OH$$ (18)

$$O_3 + h\nu \xrightarrow{O(^1D), H_2O} 2 \ OH \tag{19}$$

$$terpene + \begin{matrix} OH \ | \\ O_3 \end{matrix} \rightarrow \alpha_1 \ RO_2 + \alpha_2 \ Product, \qquad (\alpha_1 + \alpha_2 = 1) \tag{20}$$

The observed minimum at night does not result from gas phase reactions, but from deposition (Garland and Derwent 1979, Roberts et al. 1985). Of course, there can be deposition at day, too. At night, however, when atmospheric stability is dominant, the losses of ozone by deposition can not or only partly be restored by vertical transport from air layers above. As mentioned above the appearance of "second order maxima" during the minimum period is not in conflict with that.

Therefore, we believe that the reasons for the daily cycle as measured at our site are well understood and in accordance to observation made elsewhere. No data but from Krapfenbauer (1988), however, are known to us, which show inverse ozone gradients above the canopy of coniferous forests. That may be related to the fact that the number of stations is very limited, which make contineous, but nevertheless in time and height highly resolved measurements within and above mature spruce stands.

To generate this behavior, there must be either a increased destruction of ozone well above the canopy or an additional production close to (or within) it. The NO_x measurements during the period June 27-30 give no hints to an explanation, because there was nothing extraordinary at all, and no significant differences in the NO_x concentrations between the different heights. Due to a malfunction the seperation of NO and NO_2 was not possible that time, which would lead to a better understanding.

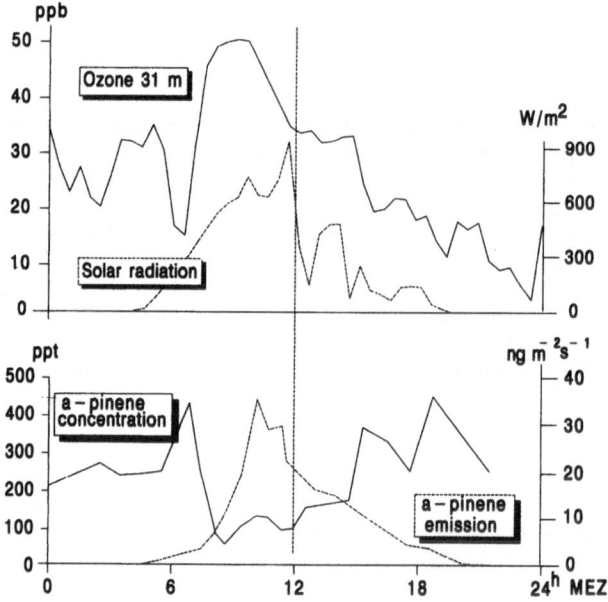

Figure 5. Diurnal cycle of ozone at 31 m, solar radiation, and α-pinene emission and concentration at 22 m on June 10, 1988. 30-min averages, terpene measurements not equally distant.

On the other hand, there are some anthropogenic and biogenic substances, which can be modified or destroyed by OH radicals primarily originating from the photolysis of ozone (see Eq. (19)). If hydrocarbons are present, among others peroxiradicals HO_2 or RO_2 originate from those reactions. According to Eq. (18), they transform the reducing NO into the most important ozone precursor NO_2 and re-generate simultaneously the catalyst OH (for details see Fabian 1987, Schurath 1988). Hydrocarbons are emitted from all trees, though in different composition and quantity. Estimates about local and global emissions of terpenes was given by Kohlmaier et al. (1983). A comprehensive review of the cycle of terpenes in the atmosphere was presented by Graedel (1979). Very different reaction pathway are possible depending on the terpene species and wether it reacts with O_3 or OH, but so far there is no fully clearness about their importance in producing ozone. A short review of the conflicting literature hereto is published by Kreuzig (1987). For model calculations, the chemical degradation pathway of the various terpenes may be simply parameterized by Eq. (20) (Hov et al. 1983).

Because our own measurements are insufficent to verify or reject this hypothesis, we started a close research cooperation with other disciplines. On June 10, 1988 terpene emission from needles and ambient concentration in the 22 m level were measured by the Institute for Biology of the Technical University Munich (Steinbrecher et al. 1988). Six monoterpenes could been identified, four of them (α-pinen, camphene, β-pinen, limonene) were emitted in appreciable amounts. As an example the diurnal cycle of emission and concentration of α-pinen are given in Fig. 5, in addition ozone at the 31 m level and solar radiation (unfortunately, ozone was measured only in one level during this day, because this joint field experiment was merely designed for testing methods and equipment). The depletion in the terpene concentration occurs, just when the emission reaches its maximum. This remarkable phase difference suggests a very quick reaction, when solar radiation is present. Also it should be notified that the ozone maximum takes place before the maximum of solar radiation and, therefore, fits better to the period of mimimum terpen concentration.

From this still limited state of knowledge, however, it is not allow to draw the conclusion that there is a causative relation between the behavior of ozone and terpenes.

Therefore, and due to the lack of NO and NO_2 measurements the ozone had to be treated as a quasi-inert gas in the model.

First, roughness length z_o and zero-displacement d as well as the surface layer parameters described in section 5 were estimated from vertical profile data sets of wind velocity, temeperature, humidity, and trace gases by least squares methods (Kramm 1989b). These methods are based on the equation set (4) to (11).

From the period June 28, 14:00 - June 30, 24:00, 116 profile data sets (30-min averages) were closer examined, of which only 33 % were suitable for evaluation, because most of the time very stable conditions with low wind velocities and temperature inversions at night were dominant due to the high pressure situation.

Typical results from the remaining data are given in detail in Table 1. The mean aerodynamic stand characteristics calculated from that table are z_o = 2.90 m, d= 20.21 m, and z_r = z_o+d = 23.11 m, which fit well to the often used approximations d = 0.7 - 0.8 H and z_o = 0.1 H (H is the average tree height of 28 m). The deposition velocities of ozone calculated for the lowest observation level (31 m) show considerable differences. The values reach from -8.9 cm/s to +1.8 cm/s. As was clear from the observations, also the calculation leads for most of the time to an exhalation of ozone (negative sign).

Figure 6 illustrates two cases with opposite sign of the deposition velocity and gives a comparison of measured and calculated profiles of wind speed, potential air temperature, specific humidity, and ozone concentration. The calculated least squares fits match the observed values very well.

32

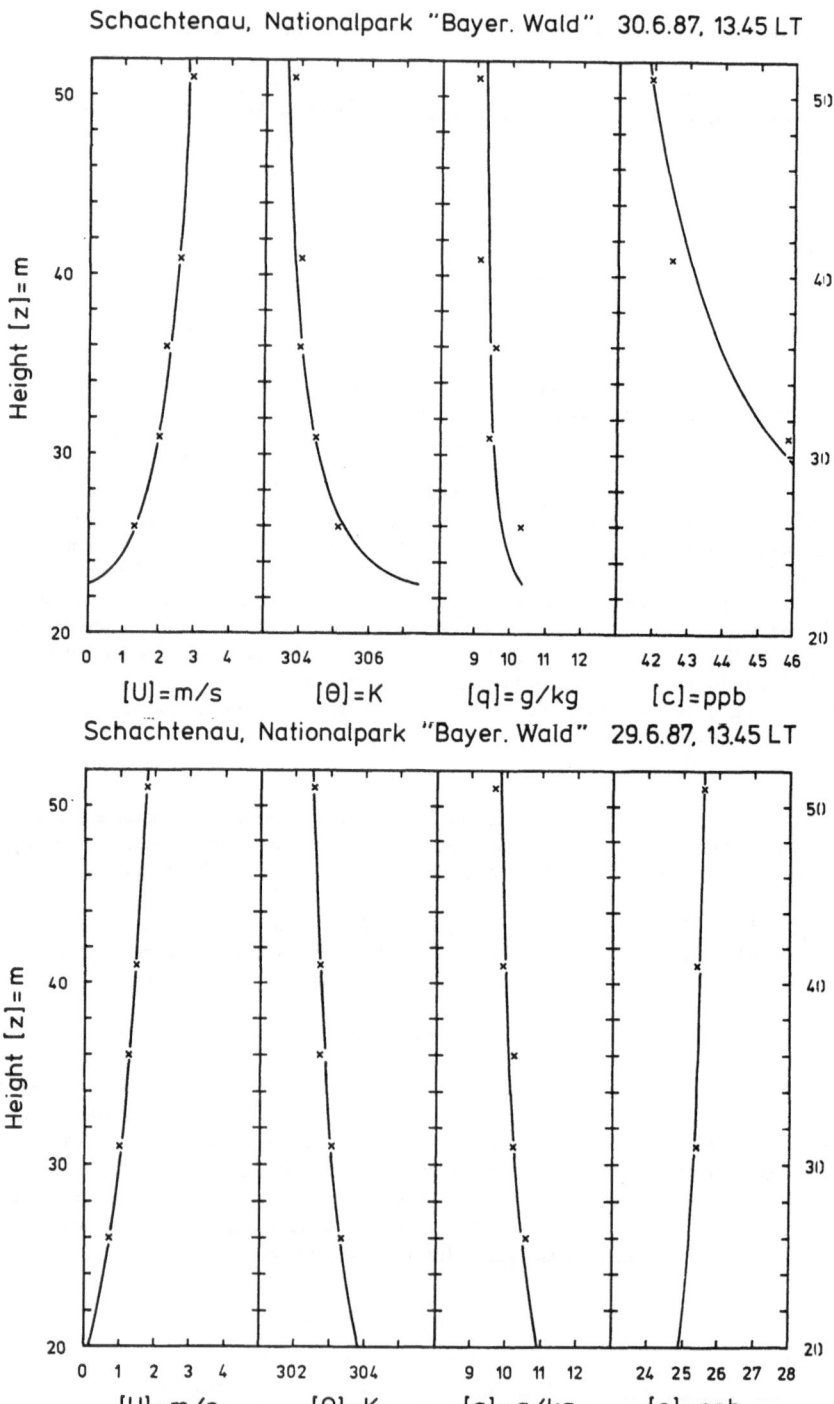

Figure 6. Typical vertical profiles of wind velocity (U), potential temperature (Θ), specific humidity (q) and ozone (c). The crosses represent the observed values and the solid lines the calculated profiles.

Day	Time	z_0 (m)	d (m)	u_* (cm/s)	θ_* (K)	q_* (g/kg)	c_* (ppb)	$v_{d,31m}$ (cm/s)	s_U (cm/s)	s_θ (K)	s_q (g/kg)	s_c (ppb)	D (deg)
28.6.87	14:15	0.06	25.59	19.3	-0.4353	-0.2577	-4.733	-2.6	24.8	0.37	0.27	0.9	183
	14:45	0.46	24.43	30.2	-0.8864	-0.5012	-5.347	-4.5	13.4	0.46	0.22	1.0	180
	15:15	1.02	24.01	39.5	-0.8474	-0.4728	-4.626	-5.2	20.1	0.53	0.25	1.5	240
	15:45	8.14	15.76	76.5	-0.6521	-0.2735	-1.992	-4.6	9.0	0.15	0.20	1.1	260
	16:15	10.33	13.23	99.2	-0.5260	-0.2780	-2.217	-6.5	8.7	0.16	0.16	1.0	254
	16:45	2.08	22.30	48.2	-0.8776	-0.5394	-3.928	-5.4	17.2	0.36	0.21	1.2	230
	17:15	14.08	9.84	115.7	-0.4673	-0.3531	-1.830	-6.0	8.6	0.15	0.14	0.6	260
	17:45	4.71	19.98	43.3	-0.5087	-0.0200	-3.482	-4.2	12.1	0.22	0.25	1.1	261
	18:45	2.98	22.51	42.7	-0.3325	-0.0706	-1.499	-1.8	27.1	0.42	0.21	1.1	263
29.6.87	10:45	2.31	22.26	39.8	-0.9826	-0.6206	-1.550	-2.1	10.5	0.20	0.14	0.7	220
	13:45	9.01	9.92	79.8	-1.1160	-0.9297	0.573	1.8	4.4	0.08	0.12	0.1	158
	14:15	0.20	25.11	21.7	-1.0100	-0.3800	-2.702	-2.2	11.5	0.44	0.18	0.4	163
	14:45	2.28	21.19	40.4	-0.7499	-0.4089	-3.353	-4.7	12.0	0.13	0.22	1.1	156
	18:45	1.44	23.74	25.6	-0.1416	0.0657	1.295	1.3	26.3	0.18	0.36	0.8	83
30.6.87	11:15	0.79	22.52	41.3	-0.6949	-0.0858	-4.216	-4.3	7.9	0.15	0.35	0.6	175
	11:45	1.26	22.07	49.7	-0.7213	-0.2276	-5.054	-6.1	11.2	0.10	0.30	1.1	147
	12:15	1.25	21.80	53.5	-1.9230	-0.5247	-6.537	-6.5	11.0	0.32	0.23	0.6	179
	13:15	0.82	22.28	43.6	-0.8375	-0.2290	-2.341	-2.5	17.0	0.12	0.35	0.4	182
	13:45	1.25	21.57	54.9	-1.0850	-0.3146	-4.473	-5.4	8.6	0.14	0.32	0.5	182
	14:15	0.11	25.19	26.0	-0.5923	-0.1276	-6.851	-3.7	17.3	0.27	0.39	0.1	175
	14:45	0.66	23.22	41.8	-1.3020	-0.5430	-10.240	-8.9	11.0	0.31	0.28	0.6	179
	15:45	0.35	24.50	35.5	-0.5679	-0.2749	-7.659	-5.6	19.2	0.20	0.29	1.1	179
	17:45	1.14	22.92	40.9	-0.6716	-0.1997	-7.974	-6.8	12.8	0.34	0.24	1.7	181

Table I. Calculated values of the surface layer parameters z_0, d, u_*, Θ_*, q_*, c_*, $v_{d,31m}$. D is the wind direction and s_x the "badness of fit" defined by $s_x = (1/N \sum_{i=1,N} (x_{M,i} - x_{c,i})^2)^{1/2}$ where $x_{M,i}$ is the observed value at the height z_i and $x_{C,i}$ the corresponding calculated value.

7. Conclusion

Further and more detailed measurements are required, for which the EUROTRAC subproject BIATEX (Biosphere/Atmosphere Exchange of Pollutants) provides the frame. Moreover, in June 1989 a two week field experiment will be carried out to investigate simultaneously at different levels diurnal cycles of terpene emission and concentration, H_2O_2, organic peroxides, NO, NO_2, O_3, PAN and reaction products of the terpene/ozone complex.

Those experiments hopefully provide data sets, which allow the incorporation of chemical reactions among those trace gases into such a model. Fluxes and deposition velocities have to be measured by other methods, e.g. by eddy correlation, to be compared to results from the gradient approach.

Acknowledgement

The authors would like to express their thank to all federal and state agencies, without their support this work would not have been possible. The experimental part is financed by the Federal Ministry of Research and Technology (contract no. 0339113B), the State Ministry of Environmental Protection (8272-62-7923), and the State Ministry of Food, Agriculture and Forestry (M23). We further thank the National Park Service for continous technical and logistic assistance.

References

Chamberlain, A.C. (1968) 'Transport of gases to and from surfaces with bluff and wave-like roughness elements', Quart. J. R. Met. Soc. 94, 318-332.

Enders, G., Teichmann, U. (1986) 'GASDEP - Gaseous deposition measurements of SO_2, NO_x, and O_3 to a Spruce Stand', in: H.-W. Georgii (ed.), Atmospheric Pollutants in Forest Areas, Reidel Publ., Dordrecht, 13-24.

Fabian, P., (1987) 'Photochemischer Smog und seine Einwirkung auf die Biosphäre', Forstw. Cbl. 106, 223-234.

Garland, J.A. (1978) 'Dry and wet removal of sulphur from the atmosphere', Atmos. Environ. 12, 349-362.

Garland, J., Derwent R. (1979) 'Destruction at the ground and the diurnal cycle of concentration of ozone and other gases', Quart.J.R.Met.Soc. 105, 169-183.

Gietl. G., Rall. A. (1986) 'Bulk deposition into the catchment "Große Ohe". Results of neighbouring sites in the open and under spruce at different altitudes', in: H.-W. Georgii (ed.), Atmospheric Pollutants in Forest Areas, Reidel Publ., Dordrecht, 79-88.

Hov, O., Schjoldager, J., Wathne, B.M. (1983) ' Measurement and modeling of the concentrations of terpenes in coniferous forest air', J. Geophys. Res. 88, 10,679-10,688.

Graedel, T.E. (1979) 'Terpenoids in the atmosphere', Geophys. Space Phys. 17, 937-947.

Kohlmaier, G., Bröhl, H, Sire, E. (1983) 'Über die mögliche lokale Wechselwirkung anthropogener Schadstoffe mit den Terpen-Emissionen von Waldökosystemen', Allg. Forst- u. J.-Ztg. 154, 170-174.

Kramm, G., Herbert, F. (1984) 'Ein numerisches Modell zur Deposition von Schadstoffen in der bodennahen Luftschicht', in: H. Reuter (ed.), Probleme der Umwelt- und Medizinmeteorologie im Gebirge, Symposium Rauris/Austria, 23.-25. Sept. 1983, Zentralanstalt f. Meteorologie u. Geodynamik, Wien, Nr. 288.

Kramm, G. (1989a) 'A numerical method for determining the dry deposition of atmospheric trace gases', Boundary-Layer Meteorology (accepted for publishing).

Kramm, G. (1989b) 'The estimation of the surface layer parameters from wind velocity, temperature and humidity profiles by least squares methods', Boundary-Layer Meteorology (accepted for publishing).

Krapfenbauer, A. (1988) 'Ozontagesgänge in 400 und 700 m Höhe am Westhang des Rosaliengebirges', in: Verteilung und Wirkung von Photooxidantien im Alpenraum, GSF-Ber. 17/88, 305-314.

Kreuzig, R. (1987) 'Zum Vorkommen flüchtiger Kohlenwasserstoffe anthropogenen und biogenen Ursprungs in unterschiedlich belasteten Waldstandorten', Diss. TU München.

Nikuradse, J. (1933) 'Strömungsgesetze in rauhen Rohren', Forsch. Arb. Ing.-Wes. 361.

Paulson, C.A. (1970) 'The mathematical representation of wind speed and temperature profiles in the unstable atmospheric surface layer', J. Appl. Meteor. 9, 857-861.

Owen, P.R., Thomson, W.R. (1963) 'Heat Transfer Across Rough Surfaces', J. Fluid Mech. 15, 321-334.

Ripperton, L., White, O., Jeffries, H. (1967) 'Gas phase ozone-pinen-reactions', Div. of Water, Air and Waste Chemistry, 147th National Meeting, Am. Chem. Soc., Chicago Ill., 54-56

Roberts, J., Hahn, C., Fehsenfeld, F., Warnock, J., Albritton, D., Sievers, R. (1985) 'Monoterpene hydrocarbons in the nigthtime troposphere', Environ. Sc. Technol. 19 (4), 364-369.

Schurath, U. (1988) 'Bildung von Photooxidantien durch homogene Transformation von Schadstoffen', in: Verteilung und Wirkung von Photooxidantien im Alpenraum, GSF-Ber. 17/88, 136-151.

Steinbrecher, R., Schönwitz, R., Ziegler, H. (1988) 'Emission of monoterpenes from needles of picea abies (L.) Karst. under field conditions', in: Abstracts 19th Int. Sympos. on Essential Oils and natural substrates, Dübendorf, 1988, L10.

Teichmann, U., Enders, G. (1988) 'Gaseous deposition of SO_2, NO_x, and O_3 to a spruce stand in the National Park "Bayer. Wald"', in: M.H. Unsworth and D. Fowler (eds.), Acid Deposition at High Eleveation Sites, Kluwer Acad. Publ., Dordrecht, 583-592.

Winkler, P. (1980) 'Störungen der nächtlichen Grenzschicht', Meteorol. Rdsch. 33, 90-94

OZONE IN FOREST STANDS - EXAMINATIONS TO ITS OCCURENCE AND DEGRADATION

G. BAUMBACH, K. BAUMANN
Institut für Verfahrenstechnik und Dampfkesselwesen (IVD)
Abteilung Reinhaltung der Luft, Universität Stuttgart
Pfaffenwaldring 23, D-7000 Stuttgart 80

Dedicated to Prof. Dr.-Ing. R. Quack in memory of his 80th birthday

ABSTRACT. Horizontal and vertical profiles of ozone (O_3) as well as diurnal variations of O_3 concentrations are investigated at different forest areas in Baden-Württemberg based on continuous measurements at or near ground level. The influence of nitrogen oxides and light intensity on the ozone concentration is discussed in this paper. Ozone concentrations are depressed in the vicinity of anthropogenic emission sources like urban areas or meteorologically influenced valleys. These interrelationships are studied from aircraft measurements carried out above the river Rhine valley and the highlands of the Black Forest. At night, when no photochemical ozone production takes place, NO sources in urban areas cause an immediate ozone reduction, while ozone levels in the forests decrease only slightly. Vertical concentration gradients are determined in forest stands. At ground level inside the forest stand, the degradation of ozone is observed during night time which may be caused by NO exhalation from the soil. During sunny day hours, especially in summer, ozone is produced above the tree canopy.

1. MEASURING SITES AND SET UP OF MEASURING STATIONS

The department Reinhaltung der Luft of IVD carries out air pollution measurements in forest areas of Baden-Württemberg for several years. The measuring sites are located in the northern Black Forest close to the town Freudenstadt (in the following context named "Schöllkopf", 840 m asl) and in the forest Schönbuch between the cities of Stuttgart and Tübingen (in the following context named "Betzenberg", 500 m asl) as shown in Figure 1. On top of "Hornisgrinde", the highest mountain of the northern Black Forest (1120 m asl) ozone measurements were taken part of the time at a measuring station of the electricity authority Badenwerk AG. Results of the measurements taken by the Umweltbundesamt on top of the "Schauinsland", the highest mountain of the southern Black Forest (1240 m asl) are evaluated for comparison. In the area between Freudenstadt, Hornisgrinde and the river Rhine valley air craft measurements were taken in cooperation with the Deutsche Forschungs- und Versuchsanstalt für Luft und Raumfahrt (DFVLR). Part of these results shall be presented here also.

It is to note, that the "neuartige Waldschäden" observed in the Black Forest are still significantly high, whereas the Schönbuch shows a lower grade of forest desease. At the IVD measuring stations Schöllkopf and Betzenberg meteorological parameters and air pollutant gradients as well as absolute concentrations are continuously measured in forest

H.-W. Georgii (ed.), Mechanisms and Effects of Pollutant-Transfer into Forests, 37–44.
© *1989 by Kluwer Academic Publishers.*

stands /1/. For this study results of ozone measurements are of special interest.

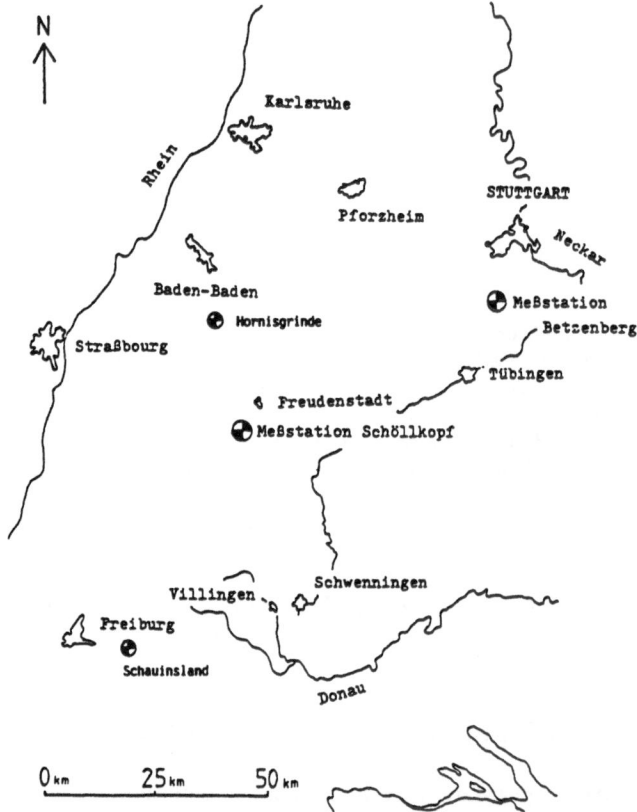

Figure 1.Locations of forest measuring stations in Baden-Württemberg

2. OZONE LEVELS AT DIFFERENTLY LOCATED FOREST AREAS

The evaluated monthly averages and maximum half-hour mean values of ozone concentrations registered at both IVD measuring stations show significant annual variations. The summer ozone levels always exceed those measured in winter, whereas the levels on Schöllkopf - receptor point at 860 m asl, lie distinctly above those detected on Betzenberg at 520 m asl.

In Figure 2, the relative excessive frequencies of ozone and nitrogen oxide (NO) are plotted versus the respective concentrations measured at Betzenberg and Schöllkopf for a quantitative comparison of all data. This presentation is based on all half-hour mean values of the hitherto measuring period. On the graphs' ordinate it can be read how often (percent of total amount of registrations) a certain concentration limit - to be determined on the abscissa - is exceeded. For instance 16 % of all values measured on Schöllkopf exceed the ozone MIK-value of 120 µg/m^3 /2/, a threshold for the safety of human health, whereas on Betzenberg this amounts only to 6 %. The difference is even bigger for 50 µg/m^3: which is exceeded by 81 % of all measured ozone values on Schöllkopf and only by 49 % on Betzenberg. The pollution load by NO is reverse: the concentration levels on Schöllkopf are much lower compared to those on Betzenberg, which is directly influenced by the densly populated area of Stuttgart whose NO emissions can reach the Betzenberg /3/.

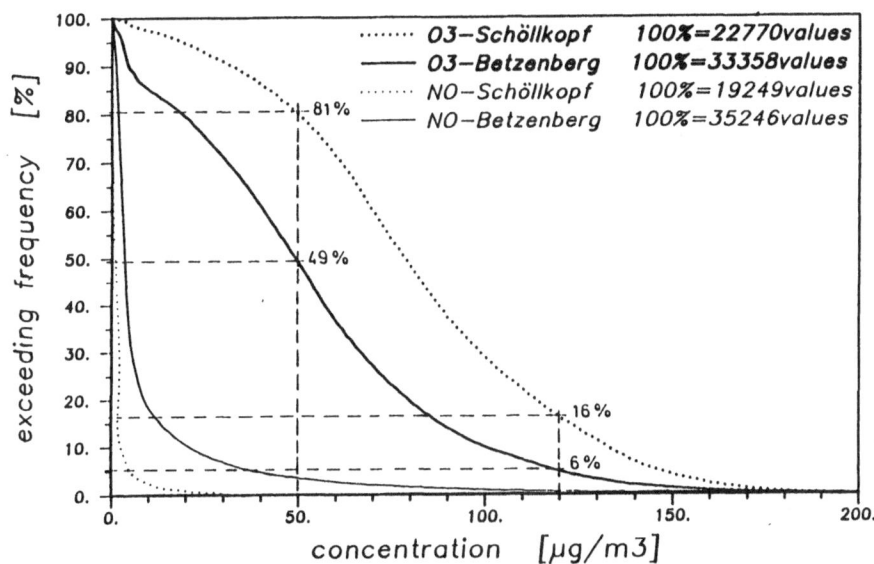

Figure 2. Excessive frequencies of NO and O_3 as percentage of their respective total of measured half-hour mean values for Schöllkopf (860 m asl) and for Betzenberg (520 m asl). Measuring Period: Schöllkopf 3.85 to 6.88
                   Betzenberg 1.84 to 6.88

3. ATMOSPHERIC BEHAVIOUR OF OZONE IN FOREST AREAS

3.1 Average Diurnal Variations

The evaluation of average diurnal cycles is well suited to characterise and to compare the location specific behaviour of ozone for a sufficient statistical amount of measured data. Seldomly appearing special weather situations may cause certain divergencies of "typical" ozone cycles. Since such situations show irregularities, they are not suitable to compare average pollution loads of different locations.

Average diurnal variations allow critical judgement whether photochemical ozone production or degradation is the ruling process for the ozone level detected at a certain location or not. Photochemical processes and atmospheric mass transport are subjected to continuous fallwinds and valley breezes. Average daily variations of ozone measured on "Zugspitze" (2963 m asl) and at the neighbouring valley town Garmisch-Partenkirchen (730 m asl) - depicted in Figure 3 as published by Kanter, Reiter and Munzert /4/, typically demonstrate the variation of ozone concentrations with daytime - meaning daily cycle of solar radiation and upwind-fallwind-system. A relatively high ozone level is detected on

40

Zugspitze with very little daily variation, while Garmisch-Partenkirchen shows a distinct daily ozone course: at night, primary air pollutants like nitrogen oxides emitted by motor traffic at the valley bottom remain in the ground inversion layer where ozone is largely degraded. After sunrise, solar radiation induces photochemical ozone production, which is fed by primary air pollutants of the valley ground. The midday ozone maximum of Garmisch exceeds the relatively high ozone level of Zugspitze where influences of air mass transportations connected with photochemical effects are barely detectable. On top of "Wank" (with 1780 m asl elevated between Garmisch und Zugspitze) almost no daily variations can be registered, but the influences on the ozone production and degradation are stronger than on top of Zugspitze.

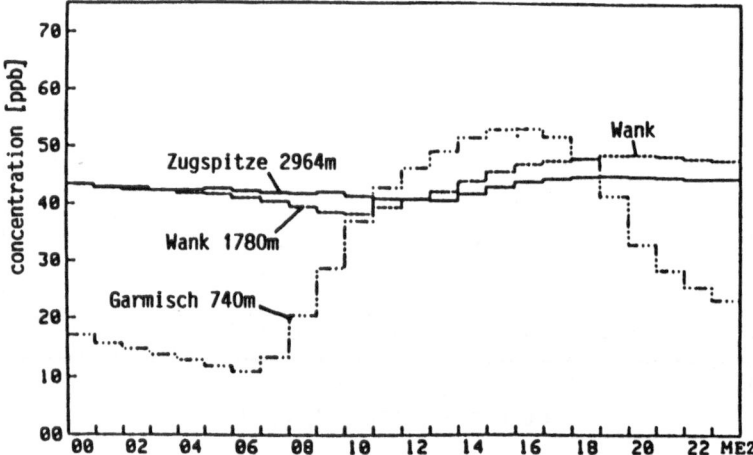

Figure 3. Average diurnal variations of ozone concentration on sunny days of the summer months June to September (1977 to 1980) at the measuring stations Zugspitze, Wank and Garmisch, from /4/.

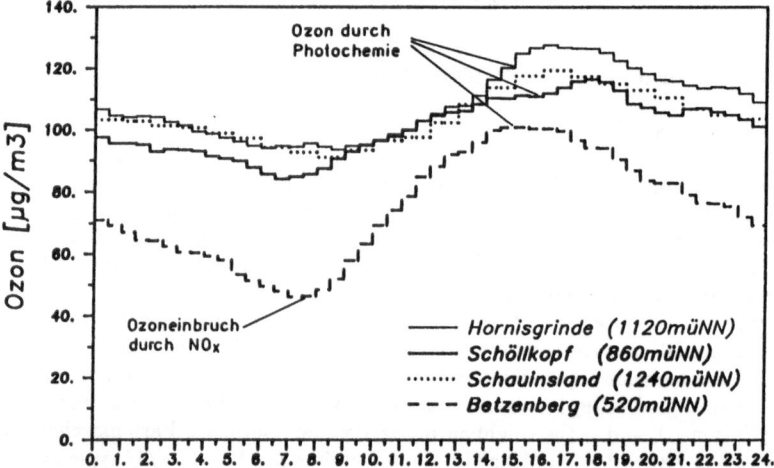

Figure 4. Average diurnal variations of ozone concentration measured at different locations of Baden-Württemberg.

Looking at the average daily cycles of ozone measured at different forest areas of Baden-Württemberg as presented in Figure 4, it can be seen, that Betzenberg is most strongly influenced by ozone production and degradation processes, which is mainly caused by the nearby urban area of Stuttgart (as reported earlier /3/). It is evident, that even in the highlands of the Black Forest - namely Hornisgrinde, Schauinsland and Schöllkopf - an ozone production takes place with its maximum between 4.00 and 5.00 pm. This ozone probably originates in air masses from neighbouring areas fed by emission sources, which may be the river Rhine valley for Hornisgrinde and Schauinsland. Precursor substances cause the photochemical ozone production during the air mass transport to the highlands. Since this pathway is longer than the distance between Stuttgart and Betzenberg, the average ozone maximum appears later in the afternoon.

All three Black Forest stations show little ozone degradation at night, which can be explained as follows: nightly ground inversions prevent transport of degradative substances, e.g. nitrogen oxides, to the highlands of the Black Forest. During stable sunny weather periods usually another inversion layer ranges above 1500 m asl and prevents air exchange from below, so that the air enclosed by the two inversion layers contains "undegradable" ozone. Paffrath named this layer "ozone reservoir" /5/. The highlands of the Black Forest are covered by this ozone reservoir, and that is why relatively high ozone concentrations can be registered there.

3.2 Aircraft Measurements

The above mentioned explanations has been proven right by aircraft measurements carried out by the DFVLR in summer 1988 /6/. Vertical ozone gradient profiles has been measured above the river Rhine valley close to Hornisgrinde at different daytimes on 26 July 1988, which was a typical sunny summer day. Low winds of SW directions prevailed at high elevations. But the Rhine valley shows its own typical meteorological system /7/.

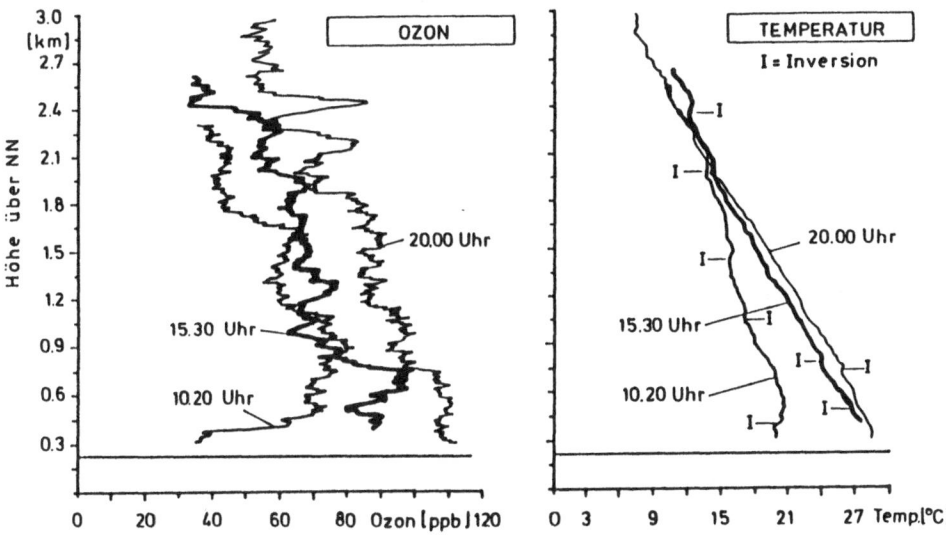

Figure 5. Vertical ozone and temperature gradient profiles above the river Rhine valley taken at different daytimes on 26 July 1988 by aircraft measurements /6/.

The positions of inversion layers were exactly determined by aircraft measurement of vertical temperature gradient profiles, which are plotted in Figure 5 for different daytimes. Below the morning ground inversion at the valley bottom up to 350 m asl, a distinct ozone degradation by primary air pollutants can be observed. At this time the high altitude inversion lies at approx. 1400 m asl. Between both inversion layers the above mentioned ozone reservoir is noticeable. It can be seen, that the barrier layers move up during the day, and that an enormous increase of ozone concentration takes place within these layers.

Aircraft measurements can describe only isolated situations, which should be statistically assured by more flights. The results mainly correspond to the explanations derived by the average diurnal variations.

3.3 Do Forests Act as Ozone Sink?

Do forests represent a more or less intensive sink for ozone, or can they possibly act as ozone sources, when connected with nitrogen oxides at certain daytimes?

Many attempts described in literature to determine the ozone flux into forest stands failed, because the situations of the various field experiments did not match all the theoretically necessary conditions. Especially ozone is subjected to chemical reactions in the air of the forests which overlap the mass transport to the trees. Simultaneous measurements of ozone and nitrogen oxide above and below the tree canopy of the Betzenberg station point to such reactions, see Figure 6.

Average daily cycles of ozone (top), NO and NO_2 (bottom) are depicted only for sunny days of July and August 1987. One can see distinctly, that at night, the average NO-concentrations inside the forest stand exceed those registered above the tree canopy. Since no traffic road is in the vicinity of this measuring station, this NO must be exhaled by the forest soil. This values lie far below the average level reached above the canopy of midday, but this night time NO-excess causes an ozone degradation inside the forest stand due to the following photostationary balance:

$$NO + O_3 \xrightleftharpoons[\text{light}]{\text{dark}} NO_2 + O_2$$

That is why relatively big ozone differences occur between the air above and below the canopy during night hours. This nightly differences are even bigger than the ones during daylight hours at an absolutely higher level. Besides NO, there may exist other gases which can cause such an ozone degradation inside the forest stand.

NO_2 photolysis activated by solar radiation is the reason for a negative vertical NO_2 gradient during the morning hours, when sun rays hit the tree canopy. That is when ozone strongly increases while a positive vertical NO gradient exists, so that the photostationary balance is shifted from NO_2 to NO plus O_3 due to solar radiation.

The role of hydrocarbons, especially terpenes exhaled by forest trees, could not yet be investigated in this respect.

This investigations showed, that at remote forest areas the night time ozone reduction is low. It can be stated, that forest areas in general do not represent a major ozone sink, but areas with sources of anthropogenic air pollutants do.

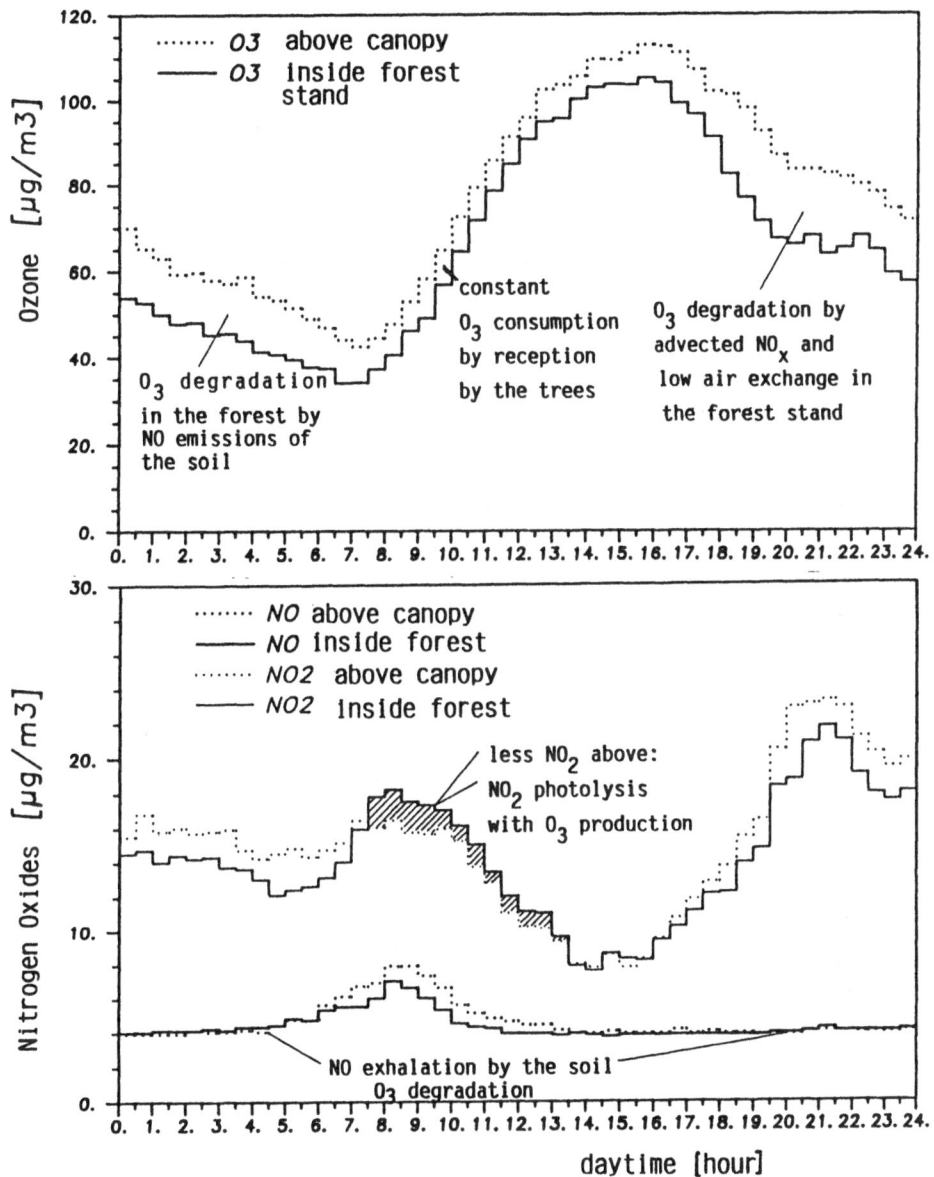

Figure 6. Average diurnal variations of ozone (top) and NO$_x$ (bottom) measured above and below the tree canopy on sunny days in July and August 1987 at Betzenberg station - Schönbuch forest.

44

ACKNOWLEDGEMENTS

This report is generally based on a research program sponsored by the German government through the Bundeminister für Forschung und Technologie (BMFT), record no. 339112. The ambitious help by Andreas Rochowiak, who maintained the equipment of the measuring stations was greatfully appreciated.

REFERENCES

/ 1/   Baumbach, G., K. Baumann and F. Dröscher (1987): Luftverunreinigungen in Wäldern. Institut für Verfahrenstechnik und Dampfkesselwesen der Universität Stuttgart. Report no. 5

/ 2/   Verein Deutscher Ingenieure: Richtlinie VDI 2310 Bl.15, Maximum Immission Value Referring to Human Health - Maximum Immission Concentrations for Ozone (and Photochemical Oxidants), Beuth Verlag Berlin und Köln, April 1987

/ 3/   Baumbach, G. (1986): Occurence of Gaseous Pollutants in Forest Stands. In H -W. Georgii (ed.): Atmospheric Pollutants in Forest Areas. D. Reidel Publishing Co., Dordrecht, pp 177 - 187

/ 4/   Kanter, H.G., R. Reiter and K.-H. Munzert (1982): Untersuchungen zur Frage der photochemischen Produktion von Ozon in Reinluftgebieten und ihrer vertikalen Verteilung. Forschungsbericht 104 02 800, Umweltbundesamt Berlin

/ 5/   Paffrath, D. (1988): Untersuchungen über die Verteilung und Bildung von Ozon im Alpenbereich aus Flugzeugmessungen. In: Verteilung und Wirkung von Photooxidantien im Alpenraum. Gesellschaft für Strahlen- und Umweltforschung, München. GSF-Bericht 17/88

/ 6/   Paffrath, D. and F. Rösler, DFVLR Oberpfaffenhofen, and Baumbach, G., K. Baumann, University of Stuttgart (1988): Aircraft Measurements of Air Pollutants in the Black Forest Area. Unpublished results

/ 7/   Malberg, H., G. Bökens and G. Frattesi (1980): Mittlere geostrophische und beobachtete Strömungsverhältnisse im Oberrheingraben. Annal. Meteorol. (N.F.) 16, pp 85 - 89

ANNUAL VARIATION OF WET DEPOSITION AT NON-URBAN SITES IN CENTRAL EUROPE

W. FRICKE
Umweltbundesamt
- Pilotstation -
Frankfurter Str. 135
D-6050 Offenbach/Main
Federal Republic of Germany

ABSTRACT. Precipitation chemistry data from a network operated by the Federal Environmental Agency in the FRG are used for discussing the annual variation of hydrogen ion, sulfate, nitrate, and ammonium deposition in Central Europe from 1982 to 1988. Except for the coastal station, deposition rates are generally highest in late spring, while the precipitation amount does not show an evident annual variation. Differences in deposition between low level sites and the mountain station are most pronounced during the first half year. Later on the mountain site deposition rates decrease to the amount which is deposited at the low level stations, despite continuing high amounts of precipitation.

1. INTRODUCTION

The German Federal Environmental Agency runs a network of currently 15 air pollution monitoring stations in rural areas within the Federal Republic of Germany. Daily samples of precipitation have been collected and analysed at five of these stations since 1982. In the following, results from Westerland, Langenbrügge (Waldhof), Deuselbach, and Schauinsland will be discussed.

Westerland is located on the island of Sylt in the North Sea, close to the German-Danish border, Langenbrügge is situated in the North German Plain. Deuselbach is located in the Mid-German Highlands on the southwestern part of the Rhine Plateau at 480 m above sea level. Schauinsland is a mountain station in the Black Forest in southwestern Germany at 1205 m above sea level.

2. SAMPLING AND ANALYSIS

Precipitation was collected by using ERNI/Eigenbrodt ARS 721 samplers. The samplers are operated as daily bulk collectors: the lids open for 24 hours starting at 08 CET. In order to cover regular weekends, four collectors are used in sequence at each site. Funnel and collector vessel are made of polyethylene. They are replaced and cleaned after every exposure, even in case of no precipitation collected. The samplers are kept dark and cool until analysed in a central laboratory within a few days.

45

H.-W. Georgii (ed.), Mechanisms and Effects of Pollutant-Transfer into Forests, 45–50.
© 1989 by Kluwer Academic Publishers.

Ion chromatography is used for the anions, ammonium is analysed with the indophenol spectrophotometric method. Deposition was calculated by using the ERNI sampler precipitation amount, which on average is about 15% less than the respective amount from the standard rain gauge. The data were not corrected for sea salt.

3. RESULTS

3.1. PRECIPITATION

If interception by fog or clouds can be neglected, wet deposition is calculated by multiplying the concentration of a trace substance in precipitation by the corresponding precipitation amount. Fig. 1 shows the median monthly precipitation amounts at the four sites. In general, there is no significant difference between the seasons because of high variability in precipitation from year to year. An exemption is Westerland, where most rain is observed from late summer to early fall. The amount of precipitation at Schauinsland mountain station is about three times that of the other sites.

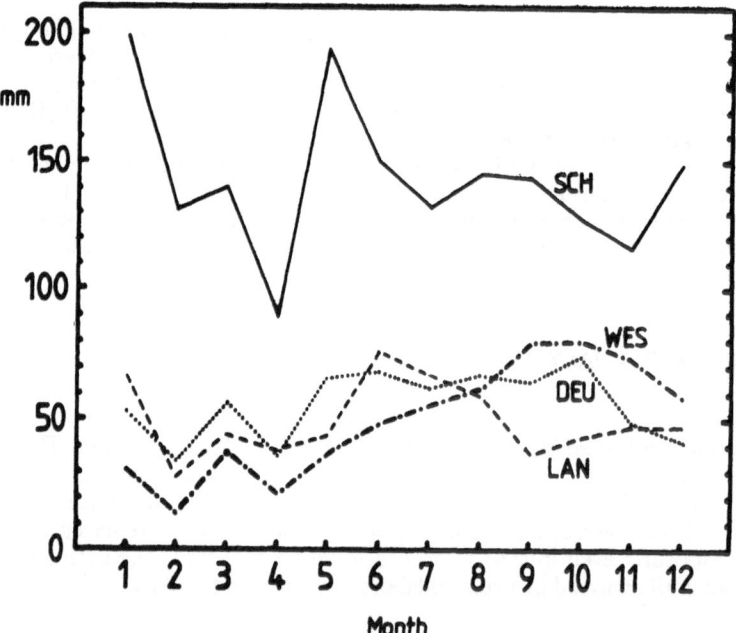

Fig. 1: Median monthly precipitation amount, 1982-1988, at Westerland (WES), Langenbrügge (LAN), Deuselbach (DEU), and Schauinsland (SCH)

3.2. H⁺-DEPOSITION

The deposition of acidity by hydrogen ions at Schauinsland shows a pronounced maximum in May, while Langenbrügge and Deuselbach peak somewhat later (Fig. 2). The coastal station Westerland, with little rain in spring, reaches a maximum between July and September. There is a marked difference in hydrogen ion deposition among the sites during the first six months: deposition at Schauinsland is about four times that of Westerland, while Langenbrügge and Deuselbach are intermediate. This spatial difference disappears almost completely during the last five months, despite the fact that precipitation amounts in those months differ from site to site by up to a factor of three.

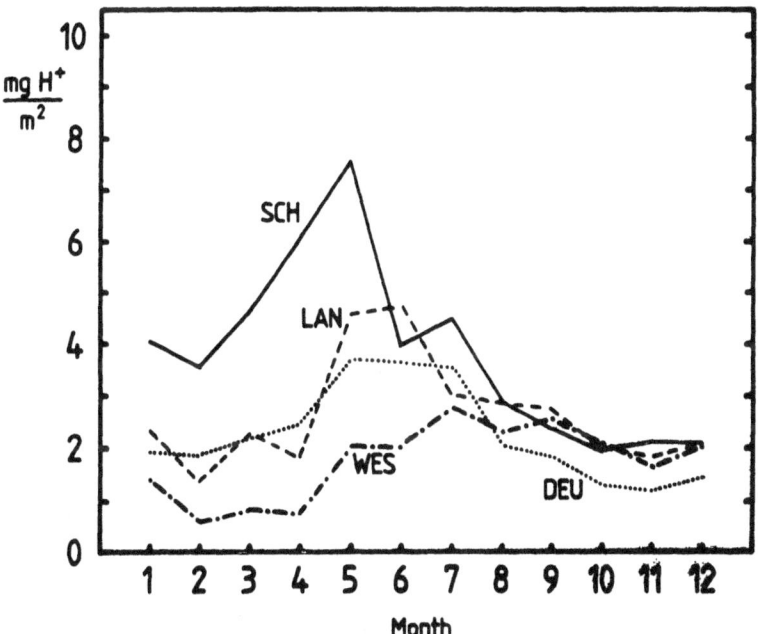

Fig. 2: Median monthly wet deposition of hydrogen ions, 1982-1988, at Westerland (WES), Langenbrügge (LAN), Deuselbach (DEU), and Schauinsland (SCH)

3.3. SULFATE DEPOSITION

Fig. 3 demonstrates the median monthly deposition of sulfate. The Westerland curve follows the precipitation amount, showing a maximum in September and October. In contrast, the Schauinsland maximum is reached in April/May; later on the deposition decreases rapidly. Langenbrügge and Deuselbach reach the maximum one month later than Schauinsland. While the Deuselbach curve drops continuously after that, a second maximum appears at Langenbrügge in August/September.

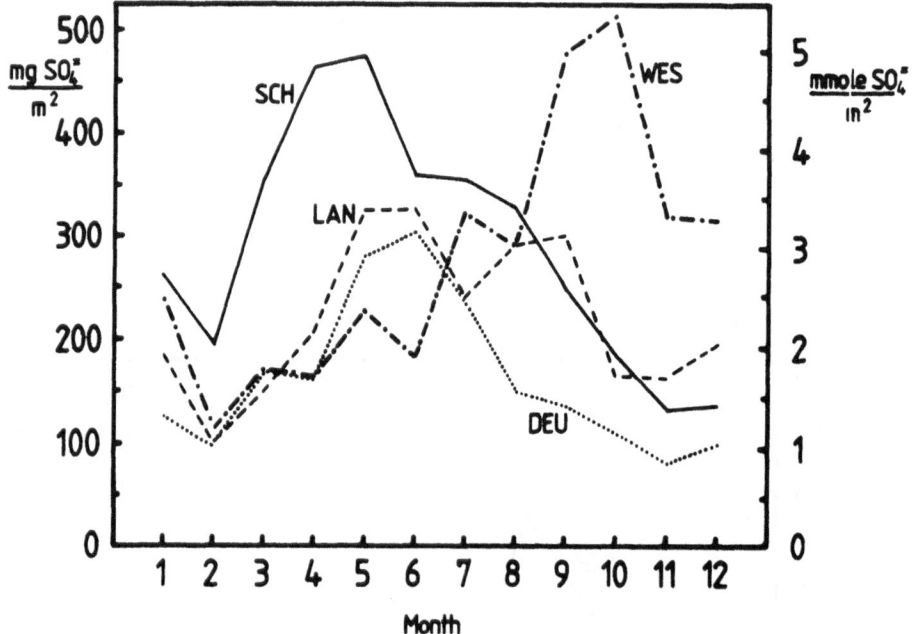

Fig. 3: Median monthly wet deposition of sulfate, 1982-1988, at Westerland (WES), Langenbrügge (LAN), Deuselbach (DEU), and Schauinsland (SCH)

3.4. NITRATE DEPOSITION

Except for Langenbrügge, nitrate deposition appears to reach its maximum about one month before sulfate (Fig. 4): at Westerland in August, at Deuselbach in May, at Schauinsland in April. From January to May nitrate deposition at Schauinsland exceeds the corresponding deposition rates at the other sites by a factor of two, while from June on it drops off to similar amounts. Nitrate deposition at Deuselbach is comparatively low from July to the end of the year, similar to sulfate and hydrogen ion deposition at that site.

3.5. AMMONIUM DEPOSITION

Fig. 5 shows the median monthly wet deposition of ammonium. At all stations except Westerland it is quite similar to nitrate: maximum values are reached at about the same time. During the first months Schauinsland receives much more ammonium than the other sites. This pattern changes at the end of the year when Schauinsland deposition drops to low Deuselbach values. At Westerland the picture is somewhat different: here, the highest deposition rate is reached in September when the rain amount peaks. From January to April ammonium deposition is almost the same at Westerland, Langenbrügge and Deuselbach.

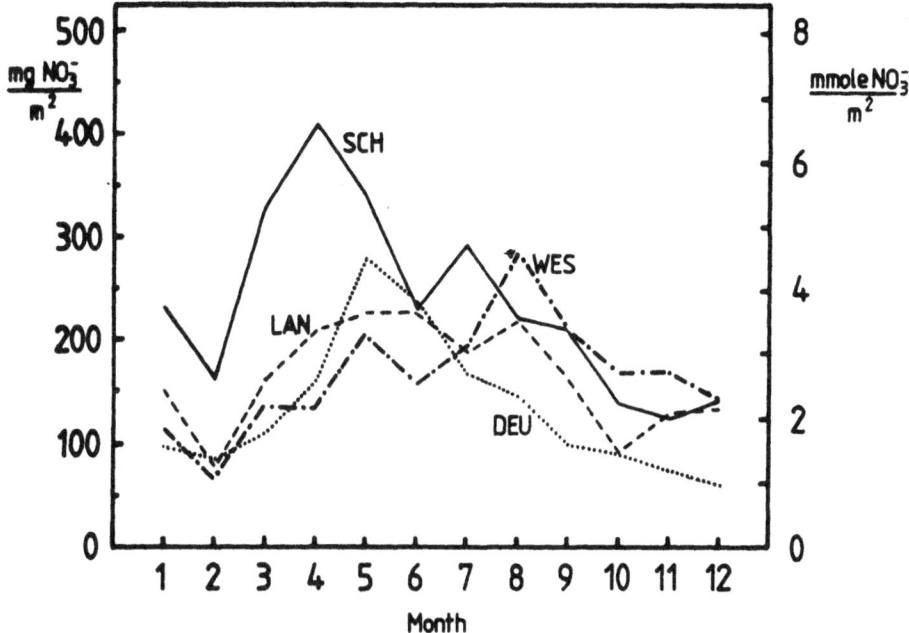

Fig. 4: Median monthly wet deposition of nitrate, 1982-1988, at Westerland (WES), Langenbrügge (LAN), Deuselbach (DEU), and Schauinsland (SCH)

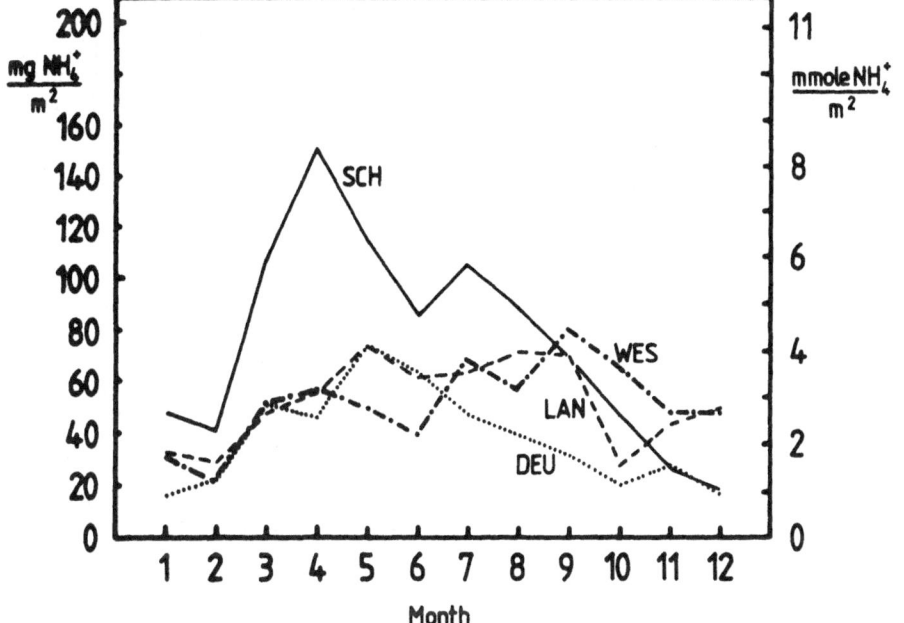

Fig. 5: Median monthly wet deposition of ammonium, 1982-1988, at Westerland (WES), Langenbrügge (LAN), Deuselbach (DEU), and Schauinsland (SCH)

4. DISCUSSION

In general, the highest deposition rates for hydrogen ions, sulfate, nitrate, and ammonium are observed in late spring, at Schauinsland mountain station about one month earlier. An exemption is the coastal station Westerland, where deposition rates are highest between July and October. This is partly due to the fact that the amount of precipitation is highest during those months. In addition, sulfate wet deposition is influenced by sea spray, especially in fall.

The observed sulfate and nitrate deposition rates can be compared with measurements obtained from a deposition network of 12 stations in the FRG, which had been established as a research project over a two year period from 1979 to 1981 (Georgii et al. 1982). The results from that network agree reasonably well with the data from Deuselbach and Langenbrügge. Differences may be due to the short period of two years, which can cause rainfall amounts not to be typical for the respective months. Ammonium ions enter cloud and rain drops as aerosol constituents or as gaseous ammonia. The predominant sources for ammonia in Central Europe are agricultural emissions: according to Buijsman et al. (1986), more than 500 kilotonnes of ammonia per year were emitted in West Germany, Belgium, and The Netherlands in the early eighties. Emission from fertilizers added another 50 kt/y in the same area. Application of manure and fertilizer usually occurs shortly before the growing season starts. More detailed analyses of the data will have to be made before the influence of these emissions on the annual variation of ammonium and other ions in precipitation can be quantified. At least the annual average concentrations of ammonium in rain at the different sites are in good agreement with results from a long-range transport model for ammonia and ammonium for Europe (Asman ans Janssen, 1987).

For the interpretation of annual variations in wet deposition of trace substances it is also important to know, to what extent the wind direction distribution during rain events at each site changes with season. This change can have a considerable effect on concentrations of trace substances in precipitation, because it relates different source regions to the receptor site. For this reason, further and more detailed investigations will have to be made in order to improve the understanding of the annual variations found.

5. REFERENCES

Asman, W.A.H. and Janssen, A.J. (1987) 'A long range transport model for ammonia and ammonium for Europe', Atmospheric Environment 21, 2099-2119.

Buijsman, E., Maas, H.F.M., and Asman, W.A.H. (1986) 'Anthropogenic NH_3 emissions in Europe', IMOU-report R-86-17, Institute for Meteorology and Oceanography, Utrecht, The Netherlands.

Georgii, H.-W., Perseke, C., and Rohbock, E. (1982) 'Feststellung der Deposition von sauren und langzeitwirksamen Luftverunreinigungen aus Belastungsgebieten', Final Report, Research Project 104 02 600, Umweltbundesamt, Berlin.

DEPOSITION BY RAIN AND SNOW - DRY DEPOSITION ON SNOW

S. GROSCH and H.-W. GEORGII
Institute of Meteorology and Geophysics
University of Frankfurt
Feldbergstraße 47
6000 Frankfurt/Main
Federal Republic of Germany

ABSTRACT. A comparative investigation of winter rain and snow is based on a sampling period from November 1986 until April 1987 at two stations at different altitudes. The average element concentrations were enhanced for snow samples. The element specific enrichment factors range from 1.1 to 4.5. They were lower for anions (1.1-1.8) than for trace metals (1.2-4.5). Generally snow/rain concentration ratios are larger at the station located 300 m higher. Higher element concentrations in snow than in rainfall samples are explained by more efficient below-cloud-scavenging of atmospheric constituents especially aerosol particles appearing for snowflakes. The total release of element masses accumulated within a snowpack during the winter season of February to April 1988 was investigated by taking consecutive snowmelt samples from a special snow deposition sampling device. Following the time distribution soluble components were characterized by increased concentrations at the beginning and decreased concentrations towards the end of the snowmelt season. Partly insoluble components showed the opposite behaviour. By calculating the total element input caused by snowmelt and by snowfall samples it was possible to approximate the amount of dry deposition on the snow surface throughout the sampling period.

1. Introduction

In deposition research usually wet and dry deposition is distinguished. Concerning the term "wet deposition" it is necessary furthermore to differentiate several types of hydrometeors like raindrops, hail, sleet, settling fogdroplets or snowflakes. Due to different characteristics of these precipitation elements the effectiveness of scavenging processes and finally the chemical composition of the resulting wet deposition is suspected to be different. The physical principals of wet deposition have been investigated in detail by Slinn et. al. (1978). Concerning below-cloud-scavenging of particles model calculations of Slinn et.al. (1978) call for a much higher collision efficiency for snowflakes compared to raindrops. Several authors report about increased ionic concentration in snow compared to rain (Bowersox and de Pena, 1980; Raynor and Hayes, 1983; Czurzwa et. al., 1988). Also organic components like PAH's and n-alkenes in snow may be enhanced by factors of 4-8 (Leuenberger et.al., 1988). Others like Summers (1977) or Reynolds (1983) observed the opposite. Topol (1986) again revealed higher NO_3^- but lower SO_4^- concentrations in snow compared to winter rain.

Snow furthermore is a most interesting and important kind of wet deposition because of its ability to accumulate element masses during the cold season. Additional dry

H.-W. Georgii (ed.), Mechanisms and Effects of Pollutant-Transfer into Forests, 51–59.

deposition will lead to an element enhancement although dry deposition on snow is known to be very limited (Whelpdale and Shaw, 1974; Dovland and Eliassen, 1976; Barrie and Walmsley, 1978; Granat and Johansson, 1983; Johansson and Granat, 1986). The sudden release of these element masses with the onset of thaw -most times along with an "acidic flush"- may cause severe impact on terrestial and especially on senible aquatic ecosystems (Gjessing et. al., 1976; Cadle et. al., 1984; Henriksen et. al., 1984). The sampling system described here allows to collect the natural snowmelt completely as well as to observe concentration changes with time.

2. Materials and methods

Both sampling sites are located in forest areas of the mountain range "Taunus" within a distance of about 20 km of the city of Frankfurt. They are only 3 km apart but differ in altitude by 300 m. Precipitation samples on a daily basis were taken since 1983 by means of an automatic wet and dry deposition sampling device. Using a high glass vessel as collector (Bergerhoff sampling system) samples of dry deposition at a time resolution of 14 days were accomplished. To study the deposition by accumulating snow a special sampling device called "snow-tunnel" was constructed and installed at the station "Kl. Feldberg". It consists of a plastic tunnel with two cylindric sampling compartments mounted perpendicual to both sides at one end of the tunnel opposite to the entrance (fig. 1). Each of these cylindric compartments holds one snow sampler consisting of a huge polyethylene bag at one side mounted to a flat teflon funnel and at the other side mounted to a short fixed cylindrical frill that defines the sampling area to be 0.16 m². The tunnel is buried into the ground so that the upper edge of the cylindric compartments ends at ground level. The advantage of this collector is that it guarantees a most realistic snow sampling right at the ground where the deposition naturally takes place. The sampling area is always adapted to changing conditions by lifting the frill with the flexible bag to the growing actual snow hight. The most important detail is that the conditions for the appearance of dry deposition on the sampling area is always comparable to the vicinity.

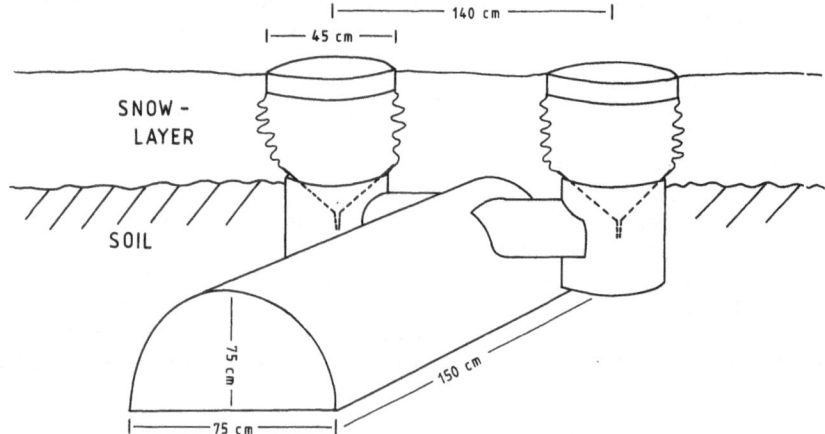

Fig. 1: Snow-deposition-sampler ("snow-tunnel")

Thereby artefacts concerning the amount of dry deposition caused by the sampling device itself should be eliminated. During thaw consecutive samples of meltwater can be received. Besides total conductivity and pH all samples have been analysed for SO_4^-, NO_3^-, Cl^-, NH_4^+, Na^+ and K^+ by IC as well as Fe, Mn, Pb and Cd by G-AAS. For detailed information on the chemical procedure refer to Georgii et. al. (1986). All results concerning SO_4^- and NO_3^- are calculated as S and N respectivly.

3. Results

3.1 THE CHEMICAL COMPOSITION OF RAIN COMPARED TO SNOW

Based on the results from precipitation sampling during the winter period of November 1986 to April 1987 a comparative investigation on the chemical composition of rain and snow was carried out. The precipitation events were seperated in two groups: those collected as rain and those collected as snow. To characterize in general the different kind of precipitation types volume-weighted mean concentrations of the chemical components were calculated. Following these data higher trace element concentrations in snow- than in rainfall samples were observed. As an example fig. 2 shows the results for the anions SO_4^-, NO_3^- and Cl^-. The concentration ratio snow/rain is generally > 1 although -especially at the station "Königstein"- the difference is not strong pronounced. Cl^- at this station even shows an opposite behaviour. Anyway the general trend is more evident for the heavy metals with snow/rain concentration ratios from 1.2 to 4.5 (fig. 3). Obviously this behaviour is more distinct at the station "Kl. Feldberg" at higher altitude.

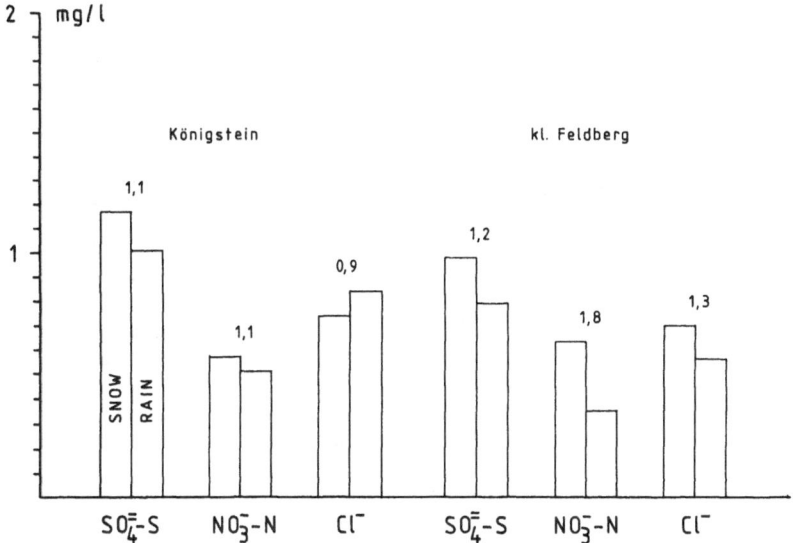

fig. 2: Mean concentrations of anions in snow- and rainfall-samples at two stations
    Figures above the bars: concentration ratio snow/rain
    (Period: Nov. 1986 - April 1987)

54

The enhanced frequency of high relative humidity at this mountain station may cause small particles to grow by adsorption of water molecules up to a size range where scavenging processes are more effective (Zinder et. al., 1988)

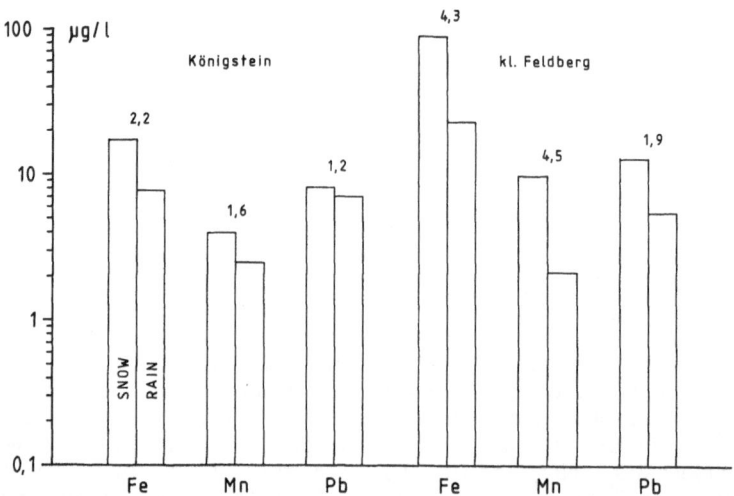

fig. 3: Mean concentrations of trace-metals in snow- and rainfall-samples at two stations
Figures above the bars: concentration ratio snow/rain
(Period: Nov. 1986 - April 1987)

3.2 RESULTS OF THE "SNOW-TUNNEL" EXPERIMENT

The experiment lasted from the beginning of February until the middle of April 1988. As a first result the total volume of precipitation collected with the wet/dry deposition sampler was compared to the total amount of meltwater. A rather small deficite of only 10 % was found for the wet/dry deposition sampler. Due to the response time of the automatically opening deposition sampler and due to principal sampling difficulties of snow even higher losses were expected. As a general view the average element concentrations (volume-weighted) for snowfall- and snowmelt samples are shown in fig. 4 and fig. 5. A distinct concentration increase for snowmelt samples can be realized for all compounds analysed except Cd. The element specific concentration enrichment ranges from 10 % (NH_4^+) to 70 % (Mn). Losses of Cd during the accumulation phase of the snowcover does not seem to be very likely. On the other hand the mean Cd concentration of precipitation appears to be somewhat enhanced compared to a longterm average concentration of < 1 µg/l usually found at this station. Therefore contamination of some precipitation samples might be responsible for the opposite behaviour of Cd.

The most detailed and comprehensive information is provided by plotting the time distribution of the concentration of individual samples both for precipitation and snowmelt along with the arising cumulative deposition. As an example this is demonstrated for the anion NO_3^- (fig. 6). The upper part represents the precipitation samples the lower part the snowmelt samples. The bars stand for the concentration of individual samples whereas the solid lines show the resulting cumulative deposition.

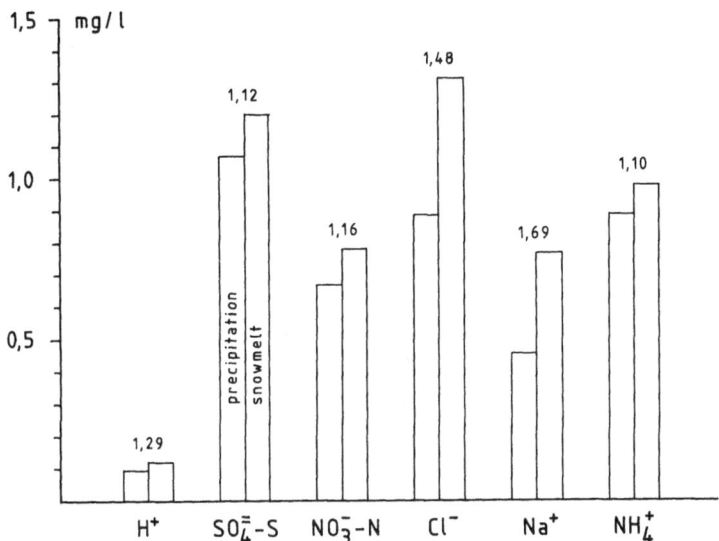

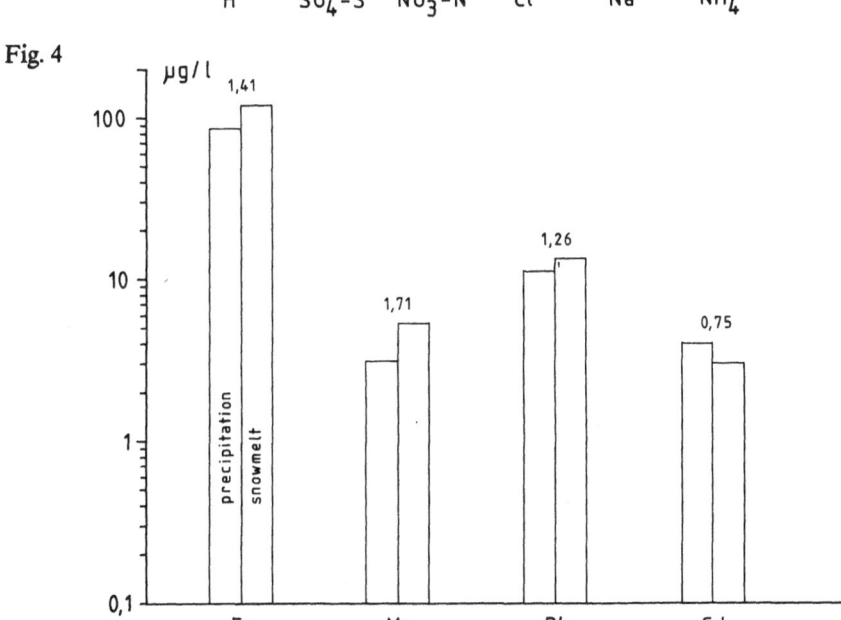

Fig. 4 and 5: Mean concentrations of anions/cations and trace metals in snowfall- and snowmelt-samples.
Figures above the bars: concentration ratio snowmelt/precipitation
(Station: "Kl.Feldberg"; Period: Feb.-April 1988)

56

Besides the demonstration of a rather large concentration variation for individual precipitation events two main results may be derived from this kind of presentation:
1. The cumulative deposition clearly shows a rather sudden release of accumulated trace substances with the onset of thaw.
2. The snowmelt samples of highly soluble compounds -like NO_3^-- bear highest concentrations generally during the initial phase of thaw while to the end the concentration decreases. An opposite behaviour can be realized for partly insoluble compounds. As an example fig. 7 shows the Fe concentrations tending to increase towards the end of the thaw periode.

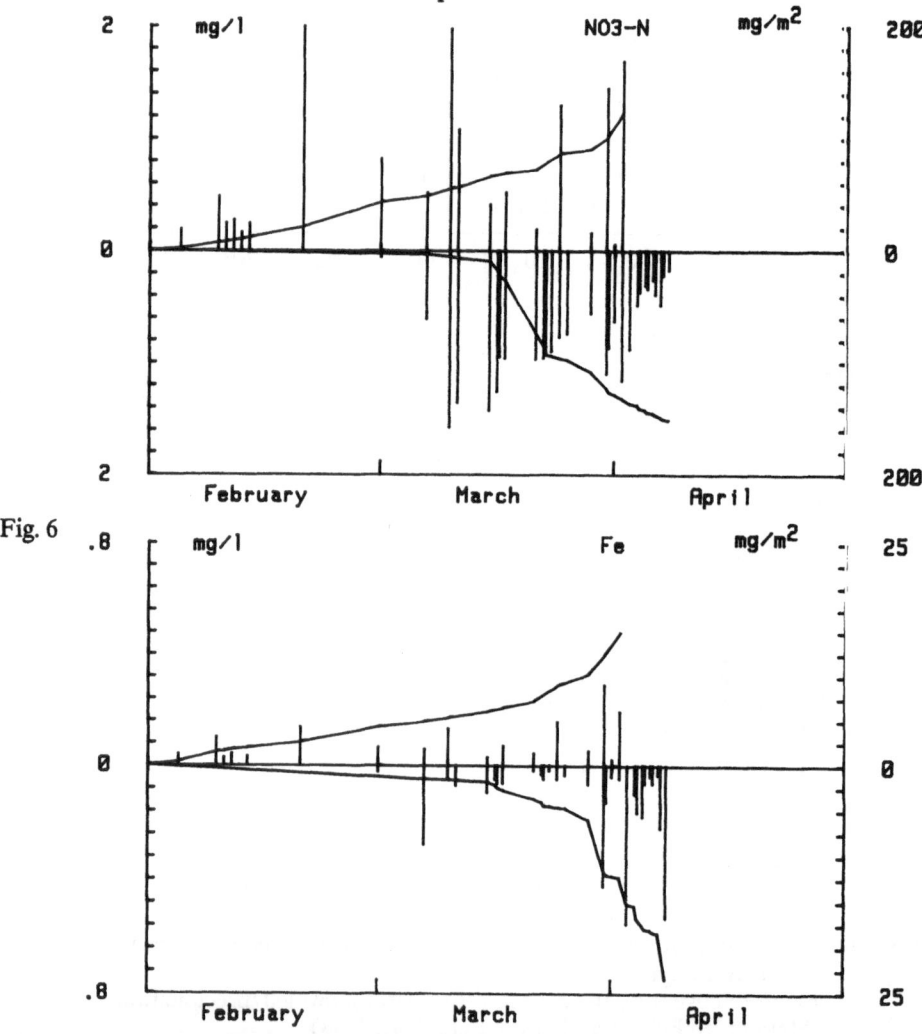

Fig. 6

Fig. 6 and 7: Concentration of individual snowfall- and snowmelt-samples (bars) as well as the temporal course of the cumulative deposition (solid line). The upper part of each figure represents the snowfall- the lower part the snowmelt events

From fig. 6 and 7 it is obvious that the cumulative deposition at the end of the research period is higher for snowmelt- than for precipitation samples. This must be caused by the additional amount of dry deposition occuring on the snow surface. By calculating the difference of deposition by snowmelt and precipitation it is possible to estimate respective dry deposition rates (tab. 1).

Although not directly comparable because of different sampling and estimating procedures but as another approximate figure mean dry deposition rates derived from the Bergerhoff sampling system (1987/1988) were also listed in tab. 1. The deposition rates for snow range from 0.009 mg/m² · d (Mn) to 1.16 mg/m² · d (SO₄⁻). Anyway these figures generally must be judged as rather small because of poor acceptor properties. Due to very stable thermodynamic conditions close to the snow-surface (~25 cm) the transport of gases and small particles towards the ground is restricted. The deposition rates for Na⁺ and Cl⁻ on snow are expected to be somewhat enhancend caused by the use of NaCl on the streets during the wintertime.

Tab. 1: dry deposition [1]: deposition by snowmelt minus deposition by snowfall
(Feb.-April 1988)
dry deposition [2]: mean dry deposition (1987/1988) derived from the Bergerhoff sampling system

component	dry deposition [1]		dry deposition [2]
	mg/m²	mg/m² · d	mg/m² · d
SO_4^*-S	60,18	0,94	1,16
NO_3^--N	36,36	0,57	0,62
Cl^-	94,79	1,48	0,58
Fe	9,83	0,154	0,193
Mn	0,562	0,009	0,009
Pb	0,698	0,011	0,009
Na^+	67,17	1,05	0,81
NH_4^+	39,01	0,69	0,36

4. Conclusions

Cloud hydrometeors during the wintertime are generated exclusively as ice-crystals regardless for the subsequent precipitation reaching the ground as rain or snow. Therefore condensation and in-cloud-scavenging processes are not expected to be different for rain and snow. For several reasons below-cloud-scavenging is believed to be more effective for snow than for rain.

1. A lower velocity of fall causes a longer residence time of snowflakes in the atmosphere and therefore a better chance for interactive processes between the precipitation elements and other atmoperic constituents like gases and particles.
2. The filigrane structure of snowflakes permits a better scavenging compared to raindrops.

3. An additional increase of residence time and distance of fall is caused by their specific kind of swinging motion as well as by the possibility of sideward and upward deflections under conditions of strong winds.

The reason mentioned last is believed to be responsible for the more pronounced concentration difference at the higher elevated station "Kl. Feldberg". Occasions of higher wind speeds are rather common at this station.

By means of the "snow-tunnel" it was possible to follow the temporal distribution of the deposition initialized by thaw. During the onset of thaw a redistribution and downward transport of soluble components within the snowlayer leads to higher concentrations in the beginning and decreasing concentrations towards the end of the snowmelt. Partly insoluble components seem to accumulate in the shrinking snowlayer leading to an opposite concentration time distribution.

5. Acknowledgements:

This investigation was supported by the German Federal Environmental Agency (*Umweltbundesamt*) under contract no. 10402635

6. References

Barrie, L. A., J. L. Walmsley (1978) 'A study of Sulfur Dioxide Deposition Velocities to Snow in northern Canada', Atm. Environm. Vol. 12, 2321-2332.

Bowersox, V.C., R. G. de Pena (1980), 'Analysis of Precipitation Chemistry at a central Pennsylvania Site', J. geophs. Res., Vol. 85, 5614-5620.

Cadle, S. H., J. M. Dasch, N. E. Grossnickle (1984) 'Retension and Release of Chemical Species by a northern Michigan Snowpack', Water Air Soil Pollut., Vol. 22, 303-319.

Czuczwa, J. C. Leuenberger, W. Giger (1988) 'Seasonal and Temporal Changes of Organic Compounds in Rain and Snow', Atm. Environm., Vol. 22, 907-916.

Dovland, H. A. Eliassen (1976) 'Dry Deposition on a Snow Surface', Atm. Environm., Vol. 10, 783-785.

Georgii, H.-W., S. Grosch, G. Schmitt (1986) 'Feststellung der Schadstoffbelastung von Waldgebieten in der Bundesrepublik Deutschland durch trockene und nasse Deposition', Abschlußbericht Teil A, Forschungsbericht 104 02 715 im Auftrag des Umweltbundesamtes, Eigenverlag des Universitätsinst. für Meteorologie und Geophysik, Frankfurt, 247 p.

Gjessing, E. T., M. Henriksen, M. Johannesen, R. Wright (1976) 'Effects of Acid Precipitation on Freshwater Chemistry', in: *Impact of acid precipitation on forest and freshwater ecosystems in Norway*, (ed.: Braekke, F.), Sur Nedbors Virkning Pa Skog og Fisk Res. Rep., 6/76, 64-85.

Granat, L., C. Johansson (1983) 'Dry Deposition of SO_2 and NO_x in Winter',
 Atm. Environm., Vol. 17, 191-192.

Henriksen, A., O. K. Skogheim, B. O. Rosseland (1984) 'Episodic Changes in pH and
 Aluminium Speciation Kill Fish in a Norwegian Salmon River',
 Vatten 40, 255-260

Leuenberger, C., J. Czuczwa, E. Heyerdahl, W.Giger (1988) 'Aliphatic and Polycyclic
 Aromatic Hydrocarbons in Urban Rain Snow and Fog',
 Atm. Environm., Vol. 22, 695-706.

Raynor, G. S., J. V. Haynes (1983) 'Differential Rain and Snow Scavenging Efficiency
 implied by Ionic Concentration Differences in Winter Precipitation'
 in: *Precipitation Scavenging, Dry Deposition and Resuspension*,
 Vol. 1 (coordinated by H. Pruppacher, R. Simonin and W. Slinn), 249-264.,
 Elsevier, New York

Reynold, B. (1983) 'The Chemical Composition at a rural Upland Site in Mid-Wales',
 Atm. Environm., Vol. 17, 1849-1851.

Slinn, W., L. Hasse, B. Hicks, A. Hogan, D. Lal, P. Liss, K. Munnich, G. Sehmel,
 O. Vittory (1978) 'Some Aspects of the Transfer of Atmospheric Trace
 Constituents past the Air-Sea Interface', Atm. Environm., Vol. 12, 2055-2087.

Summers, P. W. (1977) 'Note on SO_2 Scavenging in Relation to Precipitation Type',
 in: *Precipitation Scavenging* (1974); (ed: R. Semonin and R. Beadle),
 CONF-741 003, 88-94, Technical Information Center Energy Research and
 Developement Administration, Springfield, VA.

Topol, L. E. (1986) 'Differences in Ionic Composition and Behavior in Winter Rain and
 Snow', Atm. Environm., Vol. 20, 347-355.

Whelpdale, D., R. Shaw (1974) 'Sulphur Dioxide Removal by Turbulent Transfer over
 Grass, Snow and Water Surfaces', Tellus 26, 196-205.

Zinder, T. Schumann, A. Waldvogel (1988), 'Aerosol and Hydrometeor Concentrations
 and their Chemical Composition During Winter Precipitation Along a Mountain
 Slope'
 - II. Enhancement of Below-Cloud-Scavenging in a Stably Stratified Atmosphere'
 Atm. Environm., Vol. 22, 2741-2750.

DEPOSITION OF GASEOUS POLLUTANTS IN A DOUGLAS FIR FOREST:
FIRST RESULTS OF THE ACIFORN PROJECT

A. W. M. VERMETTEN, P. HOFSCHREUDER
Department of Air Pollution,
Agricultural University of Wageningen
P.O. Box 8129
6700 EV Wageningen
The Netherlands

ABSTRACT. In 1985 a research programme concerning the effects of air
pollution in forest areas has started in the Netherlands. In this
contribution the results of the monitoring of gaseous pollutants in
and above two stands of Douglas fir are discussed, together with some
technical problems which had to be solved. Concentrations of most
pollutants could be measured at five heights with an accuracy better
than 5% . Levels of SO_2, NO_2, NO and O_3 were found to be quite low and
comparable to the general levels in the Dutch countryside. NH_3 reached
up to hourly values of 60 $\mu g/m^3$ at an average of 5 $\mu g/m^3$ for the summer
of 1988. For O_3 some peaks up to 240 $\mu g/m^3$ were found during clear
summer days, exceeding critical levels for damage to sensitive plant
species.

1. INTRODUCTION

Within the framework of the Dutch Priority Programme on Acidification a
large part of the research is focused on the influence of air pollution, among other stress factors, on the growth and vitality of the
Douglas fir. Research groups work together on this subject in two
stands of Douglas fir in the 'ACIFORN'-project [1]. ACIFORN stands for:
ACIdification of FORests in the Netherlands.
    The start of the project was delayed considerably by the quest for
suitable research sites. The demands of the various disciplines (tree
species, vitality, age, soil structure, size, wind fetch) were quite
different and not easy to combine with the characteristics of the Dutch
forest. In general Dutch forest stands are small and surrounded by
stands of other tree species or even small agricultural areas. In the
late thirties a large part of the Dutch soil was disturbed by digging
in order to improve the soil structure. Finally two stands of Douglas
fir, meeting most of the requirements, were selected and in 1987 all
research groups started to work at the sites. The monitoring of air
pollution levels, meteorology, hydrology, soil physics and chemistry,
tree growth and physiology will proceed until the end of 1989. Some
experiments, e.g. the soil irrigation and fertilization plots, will be

61

H.-W. Georgii (ed.), Mechanisms and Effects of Pollutant-Transfer into Forests, 61–68.
© 1989 by Kluwer Academic Publishers.

continued over a longer period.

In 1985, at the previous Oberursel colloquium, we presented our plans for the air pollution measurements [2]. Most of our ideas about the experimental set-up still hold, but some have changed a great deal and a substantial amount of additional equipment is installed.

2. MATERIALS AND METHODS

2.1. Research sites

In spring 1986 two stands of Douglas fir (Pseudotsuga Menziesii) were selected by the ACIFORN research group. Both rather small stands are situated about 20 km to the west of the city of Apeldoorn, at the northern Veluwe, a forested area in the central part of the Nether-lands.

Speulderbos is a 28 years old stand, embedded in a large forested area, 2 km to the northwest of the village of Garderen, at an altitude of 50 m a.s.l., at the top of an ice-pushed ridge. Over the whole area (2.5 ha) the canopy is well closed and reaches up to an average height of 18 - 20 m; most of the needle mass is concentrated between 10 and 15 m. The one-sided LAI is calculated by structure analysis and needle counting and amounts to about 6.5 [5]. The soil surface is gently undulating; the soil profile can be classified as an orthic podzol/ luvisol [6]. It is rather heterogeneous on a small scale and composited of mainly coarse sand, mixed with layers of loam and clay. The drainage is very good, with the watertable at a depth of approximately 40 m. The mean annual precipitation is 800 mm, which holds for the other site too. Air pollution levels can be characterized as low to moderate, with incidentally high levels for nitrogen oxides and ammonia, due to con-tributions from nearby sources. The distance to the nearest highway and an area with intensive farming is about 6 km.

Kootwijk is a smaller (1.2 ha) and older stand (36 years), si-tuated 1 km to the west of the small village of Kootwijk and 12 km southeast of Garderen, at an altitude of 25 m a.s.l., in a rather inhomogeneous forest area. The height of the stand is 16 - 18 m at a LAI of 4 [5]. The soil is flat and mainly consists of coversand. The soil profile is a leptic podzol and is very well drained, with the watertable at a depth of 5 m [6]. Air pollution levels should be compa-rable to the other site, though sources of ammonia and nitroxides are somewhat nearer.

2.2. Air pollution monitoring.

Concentrations of SO_2, NO_x, O_3 and CO_2 are measured continuously at both sites, at heights of 5, 10, 15, 20 and 30 m. A large pump pulls the air down through Teflon tubing and filter inlets to the analysers (table 1) in a small housing at the foot of the 30 m tower. Every 5.5 minutes the analysers are switched to another level by Teflon valves. To avoid wall reactions the flow is kept high (residence time: 3 to 5 seconds) and the tubing is slightly heated. The $5 \mu m$ pore size filters are changed at least once a week to avoid gas-particle interactions and clogging of the filters. The price we have to pay for these precautions is a pressure drop of 20 - 50 hPa at the end of the tubing, and consequently a signal loss of 2 - 5% for the analysers. In laboratory experiments, which are repeated twice a year, these signal reductions are determined for each analyser up to 160 hPa. At the field sites the pressure drop is recorded together with the concentration data, so a correction can be applied during signal processing. The CO_2 analyser turned out to be extremely sensitive to pressure differences, due to a weak internal pump and leakage. Addition of an extra pump at the inlet solved this problem, as no losses could be detected.

Losses due to wall reactions or other processes are evaluated during field and laboratory experiments with the help of calibration gases. No losses above 2% could be detected, even with air passing through dirty filters and 80 m of Teflon (FEP, diameter 1 cm) tubing. The experiments will be repeated once or twice; NH_3 and O_3 still have to be tested under field conditions, since calibration gases were not available yet.

TABLE 1. Gas measuring techniques

Component	Analyser	Principle and remarks
SO_2	Thermo Electron 43W,43A	Pulsed fluorescence
NO,NO_2	Monitor Labs 8840	Chemoluminescence
O_3	Bendix 8002	Chemoluminescence with ethylene
NH_3	Monitor Labs 8840 + stainless steel converter	Chemoluminescence after conversion to NO_x
CO_2	ADC 225 Mk 3	Infraredabsorption

The measuring technique we intended to use for ammonia, preconcentration on tungsten oxide followed by thermal desorption, turned out to be to unsuitable for continuous measurements at five levels. Due to absorption in Teflon valves and tubing, necessary for the switching between levels, 20 - 80 % of the ammonia was lost before it got to the preconcentration tubes. We still believe that a single preconcentration tube, without any entrance tubing, will work in a proper way. However, for the ammonia measurements in the project an other approach was chosen: ammonia is converted to NOx by a catalist reaction on stainless steel at 800°C and the difference with the NOx signal is calculated. The technique cannot be used at high NOx levels and is probably not totally selective: amines are converted to NOx as well.

NH_3 is measured at one level (30 m), as the stainless steel converter is not fast enough to follow the switches in height and the damping of the signal may lead to false interpretation. From a comparison with denuder measurements during the summer of 1988 we have got the impression that with our method long term averages and the daily course of the NH_3 concentration are reproduced well. As we already

expected, the system did not respond to fast changes in time. A special
NH_3 comparison experiment is planned for the summer of 1989. For the
1989 measurements we are still looking for a monitoring technique able
to measure ammonia concentrations with an accuracy of 5%, which is
needed to establish gradients.

In addition to the continuous measurements we took some 4-hourly
samples with denuder-filterpacks to determine concentration levels of
NH_3, HNO_3, HCl, NH_4^+, NO_3^-, SO_4^{--} and Cl^-.

2.3. Meteorological data

Wind speed and direction are measured at intervals of 5 m up to 30 m;
temperature and relative humidity at the same heights, except for 25 m.
Global radiation is measured at 30 m, net radiation at 25 m. Most
instruments have a sampling period of 10 seconds; 15 minute averages,
minimum and maximum values are calculated by the datalogger and stored
on a central computer. The distribution of wind direction is stored as
a frequency table in 64 sectors. Up to now most instruments have func-
tioned properly, except for the capacitive humidity sensors (Rotronic),
which broke down regularly. At the Kootwijk site some gaps in the data
occur due to malfunctioning of the datalogger.

For the wind- and temperature sensors an individual calibration is
available; for most purposes default (average) values can be taken. If
the gradients of windspeed and temperature have to be used for deposition estimates, an accuracy of less then 0.1°C is demanded and a correction for the temperature sensors has to be applied.

During a few months each year fast flux measurements of momentum,
sensible and latent heat are carried out by KNMI and TNO, to get a
better view on transport phenomena above the forest [3]. Another new
project is the continuous monitoring of canopy wetness by a radar
scanning technique [4]. With the results of these projects and the
gradient measurements of SO_2 and O_3, we will try to relate canopy
parameters as e.g. leaf wetness to the deposition process.

2.4. Data storage and handling

After conversion to physical units all data are stored on a central
computer, a micro PDP11. At the field sites inspection of the data is
possible by means of a snapshot program. Every one or two weeks data
are transferred by tape to Wageningen for further processing on a VAX
mainframe computer. In a database environment the data are screened and
prepared for graphic presentation and further evaluation. As the time
available for interpretation is limited when the experiment is still
running, a considerable part of the data is still in its raw shape.

3. RESULTS

3.1. Air pollution monitoring

At this moment the concentration data for one of the sites, Speulderbos have been evaluated for the period April 1988 until October 1988. In table 2 and 3 monthly averages and maximum values at the height of 30 m are presented for this period. To show the small size of the gradients above the canopy, the average profiles for the month May 1988 are presented in table 4. Ozone shows the largest gradients, as we would expect for a reactive gas. For SO_2 we have found particularly large gradients in June and October 1988, both months with a lot of rain. This phenomenon could be related to canopy wetness. Further statistical analysis of the data has to confirm this hypothesis.

TABLE 2. Monthly averaged concentrations at 30 m height in ppbv. Speulderbos, 1988. Ox(idant) = NO_2 + O_3.

	NO_x	NO	NO_2	SO_2	O_3	NH_3	O_x
April	12.7	1.1	11.5	8.0	36.3	11.4	47.9
May	9.1	1.4	7.7	7.9	45.2	8.3	53.3
June	5.5	0.7	4.8	3.3	35.3	7.5	39.9
July	9.9	1.5	8.4	4.8	25.0	7.5	33.4
August	12.8	1.8	11.0	4.7	28.8	10.1	39.8
September	14.9	2.4	12.5	4.9	24.7	7.7	37.1
October	25.9	9.0	16.9	7.5	13.2	4.4	30.1

TABLE 3. Monthly maximum concentration values at 30 m in ppbv. Speulderbos, 1988.

	NO_x	NO	NO_2	SO_2	O_3	NH_3	O_x
April	55.4	27.8	47.9	30.3	88.3	70.2	99.3
May	48.3	14.5	37.1	30.1	98.3	69.7	111.3
June	37.8	13.0	36.3	23.6	89.6	26.4	94.8
July	44.5	26.1	26.5	22.5	73.9	46.9	75.2
August	54.1	21.0	40.0	20.2	122.1	60.4	131.4
September	87.5	50.2	45.3	21.2	74.8	92.8	92.5
October	117.0	88.4	50.0	34.0	52.0	28.0	- - -

The daily courses of NOx and O3 are, if presented as a monthly average, quite characteristic and easy to explain. O3 shows a maximum in the late afternoon, following the radiation and temperature course. The concentrations of NOx and NO are determined by traffic emissions and the reaction with O_3, so a maximum occurs in the rush hours. Of course, boundary layer mixing is an important process for all species, as is illustrated by the daily pattern for SO_2, which often displays a peak at the end of the morning due to downmixing [7]. Even the monthly averaged daily course shows this feature. The concentration levels of NH_3 are often determined by the spreading of manure on arable land at a distance of more than 1.5 km of our measuring sites and show no distinct daily course. In an earlier stage of the project we published graphic data for some individual days in 1987 and 1988 [7].

TABLE 4. Concentration profiles Speulderbos May 1988 (ppbv)

H(m)	5	10	15	20	30
SO_2	7.0	7.3	7.7	7.8	7.9
NH_{3*}	8.3	7.7	7.6	8.0	8.3
O_3	39.5	41.3	43.6	44.8	45.2
NO_x	9.0	8.9	9.1	9.1	9.1
NO_2	8.0	7.9	7.9	7.8	7.7
NO	1.0	1.0	1.2	1.3	1.4
O_x	47.8	49.5	51.8	52.9	53.3

* NH_3 gradients may be smoothed by the slow response of the measuring system.

3.2. Campaigns

The concentrations of some other important gases were determined by taking 4-hour samples with denuder-filterpacks. In table 5 the results for the 30 m level are shown. Though the concentrations of acid gases are low, they may be an important contribution to the acid deposition, because the vegetation might act as a perfect sink for these reactive gases.

TABLE 5. Concentrations of other gases, from denuder-filter pack samples at an height of 30 m, in $\mu g/m^3$. Speulderbos, Nov'87 - Nov'88. Averages of ± 20 4-hour periods.

	Average	Minimum	Maximum
NH_3	5.2	0.9	12.9
HNO_3	1.0	0.0	9.3
HCl	1.1	0.2	2.7

3.3. Deposition estimates

An analysis of the gradient data for SO_2 and O_3 for a couple of days, using the diabatic correction functions of Dyer and Hicks [8], showed deposition velocities of 0 - 2 cm/sec, values consistent with those cited in literature. These calculations will be repeated as soon as we have more insight in the validity of the gradient approach over our particular forest. We expect the flux-profile relationships to be disturbed by transitions in surface roughness and forest height at short distances of our field sites. Another important issue is the representativity of these point measurements for the stand or even the surrounding forest area.

Another way of estimating dry deposition is taking the difference between the yearly throughfall (16 samplers below the canopy) and rain-fall deposition (wet-only sampler). In table 6 the througfall and rainfall data are given for the Speulderbos site. The throughfall deposition of NH_4 and SO_4 is low for this site, when compared to data from other Dutch forests; NH_4 amounts can be up to 10.000 moles/ha/a in certain areas. If we assume that the only change in concentration for SO_4 and NH_4 is caused by the dry deposition of SO_2 and NH_3 (no aerosol deposition) and that the uptake by the stomata is much lower, we can calculate a yearly average deposition velocity for these gases from the Speulderbos data. For NH_3 the resulting deposition velocity is 1.2 cm/s and for SO_2 0.8 cm/sec. Bringing back to mind the assumptions we have made during the calculation, we estimate the uncertainty in these figures to be about 50 % .

TABLE 6. Deposition in throughfall and rainfall (mol_c/a,ha) Speulderbos, Nov'87 - Nov'88.

	Rainfall	Throughfall	Difference
SO_4^{--}	876	2136	1260
NH_4^{+}	979	2102	1123

4. DISCUSSION

Considering the results of the pollutant concentration monitoring, we may state that with our set-up we are able to determine the concentra-tions of all relevant pollutants within 5 % . Though this is precise enough for the description of the 'pollution climate' at our sites, we have to improve upon these results to allow us to make deposition estimates using micrometeorological methods. Gradients are typically in the range of 1 - 10 %, and often less, e.g. for NO_x. The switching between the measuring heights frees us from zero shifts and other systematic errors related to the use of more then one monitor, but we have to be careful when concentrations are changing fast in time: there is a delay of 5.5 minutes between the two upper levels. For the SO_2 gradients the problem is solved: another monitor will be measuring the trend at the 30 m level, with the first one scanning the profile.

For NH_3 gradients we still do not have a reliable method, except for wet chemical methods as e.g. denuders. For dry deposition estimates we will have to rely on the concept of the flux as the product of an average concentration times a deposition velocity.

Another problem is the representatitivity of our point measure-ments. From the additional flux measurements by TNO/KNMI we will get an idea about the variation in evaporation, heat and momemtum exchange, but extrapolation to a larger area or even the Dutch forests in general will be quite difficult. Our hope is to establish a link between micro-meteorological and ecological (i.e. with throughfall samplers) deposi-tion estimates. The latter are more often applied and can give us some information about the spatial variation in the deposition of pollutants.

A final remark which has to be made is that a thorough interpretation of the data will take approximately the same amount of time as the period they cover. We will try however to get most of the data available within one year from now.

5. ACKNOWLEDGEMENT

We are much obliged to ECN (Petten) for the preparation and analysis of the denuder-filterpacks, and to Rien van der Maas (Agricultural University of Wageningen) for supplying the throughfall and rainfall data.

6. REFERENCES

[1] Evers P. et al. (1988) 'The ACIFORN-project' In: Mathy P.,(ed.) 'Air Pollution and Ecosystems.' Proceedings COST 612 symposium, Grenoble, 18-22 mei 1987, 887-909. Reidel, Dordrecht.

[2] Vermetten A.W.M., Hofschreuder P., Harssema H. (1986) 'Dry deposition of gaseous pollutants in a Douglas fir forest.' In: H.W. Georgii (ed.), Atmospheric pollutants in forest areas, 1-11. Reidel Dordrecht.

[3] Duyzer J.H., Meyer G.M., Aalst R.M. van, (1986)(in Dutch) 'Metingen van droge depositie van luchtverontreiniging' TNO-Rapport R86/290.

[4] Bouten W., Schaap M., Hakkaart P. (1989) 'Monitoring and modelling rainfall interception and canopy wetness.' Contribution COST-612 Workshop on Monitoring Air Pollution and Forest Ecosystem Research, Bilthoven, February 20-21.

[5] Evers P.W.(ed.)(1988) (in Dutch) 'Koppeling van ecofysiologische parameters van luchtverontreinigingsinvloeden aan biometrie in de ACIFORN Douglasopstanden.' Rijksinstituut voor onderzoek in de bos- en landschapsbouw "de Dorschkamp", Wageningen. Rapport 512.

[6] Tiktak A., Konsten C.J.M., Maas R.v.d., Bouten W. (1988) 'Soil chemistry and physics of two Douglas-fir stands affected by acid atmospheric deposition on the Veluwe, the Netherlands.', Dutch Priority Programme on Acidification 03-01. Amsterdam/Wageningen.

[7] Vermetten A.W.M. (1988)(in Dutch) 'Luchtverontreiniging in bossen. Additioneel Programma Verzuringsonderzoek, 1e fase 1985-1987 Eindverslag project 14.', Vakgroep Luchthygiëne en -verontreiniging, Landbouwuniversiteit Wageningen, rapport R-306.

[8] Dyer A.J., Hicks B.B. (1970) 'Flux-gradient relationships in the constant flux layer.' Quart. J. Roy. Met. Soc. 96, 715-721

EXPERIMENTAL ESTIMATION OF THE SO₂-DEPOSITION ON AN ARTIFICIAL SURFACE

Brigitte FÄHNRICH, Hans-Walter GEORGII
Institute of Meteorology and Geophysics
University of Frankfurt/M, Feldbergstr. 47
6000 Frankfurt/Main
Federal Republic of Germany

ABSTRACT. A simple method for direct measurement of SO_2 fluxes to an artificial surface has been developed. By employing a filter sampling technique for deposition measurements a chemiluminescence method is applied for the detection of gaseous SO_2. The quality of filter sampling is tested by investigations in a calibration chamber. Some measurements of SO_2 concentration- and deposition profiles within forest areas are presented in order to demonstrate the applicability of the developed filter sampling technique. Deposition velocities in the range of 0.11-0.86 cm/s have been derived, whereby the deposition velocity turned out not to be constant with increasing height but to depend on the effectiveness of turbulent exchange of air masses.

1. Introduction

Dry deposition is defined as transport of pollutants from the atmospheric-boundary-layer to any surface. The dry deposition of trace constituents has to be considered as a very important cleaning mechanism for the atmosphere. Close to industrial sources the contribution of dry deposition to the total SO_2 deposition rate is estimated to be about 80 percent (Kuttler 1982).

1.1. DEFINITON OF THE DEPOSITION VELOCITY

The pollutant flux generally is described as proportional to the referring concentration gradient. Assuming this gradient to be directed vertically and the concentration of a pollutant to disappear at any surface (Battelle Inst. 1982):

$$F \sim c(z) - c(z_o) \text{ and } c(z_o) = 0$$

leads to the simple relationship:

$$F \sim c(z) \text{ or } F = v_d \cdot c(z)$$

$c(z)$: concentration of the pollutant in the height z
F : pollutant flux (mass/unit area · time)
v_d : constant of proportionality, defined as deposition velocity (cm/s)

69

H.-W. Georgii (ed.), Mechanisms and Effects of Pollutant-Transfer into Forests, 69–76.
© *1989 by Kluwer Academic Publishers.*

The deposition velocity on one hand depends on the effectiveness of the pollutant transport determined by the stratification of the atmosphere and on the other hand on the physical and chemical characteristics of the surface considered. Also it is influenced by the wind speed.

The pollutant flux is counteracted by various transport resistances arranged in series (fig. 1).

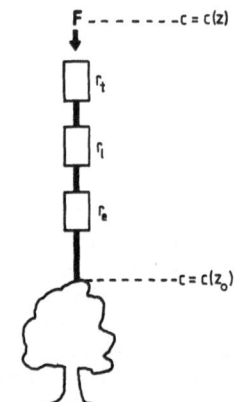

Fig. 1. Scheme of the transport resistances (Roth 1975).
r_t: turbulent transport resistance
r_i: diffusive transport resistance
r_s: surface resistance, estimated by the special properties of the surface (e.g. pH, stomata conditions)

$$r = r_t + r_i + r_s$$

The sum of r_t and r_i is defined as atmospheric resistance r .

The following equation describes the relation between deposition velocity and transport resistances:

$$v_d(z) = 1/r = 1/r_a + r_s = F/c(z)$$

2. Measuring Method

The measuring method for SO_2 applied in this investigation has been developed by West and Gaeke (1956). It is based on the adsorption of the gaseous SO_2 in an 0.1 M sodium-tetrachloromercurate-solution (TCM: $Na_2[HgCl_4]$) and its fixation in a stable, nonvolatile disulfitomercurate-complex. West and Gaeke determined TCM to be a 100-percent sink for SO_2.

For SO_2 adsorption an appropriate filter material (Delbag Microsorban 98, 47 mm dia.) was chosen as an artificial surface. The TCM-impregnated filters are kept in selfmade filter holders - which allow deposition fluxes to both sides of the filters - and exposed for a certain period of time. Impregnation and analysis of the samples are performed by the Chemiluminescence Technique developed by Stauff and Jaeschke (1978).

2.1. ANALYTICAL METHOD

The Chemiluminescence Technique is shortly summarized in the following section:

The disulfitomercurate complex, which is formed during the SO_2 sampling, is generally stable against oxidation under neutral pH conditions. However, when treating this complex with an acid potassium permanganate solution (pH 2.5) dissociation occurs, whereby a chemiluminescence indicates the oxidation of the free bisulfite ion to sulfate.

The light yield of this chemiluminescence is proportional to the complexed SO_2 present in the sample.

For calibration of the chemiluminescence effect a sulfite standard solution is used which is diluted with TCM to provide a sufficient stability of the sulfite. A set of standards in the range of about 2-80 ng complexed SO_2/ml is achieved by further dilution of the primary standard solution with 0.1 M TCM. Prior to any measurements the chemiluminescence apparatus should have been calibrated (fig. 2).

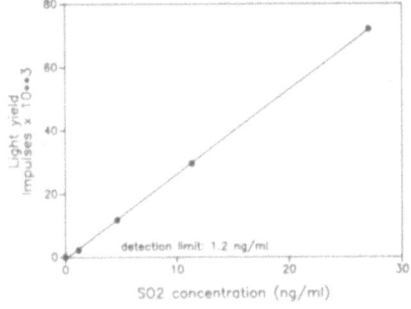

Fig. 2. Typical calibration curve for the chemiluminescence technique. Referring to the obtained detection limit an air concentration of 240 ng SO_2/m³ respectively an SO_2 deposition rate of 24 ng/m² · s is detectable.

2.2. FILTER SAMPLING

For transportation to a measuring site the filter holders are inserted in an air-tight box to protect the impregnated filters against contamination with surrounding air. Fig. 3 shows exposed filter holders.

Fig. 3. Exposed filter holders. The measurements preferably take place in forest regions in different heights at specially equiped measuring towers . Three filters will be exposed at each level.

3. Laboratory Experiments

In advance of field measurements the sampling method was tested at definite conditions by means of a calibration chamber (fig. 4).

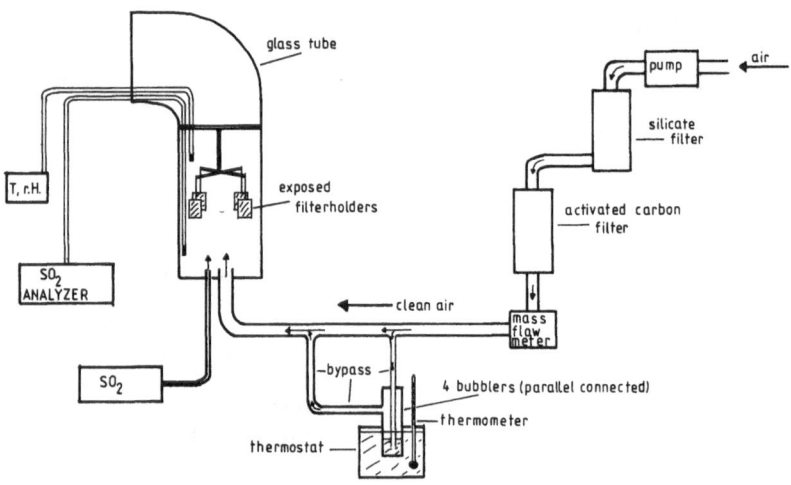

Fig. 4. Scheme of the calibration chamber.

The SO$_2$-free airstream is directed towards a glass tube (20 cm dia.) where it is homogeneously mixed with controlled rates of pure SO$_2$ (99.975 vol.%). The mass flow rate amounts to about 100 l/min. By using a flexible tension ring the filter holders including the impregnated filters are exposed in the laminary SO$_2$ flow inside of the glass tube. For individual regulation of the relative humidity the clean air stream can be directed partly through four thermostated bubblers. SO$_2$ concentration, temperature and relative humidity within the glass tube are registerated continuously.

The results of these investigations should give informations on:
-reproducibility of the sampling method
-the influence of temperature and relative humidity
-the suitable exposition period during field measurements

By simultaneous exposition of four filters under the same conditions (c=37ppb, T=24.7°C, r.H.=51%, t=50min) the reproducibility of the measurements has been examined (Table 1).

Table 1. Reproducibility.

Filter	Deposition [$10^{-2}\mu g/m^2 \cdot s$]
1	16.2
2	18.3
3	17.8
4	17.2

An average deposition rate of about $17.4 \cdot 10^{-2}\mu g/m^2 \cdot s$ with a relative standard deviation of $0.9 \cdot 10^{-2}\mu g/m^2 \cdot s$ (= 5.2%) has been obtained. Two additional measurements performed under the same conditions at two different days showed a maximum standard deviation of 18 percent.

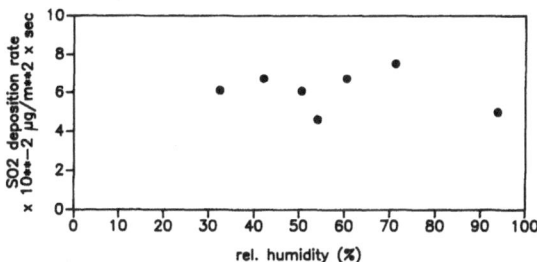

Fig. 5. Influence of the relative humidity on the deposition rate.

The relative humidity has been variied in a range of 32-94% (T=22.7°C, t=10-135min), (fig. 5) .The results have been normalized to an SO_2 concentration of 20 ppb to exclude deviations due to different concentration values during single measurements. According to these results the sampling method does not depend on the relative humidity.

No temperature dependence within the range of 19-27°C could be observed either. Due to technical reasons it was not possible to exceed this investigation to a lower temperature range.

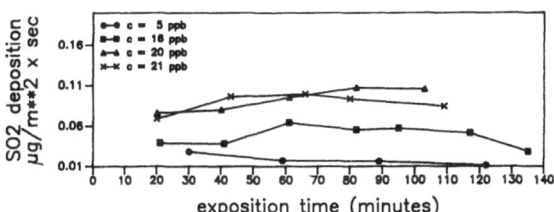

Fig 6. The SO_2 deposition rates corresponding to four different SO_2 concentrations (5-21 ppb) plotted as a function of the exposition time.

As expected an increasing deposition rate is found with increasing SO_2 concentration. In the initial phase of the exposition period up to 40 minutes the curves show a less stable behaviour. Furtheron up to 100 minutes the dependence remains rather constant, showing the filter capacity to be nearly stable. After 100 minutes exposition time a slight decrease of the curves is realized, possibly caused by drying out of the filters (fig. 6). Due to these investigations an exposition period of 50-90 minutes was found to be suitable with 70 minutes preferably.

Summarizing these results the presented filter sampling method turned out to be useful for field measurements.

4. Results

In fig. 7,8 and 9 some results of field measurements at measuring towers in forest stands are presented. The individual measurements were performed at several different levels. The dotted lines represent the upper boundary of the forest stands. The corresponding deposition velocities at each level are given in brackets. Every figure is divided in two parts, the left one for the concentration-, the right one for the deposition profile.

During the deposition measurements the actual SO_2 concentration was taken using an automated air sampling system in the respective levels.

Measuring tower I is located 30 km NW of Frankfurt/Main within a spruce stand (maintainer: Hessische Landesanstalt für Umwelt; height: 30 m). Stable atmospheric conditions involving high transport resistances were dominant at both measuring times. Deposition rate and deposition velocity on 15.12.1987 were nearly constant with increasing height especially within the stand. On 4.7.1988 a slight decrease of deposition rate and deposition velocity with height has been realized, probably caused by a lower transport resistance in the soil surface area (fig. 7).

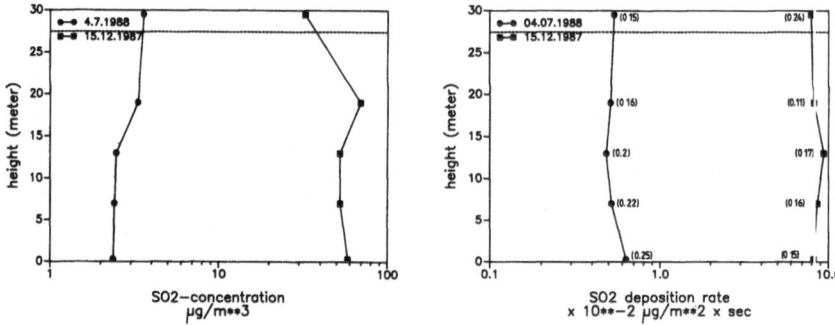

Fig. 7. Results of Measuring Tower I.

Measuring tower II is located within in a beech stand 5 km E of the city of Göttingen (maintainer: University of Göttingen; height: 42 m). At both times extremely low SO_2 concentrations (< 350 pptv) have been observed. The concentration profiles decreased strongly towards the soil surface. According to that, only in the highest measuring level deposition rates could have been estimated. Above the stand strong turbulent exchange of air masses took place with increasing wind speed which affected an increasing deposition velocity on 24.6.1988 (fig. 8).

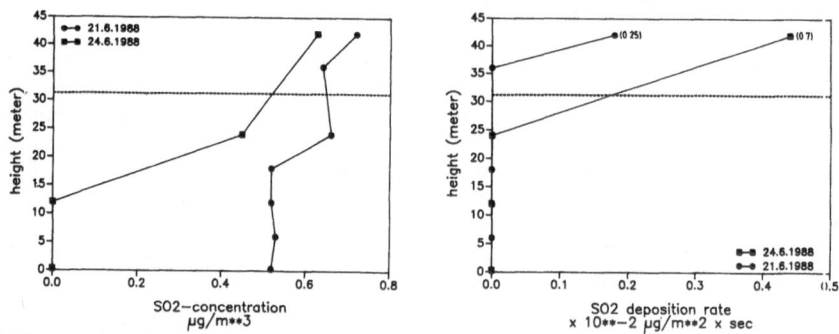

Fig. 8. Results of Measuring Tower II.

Measuring Tower III is located in the Nationalpark Bayerischer Wald also within a spruce stand (maintainer: University of Munich; height: 52 m). On 2.11.1988 concentration- and deposition rate were increasing with height. Due to turbulent exchange above the stand higher deposition velocities have been estimated. Rather stable atmospheric conditions were dominant on 3.11.1988. The prevailing winds affected an air mass transport from northerly directions (CSSR). The deposition profile has been in correspondence with the concentration profile. According to that the deposition velocity was rather constant with increasing height (fig. 9).

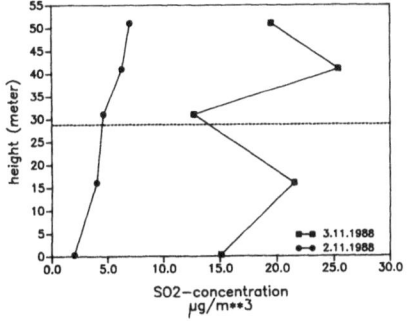 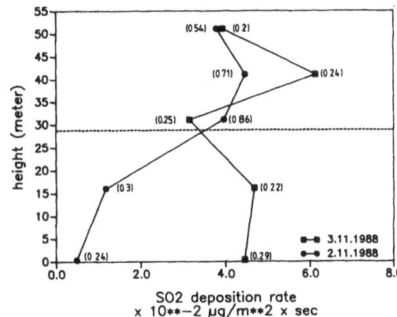

Fig. 9. Results from Measuring Tower III.

4.1. COMPARISON WITH OTHER METHODS

There are existing several methods for determination of SO_2 fluxes. The advantage of filter sampling is based on the direct measurement of the pollutant flux. Based on the assumptions described in Chapter 1.1. only one measuring height is necessary. The measurement itself is independent of meteorological data in contrast to micrometeorological techniques (e.g. Gradient Method, Eddy Correlation).

However, a comparison with natural surfaces is not easily possible. Nevertheless this technique allows to compare results obtained at different sites and under different meteorological conditions. In Table 2 some deposition velocities - estimated with different measuring methods - are listed.

Table 2. Comparison of the measured deposition velocities with results of other methods (Meyers and Baldocchi 1988).

Surface	Method	v_d (cm/s)	Author	Year
water	Gradient	0.41	Garland	1977
calcareous -soil	Gradient	1.2	Garland	1977
grass	$^{35}SO_2$-Tracer	0.8	Owers	1974
wheat	Gradient	0.74	Fowler	1976
pine	$^{35}SO_2$-Tracer	0.1-0.6	Garland	1978
oak-hickory	Eddy-Correlation	0.21-1.14	Meyers	1986
spruce	Filter sampling	0.11-0.86	Fähnrich	
beech	Filter sampling	0.25-0.7	Fähnrich	

5. Conclusion

According to the laboratory investigations, the developed filter sampling technique for direct SO$_2$ flux measurements shows the following advantages:
  -reproducibility > 80 percent
  -low detection limit
  -no significant influence of temperature and relative humidity
   on the sampling method
  -short exposition periods
The method has been successfully established for field measurements within forest areas. The obtained deposition velocities for SO$_2$ are comparable with results of other techniques. The concentration- and deposition profiles into the forest stands - especially in the crown regions - show a rather unsteady behaviour indicating a possible turbulent exchange of air masses. According to the previous results not only the SO$_2$ deposition rate itself but also the deposition velocity cannot be assumed to be constant with height. In addition there is no indication for anticipating an increase of concentration and deposition rate with height. Summarizing the results it can be verified that deposition rate and deposition velocity strongly are influenced by the stratification of the atmosphere.

6. References

Battelle Institut (1982) "Austausch von Luftverunreinigungen an der Grenzfläche Atmosphäre/Erdoberfläche (trockene Deposition, Teilprojekt 1: Deposition von Gasen", Forschungsbericht 104 02 609, Battelle Institut Frankfurt/Main

Jaeschke, W. und Stauff, J. (1978) "Die Chemilumineszenz der SO$_2$-Oxidation und ihre Anwendung in der Chemie der Atmosphäre", Ber. Bunsenges. Phys. Chem. 82, 1180-1184

Kuttler, W. (1982) "Investigations about wet deposition of pollutants in an urban ecosystem" in: H.-W. Georgii and J.W. Pankrath (eds): Deposition of atmospheric pollutants, Proc. Oberursel, November 1981, D. Reidel Publishing Company, 97-113

Meixner, F.X. and Jaeschke, W.A. (1981) "The detection of low atmospheric SO$_2$ concentration with a chemiluminescence technique", Intern. J. Environ. Anal. Chemistry Vol. 10, 51-67

Meyers, T.P. and Baldocchi, D.D. (1988) "A comparison of models for deriving dry deposition fluxes of O$_3$ and SO$_2$ to a forest canopy", Tellus 40B, 270-284

Roth, R. (1975) " Der vertikale Transport von Luftbeimengungen in der Prandtl-Schicht und die Deposition-velocity", Meteorol. Rundschau 28, 67-71

West, P.W. and Gaeke, G.C. (1956) "Fixation of sulfur dioxide as Disulfitomercurate (II) and subsequent colorimetric Estimation", Journ. of Industr. Engineering, Anal. Chemistry Vol. 28, 1816-1819

DRY DEPOSITION OF ATMOSHPERIC PARTICLES TO AN OLD SPRUCE STAND

A.WARAGHAI AND G.GRAVENHORST
Institute for Bioclimatology
Department of Forestry
Buesgenweg 1
D-3400 Goettingen

ABSTRACT. The dry deposition flux of air-borne particles to the surface of fresh needles was determined for a 110 year old spruce stand in the "Solling" area. The increase in the area density of particles deposited on the surface of one- to ten days old needles was determined by scanning electron microscope. The deposition flux was differentiated with respect to particle size and extrapolated to the whole canopy. By relating these fluxes to measured particle concentrations in the air, dry deposition velocities for a ground base area could be deduced for different particle sizes. The dry deposition velocities fall in the range of 1 cm/sec and indicate a minimum at a particle diameter of about 2-5 μm. These results support, therefore, high deposition velocities for canopies with a rough surface and a high leave area index.

INTRODUCTION

Trace substances in the atmosphere can interact with the biosphere either by changing the radiation input to the ground or by altering the matter exchange between ecosystems and the atmosphere. The exchange of gaseous components between the atmosphere and plants, and soil/water systems is important for processes such as photosynthesis, respiration, nitrogen fixation, denitrification and evapotranspiration.

The exchange of particulate matter is important in situations where wind erosion diminishes bioelements of the soil, where seeds and pollen are transported for reproduction purposes through the air, and where nutrients like phosphorous are needed to maintain soil fertility or to cause lakes to become eutotrophic via dry and wet deposition. The deposition of components incorporated in rain droplets is important when surface water systems and forest soils lose their buffer capacity by acidic input from the air. The acidification of ecosystems is, however, not only due to wet deposition via rain but also due to dry deposition of gases and particles. The dry deposition flux from the atmosphere to the plant-soil system is much more difficult to estimate than rain deposition.

The natural interface between the vegetated land surface and the atmosphere can hardly be realistically modelled in order to derive matter fluxes from concentration measurements in the air (Hicks et al; 1980). Profile methods above the canopy have, however, been made to

77

H.-W. Georgii (ed.), Mechanisms and Effects of Pollutant-Transfer into Forests, 77–86.

estimate particle deposition and resulted in higher deposition velocities than assumed for smooth surfaces (Sievering 1982; Lorenz and
Murphy 1989).

Artificial surfaces have been used to collect air borne substances.
The application of the transfer function derived from surrogate surfaces
to natural surfaces is quite uncertain. It has, therefore, been the aim
of several investigations to determine the change of substances on the
surface of the vegetation itself during a certain time to estimate dry
deposition fluxes. In these studies the dry deposition fluxes could have
been the result of both gas phase and particulate phase deposition, when
the fluxes of sulphate, nitrate, ammonium, and chloride are investigated. Furthermore, an unknown contribution by leaching from the interior of the foliage and absorption by the foliage may have complicated
the interpretation of the flux measurements of elements like K, Mg, Mn,
Ca, Na, Cl, and soluble components like SO4, NO3, and NH4.
It was also attempted to determine the depositio of artificial particles
onto natural surfaces (Jonas 1984; Brueckmann 1988). These measurements
characterize special particle size ranges which represent only a small
fraction of natural air-borne particles.

The dry deposition of natural particles to vegetation surfaces can be
estimated by throughfall measurements when leaching and absorption can
be quantified. The first estimates of dry deposition velocities for
beech and spruce stands have been made by comparing rain chemistry in
the open field and beneath the forest canopy and relating these differences to air-borne particle concentrations (Hoefken and Gravenhorst,1983). The deposition velocities found were rather high (ca. 1
cm/sec) compared to other estimates. These discrepancies could, however,
be accounted for when roughness length as an indicator for the turbulent
structure of the air flow within and above the vegetation canopy was
considered (Hoefken and Gravenhorst,1983). It can be shown by model calculations, that the higher the roughness length, the higher the particle
deposition velocities, and the smaller is the change in deposition velocity with particle size (Sehmel and Hodgson 1974; Slinn 1982).

The interpretation of throughfall chemistry to derive particle deposition velocities is based on several assumptions. It is, therefore,
necessary to estimate the flux of natural particles to vegetation
canopies by other means as well. We report here an attempt to determine
particle flux to needles in an 110 year old spruce forest which had
earlier been investigated by Hoefken and Gravenhorst (1983) for particle
deposition.

Scanning Electron Microscopy (SEM) has already been used to study
particle deposition onto vegetation surfaces (Elias and Croxdale 1980;
Davidson and Chu, 1982). These studies, however, concentrated on the deposition of preselected chemical constituents and no deposition velocities could be deduced. Previous studies conducted to determine the particle flux to foliage surfaces by SEM have suffered from the heavy load
of materials already deposited on the foliage surfaces. This made it
impossible to distinguish old and new particle deposits (Coe and
Lindberg 1987). We tried to circumvent this problem by investigating the
change of particle concentration on fresh needle surfaces in the field

and relating the increase of particle density on the needles with time to the concentration of air-borne particles.

METHODS

One- to ten days old green needles from 110 year old spruce trees were taken for our experiment. They were investigated in the top of the crown of about 31 m tall trees in the Solling region in May 1988. We tried to set a precise zero-time from which the needles were exposed to particle deposition. When needles develop from the bud state they are first covered and held together by a bud scale. This protective bud scale was removed from the needle ensemble by hand so that air borne particles could be deposited on the outer surface. The configuration of this needle ensemble as a whole influences the air movement in the vicinity of the needles. The development from one-day needle ensemble, when all needles are still held together, to individual needles occured two to three days after bud scale removal. The collection efficiency of these needles may be different than for ten days old needles.

To determine particle concentration on a needle surface by the SEM method, the needles have to be removed from the field. Hence, repeatitive measurements on the same needle are not possible. Thus, the different particle concentrations on separate needles at different times are interpreted as an average change of particle concentration on an average needle. From these differences, particle fluxes to needle surfaces were derived.

No rainfall occurred during the period when needles were exposed. The particles on the needles were measured with a scanning electron microscope. The samples were carefully protected against contamination and surface disturbance during transport to the laboratory. To prepare the samples for the SEM, they were dried at 30°C, put under vacuum (0.066 hPa), and blazed with carbon vapour for 3 seconds. The particles were classified on photographs into 6 size ranges: 0.5-1μm, 1-2μm, 2-5μm, 5-8μm, 8-10μm, and >10μm in diameter. The size of the particles was estimated as the diameter of a circle having a similar area as the projected particle. The magnification of the particles was between 300 and 6000. During the period of exposure in the field the concentration of air-borne particles in the crown of the tree was measured with nuclepore filter (0.4 micron pore diameter) for SEM observation. During the time the samples were kept under vacuum, volatile particles might have evaporated. Since we are not interested in absolute concentrations but rather in the ratio between the particle number on the needles and on the filters, this problem should not influence our results.

RESULT AND DISCUSSION

The average particle concentration on the needle surface increased with time of needle exposure for all size classes (except from 4th to 5th day for particles between 8 and 10μm diameter). An example is shown in Figure 1, where the average particle concentration on 5 needles, each needle contributing 10 observation fields, is plotted as a function of time. These needles were covered by the same bud scale before they were

exposed to the air. The concentration of all particles on the various
fields of one needle varied by about 14%, whereas the average concentration on different needles varied by about 20% (Figure 2). This reproducibility appears to be acceptable for a first approximation of particle fluxes. These results must be substantiated by further measurements. Differentiation of the curves in Figure 1 with respect to time
results in particle fluxes to the needle surfaces (Figure 3). Generally
the fluxes increase with decrease in particle size and vary from day to
day by a factor of about 3. Since the deposition flux depends on air
borne particle concentrations and deposition conditions, these variations seem reasonable.

The flux rates alone do not indicate how efficient the particles are
captured. The capture efficiency can be estimated when particle fluxes
are related to air-borne particle concentrations. Such a normalisation
can determine whether the decrease of particle flux to the needle surface with increasing particle size indicates a corresponding lower capture efficiency or not. The proportionality factor between the flux rate
to the needle surface and the concentration of air-borne particles in
the vicinity of the needles is called deposition velocity. The deposition velocity characterizes the surface of the vegetation, the air-borne
components, and the transport properties of the air. This parameter is
often used to estimate deposition rates if one assumes that these rates
depend on the air borne particle concentration. In order to estimate
matter flux from the atmosphere to forest ecosystems it is necessary to
know the flux to its atmospheric interface, integrated over all foliage
surfaces. Leaf area index (LAI), the ratio of total leaf area to the
corresponding ground area for this spruce stand, was determined to be
close to eleven (Schulze et al.1977). The total flux per unit ground
area of the spruce stand could, therefore, be one order of magnitude
higher than to a needle surface unit, provided particle concentrations
in the air and collection efficiencies of the needles are similar
throughout the canopy.

The concentration of particles in the size ranges investigated here
differed by 10-41% between the forest floor and the crown top (Hoefken
and Gravenhorst,1983). The horizontal wind velocity, that can influence
the collection efficiency of air-borne particles, decreased from the
crown top to a mean value in the crown by a factor of about 1.2 for a
beech stand near Goettingen (Kreilein, 1988). The deposition velocity
related to the needle surface can, therefore, be multiplied by a factor
of about 10 to obtain the deposition velocity for air-borne particles
for this spruce stand. Figure 4 shows derived deposition velocities for
different particle size classes. They are rather high and indicate a
minimum in mid-size ranges. These values for the deposition velocity are
compared in Figure 5 with values derived for the same spruce stand by
interpretation of throughfall chemistry (Hoefken and Gravenhorst, 1983),
and by determining the amount of radioactive Cs 137 introduced into the
forest after the Chernobyl accident via means other than rainfall
(Brueckmann, 1988). All three independent measurments result in a deposition velocity of about 1 cm/sec. For a similar spruce stand, a dry
deposition velocity in the same range was derived from throughfall
chemistry (Grosch and Schmitt 1988). These dry deposition velocities for

a rough atmosphere/earth interface with a high leaf area index are
rather high compared to other values for interfaces such as grass,
water, and snow surfaces (Figure 6). The deposition velocities for water
and short grass were measured in a windtunnel(Chamberlain, 1973) and
modelled for a 10 cm reference height (Davidson, 1989).

Deposition velocities higher than for smooth surfaces have also been
found by Gravenhorst et al.(1983), Sievering (1982), Hicks et
al.(1982), and Lorenz and Murphy (1989). The minimum for the dry deposition velocity seems to shift from small particle sizes for smooth and
bare surfaces to large sizes for rough surfaces with canopies of high
leaf area indices (Figure 6). According to these results, it could well
be that onto forest surfaces particles are more efficiently deposited
than gaseous SO_2. The transfer resistance of the canopy appears to be a
limiting factor for dry deposition of particles. A rough forest interface seems to enhance particle but not SO_2 deposition onto the foliage.
In mass balance and transport models for sulphur removal from the atmosphere to the ground, the dry deposition of gaseous SO_2 is often assumed
to be more important than of particulate sulphate. Our measurements,
however, question this assumption in inhomogenous rough forested areas
with high leaf area indices. Further verification of our results is
needed. We will, therefore, try to continue our attempt, to determine
the rate of particle deposition to spruce stand by the SEM Method.

Acknowledgement : This research work was supported by the Ministry of
Science and Technology , Bonn, through grant number : 07VTD03

References

Brueckmann,A. (1988) Radionuklidbilanz von 4 Waldoekosystemen nach
dem Reaktorunfall in Tschernobyl und eine Bestimmung der
trockenen Deposition. Diplomarbeit Forstwissenschaftlicher Fachbereich, Goettingen.

Chamberlain,A.C. (1973) The movement of particles in plant communities. in: Vegetation and Atmosphere,Vol. 1. Editor:J.L. Monteith.
London, Academic Press. 115-201.

Davidson,C.I.(1989) Mechanisms of Wet and Dry Deposition of Atmo-
spheric Contaminants to Snow Surfaces. The Enviromental Record in
Glaciers and Ice sheets, eds H. Oeschger and C.C.Langway, John
Wiley and Sons Ltd. 29-51.

Davidson,C.I.and Chu,L. (1981) Scanning electron microscope study
of ironcontaining particles on foxtail, Environ. Sci. Technol.
15: 198.

Elias,R.W.,and Croxdale,J. (1980) Investigation of the deposition
of Pb aerosols on the surface of vegetation. Sci. Total Environ.
14: 265.

Gravenhost,G. Hoefken,K.D. and Georgii,H.W. (1983) Acidic input to a beech and spruce forest, in: Acid deposition,S. Beilke, A.: Elshout (eds.), CE, Reidel Publ.Com. 161-171.

Grosch,S. and Schmitt, G. (1988) Experimental investigations on the deposition of trace elemets in forest areas K.Grefen and J.Löbel (eds), Enviromental Meteorology, 201-216. Kluwer Academic Publishers.

Hicks,B.B.,Wesely,M.L.,Couler,R.L.,Hart,R.L.,Durham,J.L.Speer,R.E.a nd Stedam D.H.(1982): An experimental study of sulfur deposition to grassland, in H.R.Pruppacher,R.G.Semonin and W.G.N. Slinn(eds), Precipitation Scavenging, Dry deposition and Resus pension, Elsevier Pub.Co.,933-950.

Hoefken,K.D. and Gravenhorst,G. (1983) Untersuchung ueber die Deposition atmosphaerischer Spurenstoffe an Buchen- und Fichten wald. In: UBA – Berichte 6/83, Teil II, Schmidt-Verlag, Berlin.

Jonas,R.(1984) Ablagerung und Bindung von Luftverunreinigungen an Vegetation und anderen atmosphaerischen Grenzflaechen, Berichte der Kernforschungsanlage Jülich,Nr.1949.

Kreilein,H.(1987) Energie und Impulsaustausch in der atm- osphaerischen Grenzschicht ueber einem Waldbestand Diplomarbeit am Institut fuer Geophysik der Univsitaet Goettingen.

Lorenz,R. and Murphy JR. (1989) Dry deposition of particles to a pine plantation.,Boundary-Layer Meteorology 46:355-366.

Schulze,E.D.,Fuchs,M.I.,and Fuchs,M. (1977) Spacial distribution of photosynthetic capacity and performance in a mountain spruce forest of northern Germany.I biomass distribution and daily CO2 uptake in different crown layers. Oecologia (Berlin) 29, 43-61.

Sehmel G.A.,and Hodgson W.H. (1974) Predicted dry deposition velo- cities. Proceeding of a symposium held at Richland, Wash. 1974, ERDA symposim series CONF 740921, 1976, 399-422.

Sievering,H. (1982) Eddy flux and profile measurements of small particle dry deposition velocity at the Boulder atmospheric Observatory,in H.R. Pruppacher,R.G. Semonin and W.G.N. Slinn(eds), Precipitation Scavenging, Dry Deposition and Resuspension, Elsevier Pub. Co., 963-977

Slinn,W.G. (1982) Prediction of particle deposition to vegetative canopies, Atmos.Env.16, 1785-1794.

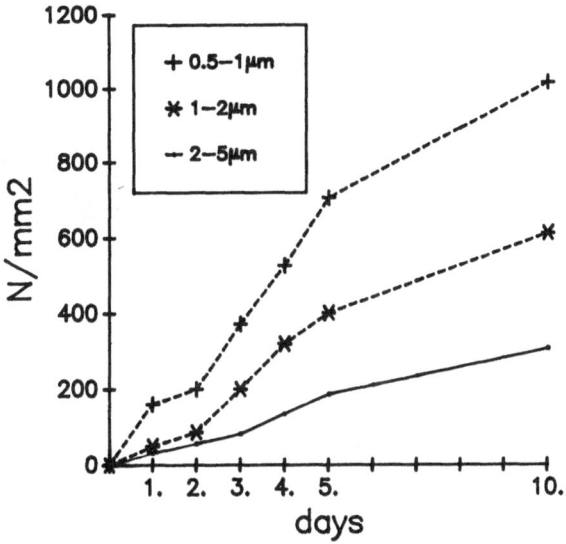

Figure 1a. Relationship between the increase of particle densities on green spruce needles with the time the were exposed to the atmosphere in field
(paramrter : particle diameter)

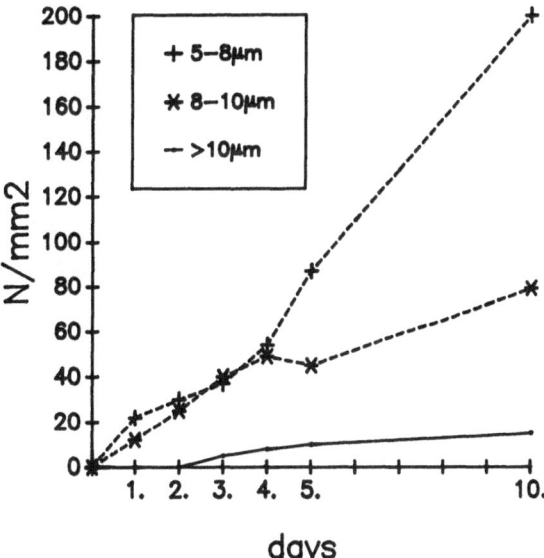

Figure 1b. Relationship between the increase of particle densities on green spruce needles with the time the needles were exposed to the atmosphere in the field.
(parameter : particle diameter)

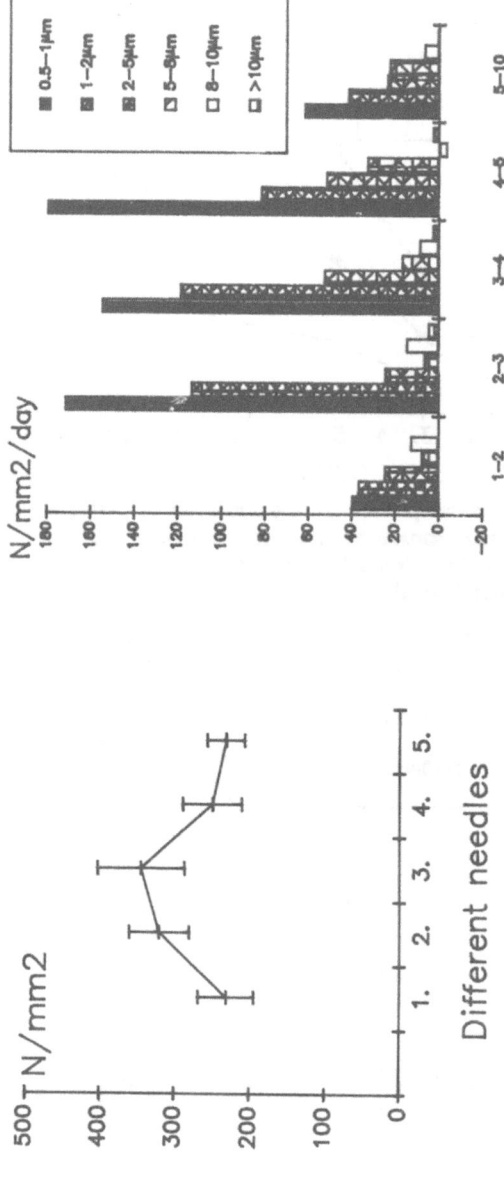

Figure 2. The number concentration of the particles (r > 0.25μm) measured by SEM on different needles after one day exposure to the atmosphere. on each needle 10 fields of max 0.1mm2 were analysed. (standard deviation: on one needle about 14%, on different needles about 20%)

Figure 3. The flux of particles to spruce needles during different time periods in May 1988 in a 110 year old tree stand. At day 0 the bud scale was removed from the bud. (no rainfall during the whole period, parameter: particle diameter)

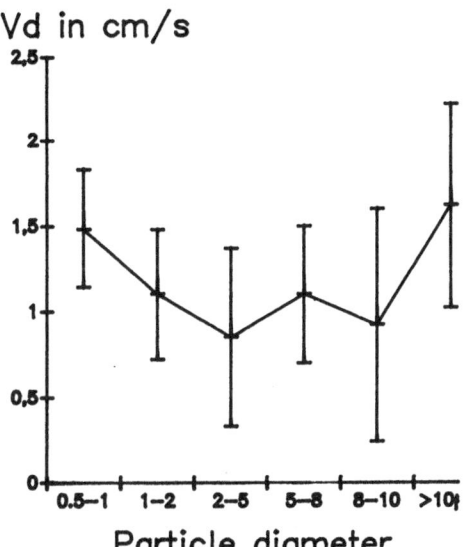

**Figure 4. Particle deposition velocities
to a spruce stand in the "Solling" area
(May 1988)**

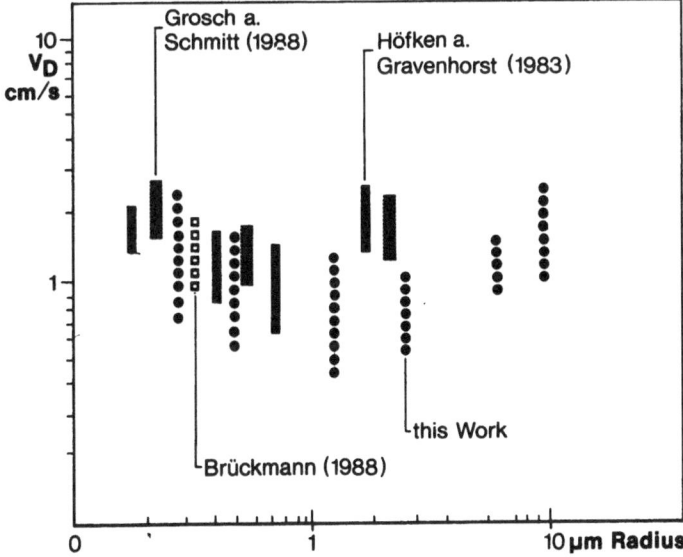

Figure 5. Deposition velocities of air borne
particles onto spruce canopies.

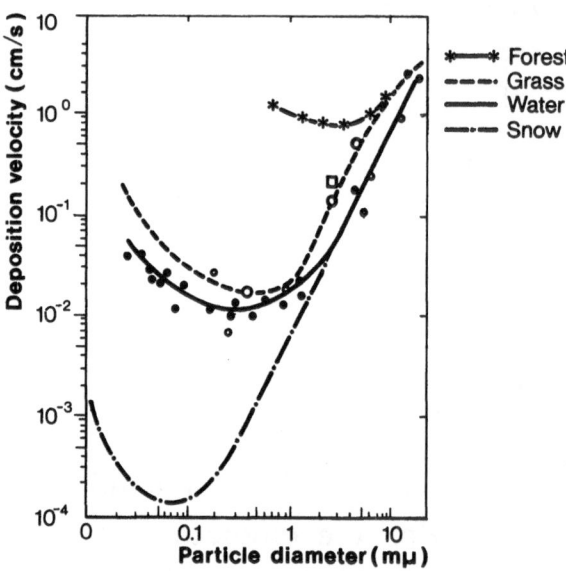

Figure 6. Deposition velocities
onto different surfaces (see text).

A CONTRIBUTION TO THE EXPERIMENTAL QUANTIFICATION OF DRY DEPOSITION TO THE CANOPY OF CONIFEROUS TREES

BARTH, S. and D. KLOCKOW
Universität Dortmund
Fachbereich Chemie
Postfach 500 500
4600 Dortmund 50

ABSTRACT.   A field method is presented, which allows to determine dry deposition rates to the canopy of a spruce tree for some atmospheric constitutents such as Ca^{2+}, Mg^{2+}, Na^+, Cl^- and total sulfur. The determination is based on the fact, that substances deposited through dry processes on needle surfaces, are washed off by subsequent rainfall and therefore are accumulated and can be measured in precipitation collected below the canopy. First results of this mass flux balance model, which directly includes the receiving surface, are discussed. The calculations carried out under neglection of source and sink terms indicate reasonable results for chloride and total sulfur, which are not subjected to considerable leaching or immobilization. For lead, which is immobilized on plant surfaces, the method presented underestimates the dry mass fluxes; for manganese, which is washed out of the needles to a great extent, it clearly overestimates them.

1.   Introduction

The deposition processes listed in table 1.1. are classified according to whether they are looked at from the viewpoint of the depositing compound, or from the viewpoint of the receiving surface.

The method presented is an attempt to determine dry deposition rates, $F_{d,i}$, for a component i to the canopy of spruce trees by experimental methods only. The results obtained are summation parameters to which the pathways B, C and D of table 1.1. contribute to an unknown extent. Their determination is based on the fact, that compounds deposited through dry or "occult" (Fowler and Leith (1985)) processes on needle surfaces, are washed off by subsequent rainfall. They are accumulated and can be measured in precipitation collected below the canopy, the so-called canopy drip.

The field measurements, to which the method is applied, are carried out at the research site Aberg in Schmallenberg-Grafschaft/Hochsauerland. They take place since August 1985, inside and outside a 30 years old spruce stand with a mean treetop height of 15 m. The results presented refer to the period until December 1987.

H.-W. Georgii (ed.), Mechanisms and Effects of Pollutant-Transfer into Forests, 87–95.
© *1989 by Kluwer Academic Publishers.*

deposition process	deposition type	
	viewpoint of depositing phase and compound	viewpoint of receiving surface
A precipitation of rain or snow with dissolved or undissolved matter	wet deposition	deposition trough precipitation
B sedimentation of particles other than rain drops or snow flakes		
C impaction of aerosol particles including fog and cloud droplets	dry deposition (incl. occult deposition)	deposition through interception
D sorption of gases on wet surfaces		

TABLE 1.1.   Classification of deposition processes (after Ulrich (1983)).

2.   Principle of the method

The mass fluxes within the canopy of a spruce tree can be described as shown in figure 2.1.

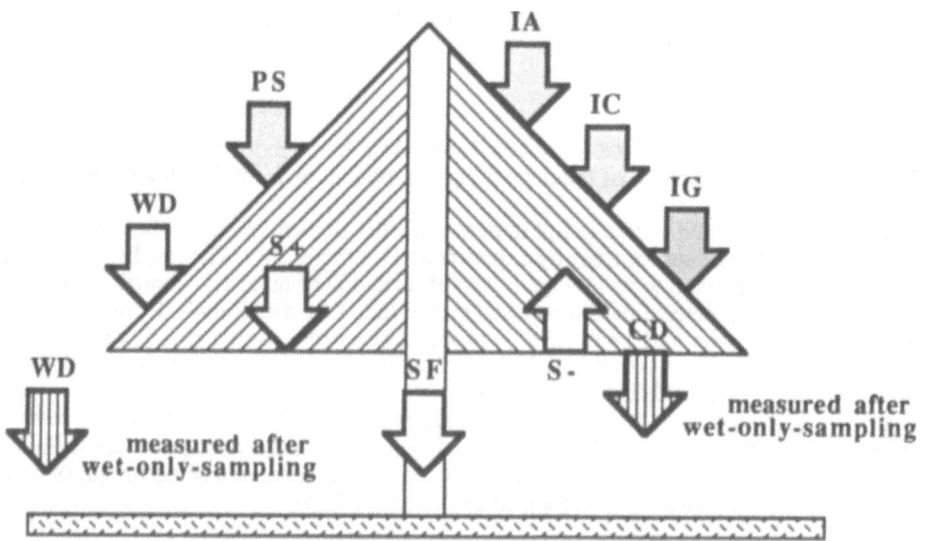

FIGURE 2.1.   Scheme of mass fluxes in a spruce tree canopy.

with CD =   canopy drip
      SF   =   stemflow
      WD =   wet deposition
      PS   =   particle sedimentation
      IA   =   impaction of aerosol particles

IC = impaction of fog and cloud droplets
IG = sorption of gases
S+ = source term (leaching of trace compounds from needles or bark in the canopy)
S- = sink term (uptake and immobilization of atmospheric trace compounds trough needles or bark in the canopy)

On the basis of this depiction the following mass flux balance for a component i in the canopy can be derived:

$$TF_i = CD_i + SF_i \tag{2.1}$$
$$\text{for spruce: } SF_i \approx 0$$

$$TF_i \approx CD_i = PD_i + ID_i + (S_i+ -S_i-) \tag{2.2}$$
$$= WD_i+PS_i + IA_i+IC_i+IG_i + (S_i+ -S_i-) \tag{2.3}$$

with TF = throughfall
PD = deposition through precipitation (see table 1.1.)
ID = deposition through intercecption (see table 1.1.)

According to table 1.1. the dry deposition rate, $F_{d,i}$, is that quantity of a substance i, which is deposited on a certain receiving surface area during a certain dry period:

$$F_{d.,i} = PS_i + IA_i+IC_i+IG_i \tag{2.4}$$

Under assumption of the mass flux balance described in equation (2.3) this leads to:

$$F_{d,i} = CD_i - WD_i - (S_i+ -S_i-) \tag{2.5}$$

The difference $CD_i - WD_i$ can be determined by making use of the measured concentrations of the component i in rain collected in the open field as well as in the canopy of the spruce:

$$F_{d,i} = \frac{c_{CD_i} * V_{CD} - c_{WD_i} * V_{CD}}{A_R * t_d} - (S_i+ -S_i-) \tag{2.6}$$

with c_{CD_i} = concentration of component i in rain in the canopy collector (μMol/l)
c_{WD_i} = concentration of component i in rain in the open field collector (μMol/l)
V_{CD} = rain volume collected in the canopy (l)
A_R = receptor surface area (m^2)
t_d = duration of dry period between two rain events (h)

The canopy collector, a wet-only sampler, is placed on a platform of 9.30 m height within the spruce stand, so that only a few spruce branches are located above the collector funnel (11.30 m height above ground). However, these few branches represent a sufficiently large receptor surface area, as could be shown during the investigations. An identical precipitation collector is placed in the open field, at a distance of approximately 40 m from the edge of the forest stand.

The receiving surface area A_R, i.e. the effective needle surface above the canopy collector, has been determined experimentally from the interception loss (154 ± 12 g H_2O per m^2) as $0,43 \pm 0,03$ m^2 above $0,06$ m^2 collector funnel opening.

The duration of the dry period between two rain events, t_d, is derived from continuous measurements of a precipitation recorder in the open field.

The source and sink terms (S+ - S-) are not determined experimentally. They take into account that some substances on the one hand are immobilized by the plant because of certain processes after deposition and therefore are no longer found in the canopy drip, and on the other hand are washed out of the plant tissue, although they were deposited neither by wet nor by dry processes. In first approximation the source and sink terms are supposed to be zero, and under this assumption dry deposition mass fluxes are calculated for various compounds employing equation (2.6).When interpreting the results it is tried, however, to take into consideration some estimates on the extent of immobilization and leaching, respectively, as known from the literature.

3. Results

3.1. DRY DEPOSITION MASS FLUX FOR HYDROGEN IONS

As was to be expected, the already mentioned various influences on dry deposition onto the canopy and on the composition of the canopy drip do not lead to one uniform value for dry deposition mass flux. There exists a wide range of mass fluxes as shown for hydrogen ions in figure 3.1.

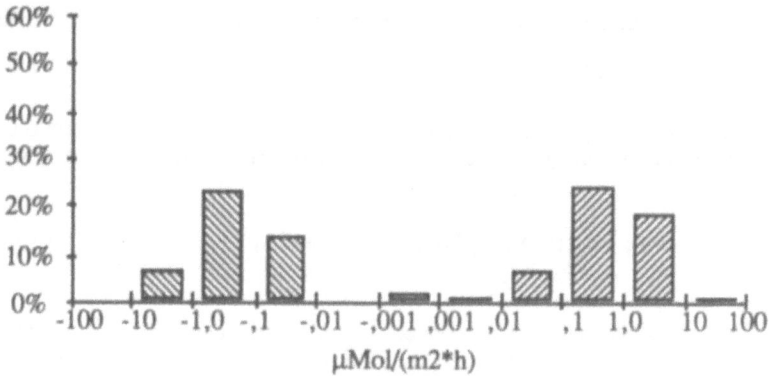

FIGURE 3.1.   Dry deposition mass fluxes for hydrogen ions; frequency distribution.

Two maxima can be seen: About half of all measurements is located in the positive, the other half in the negative range. The latter seems to indicate a negative flux of hydrogen ions, or in other words, a depletion of H+-ions in falling rain. This effect is due to chemical reactions in the canopy, which eventually lead to a source and sink term different from zero.

Hydrogen ions are deposited with acidic particles or gasous acids and by rain. After Seufert (1988) high proton concentrations maybe measured in the canopy drip especially during episodes with high SO_2 concentrations or after longer lasting moistening of needle surfaces and high SO_2 loads. A separated evaluation of the data for sommer and winter months seems to support this finding: The high positive fluxes of protons are measured particularly during the winter months. There exist also sinks for H+-ions in the canopy, which for example become effective when hydrogen ions are exchanged for calcium or magnesium ions. Consequently the buffered H+-ions are missing in throughfall (Ulrich (1983)).

3.2. DRY DEPOSITION MASS FLUX FOR CHLORIDE IONS

Chlorides are removed from the atmosphere in form of easily soluble compounds. Their main source - at least in background areas - is sea spray. They reach the receptor surfaces either as solutes in rain (snow) or as aerosol particles.

Because atmospheric chloride is neither immobilized on plant surfaces nor significantly taken up by them, the canopy represents no sink for chloride. Furthermore chlorides may be washed out of senescent leaves in autumn only to a low extent (Ulrich (1983)). For spruce, recent investigations (Mitterhuber et al. (1989)) show that leaching of anions is of little importance. Therefore the data given in figure 3.2. most likely represent a real picture of the dry chloride mass fluxes. A possible contribution of gaseous hydrochloric acid cannot be detected as such by the applied technique.

The dry mass fluxes for chloride ions vary in the range of 0,01 to 10 $\mu Mol/(m^2{*}h)$. For Na+-ions similar results are obtained.

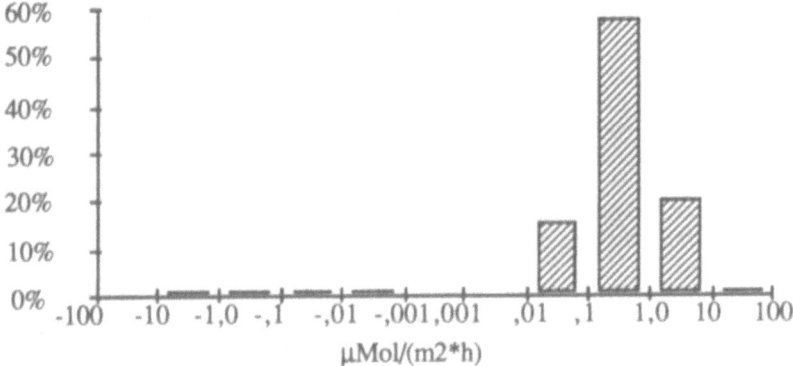

FIGURE 3.2.   Dry deposition mass fluxes for chloride ions; frequency distribution.

3.3. DRY DEPOSITION MASS FLUX FOR TOTAL SULFUR

Sulfur is an element which can be transferred in large amounts into forest stands by various deposition processes, wet as sulfate or sulfite, dry as sulfate particles or gaseous SO_2. Sulfur compounds in the form of neutral sulfates play an important role as nutrients for the ecosystem. But sulfuric acid and SO_2 as acid precursor have to be looked at as potentially damaging. Because no discrimination between species of an element is possible with the presented mass flux balance technique, only a dry mass flux of total sulfur can be given, which also includes SO_2 deposited in gaseous form and converted into sulfate on the receptor surface.

If present at high concentrations SO_2 can be taken up by the stomata and can be included into the plant metabolism (Ziegler (1975)). Conditions are different for sulfate particles: Usually they are soluble and not immobilized on needle surfaces to a significant extent.

92

Moreover, leaching of sulfate from the needles in the canopy can be ignored (Mitterhuber et al. (1989)). For spruce the neglection of the source and sink terms therefore means an underestimate of the dry mass flux of total sulfur (see figure 3.3.) for that part of SO_2, which is taken up by the plant metabolism.

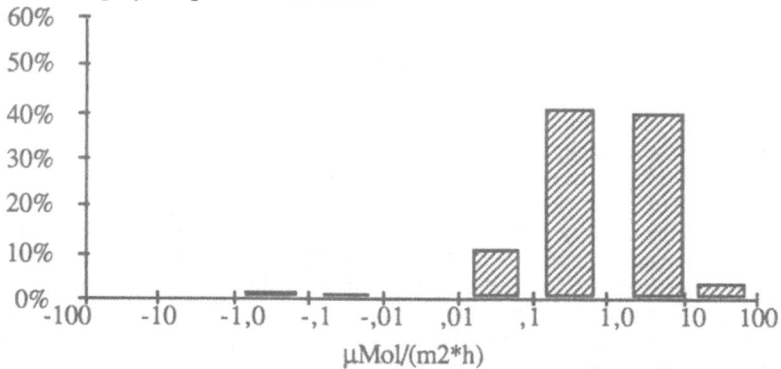

FIGURE 3.3.   Dry deposition mass fluxes for total sulfur; frequency distribution.

Examining the dry deposition mass fluxes as a function of the period between two rain events, it became obvious, that deposition rates are largest for short dry periods up to 20 h. This can possibly be explained by the fact, that the spruce twigs were still moistened during the period without rain and therefore exhibited only negligible resistance towards deposition of gaseous or particulate sulfur compounds. The results of Fowler ((1978) and (1980)) support this hypothesis by the observation, that resistances against SO_2 deposition are almost zero on wet surfaces. There exists only little information on the influence of wet surfaces on deposition of particles, but Garland (1983) and Sehmel (1980) agree that resuspension of deposited particles from wet surfaces is unlikely to happen. Therefore an accumulation of sulfate particles can be expected as a consequence ot the existance of water films on the twigs.

Total sulfur mass fluxes as shown in figure 3.3. may be compared with results obtained by Lindberg and Lovett (1983) through surrogate surface and leaf extraction methods in oak tree canopies. For sulfate these authors found dry mass fluxes of about 0,5 µMol per hour and squaremeter of effective receptor surface, which is within the maximum to be seen in figure 3.3.

3.4. DRY DEPOSITION MASS FLUX FOR CALCIUM IONS

Calcium input occurs either by rain (snow) or by dry deposition of particles. For the canopy calcium plays an important role due to cation exchange with H^+-ions, a process, by which calcium ions also reach the canopy drip. After Seufert (1988) leaching should be especially efficient at high SO_2 loads, during episodes with high SO_2 and O_3 concentrations, and during longer lasting periods with wet needle surfaces and high SO_2 concentrations at the same time. Thus the canopy represents a source of calcium, and the dry mass fluxes shown in figure 3.4. are most probably somewhat high. This statement is true for magnesium also.

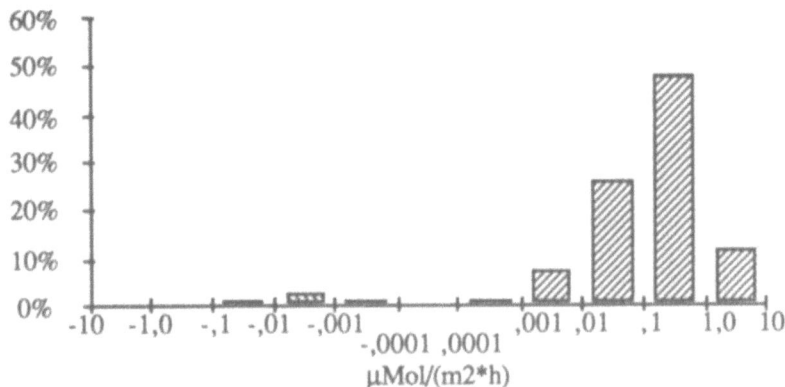

FIGURE 3.4.   Dry deposition mass fluxes for calcium ions; frequency distribution.

3.5. DRY DEPOSITION MASS FLUX FOR LEAD IONS

The heavy metal lead reaches the atmosphere mostly by combustion processes and can be transported over moderate distances as small diameter particles or as a gas (in form of alkyl-lead compounds). It can be assumed that no measurable leaching of lead takes place (Mayer (1983)). On the other hand, the canopy appears as a sink by storage of particles and precipitation of ions (immobilization) to a significant extent (Mayer (1983)). The mass fluxes shown in figure 3.5. reflect these observations because of the high fraction of negative results and therefore underestimate the real dry lead deposition rates to the spruce canopy.

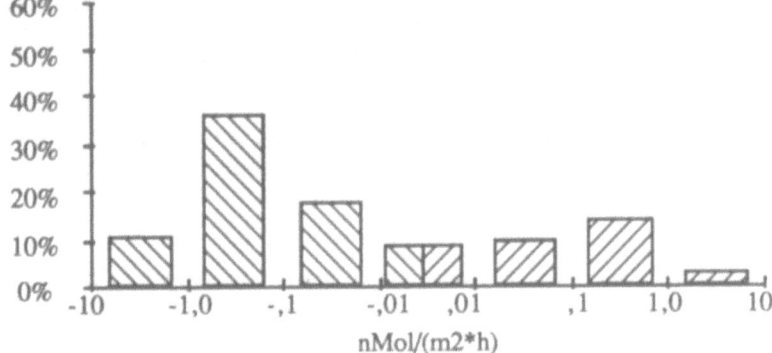

FIGURE 3.5.   Dry deposition mass fluxes for lead ions; frequency distribution.

3.6. DRY DEPOSITION MASS FLUX FOR MANGANESE IONS

Manganese concentrations found are higher by a factor of up to 100 in throughfall than in open field rain. This finding cannot be explained by dry deposition, but only by the high mobility of manganese within the plant. Large amounts of manganese are taken up by the

94

roots and are leached from the canopy extensively. The dry mass fluxes shown in figure 3.6. turn out to be much too high.

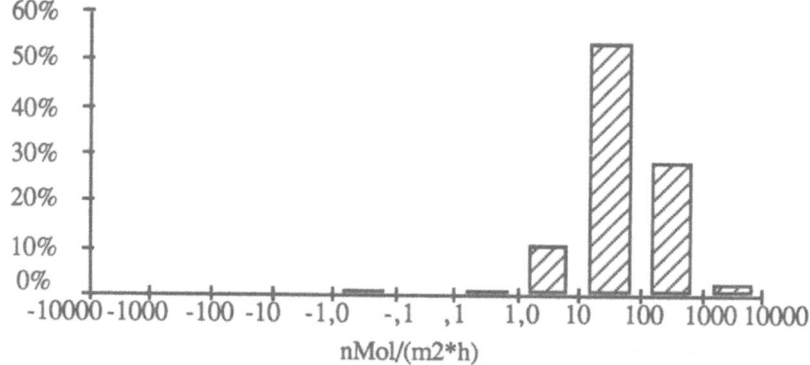

FIGURE 3.6.   Dry deposition mass fluxes for manganese ions; frequency distribution.

4.   Conclusions

The experimental method presented here offers a possibility to estimate dry deposition rates to spruce canopies of some atmospheric constituents as such as Ca^{2+}, Mg^{2+}, Na^+, Cl^- and total sulfur with moderate technical expenditure. The results obtained may vary by several orders of magnitude from one period to the other because of the influence of atmospheric (i.e. concentrations, air flow) as well as of receptor related (i.e. state of surfaces) conditions.

At the time being the dry mass fluxes shown in the figures are valid only for that spruce tree chosen for the investigations described. The method, however, is flexible enough to adapt its time (days to years) and spacial (individual trees, network) scale to the requirements of effects research.

The canopy mass balance approach is restricted to those components which can be washed off the needles by rain with high yield. Difficulties arise in case of ions which are either leached from plant tissue (e.g. K^+, Mn^{2+}) or immobilized and converted, respectively, in the canopy (e.g. H^+, NH_4^+, Pb^{2+}).

Acknowledgements. The work presented was financially supported by the German Federal Environmental Agency, Berlin, and the Commission of the European Communities, Brussels.

5.   References

Fowler, D. (1978) 'Dry deposition of SO_2 on agricultural crops', Atmos. Environ. 12, 369-373.

Fowler, D. (1980) 'Removal of sulphur and nitrogen compounds from the atmosphere in rain and by dry deposition', in Drabløs, D. and A. Tollan (eds.), Proc. Int. Conf. Ecol. Impact of Acid Precip., Sanderford, Norway, March 11-14, 22-32.

Fowler, D., and I.D. Leith (1985) 'Biophysical mechanisms in the uptake of air pollutants', Staub Reinhalt. Luft 45, 253-256.

Garland, J.A. (1983) 'Principles of dry deposition; application to acidic species and
ozone', in VDI-Berichte Nr. 500, VDI-Verlag, Düsseldorf, 83-95.

Lindberg, S.E., and G.M. Lovett (1983) ' Application of surrogate surface and leaf extraction methods to estimation of dry deposition to plant canopies', in Pruppacher,
H.R., Semonin, R.G., and W.C.N. Slinn (eds.), Precipitation Scavenging, Dry Deposition and Resuspension, Elsevier, New York-Amsterdam-Oxford, Vol. 2, 837-
848.

Mayer (1983) 'Interaction of forest canopies with atmospheric constituents: aluminium
and heavy metals', in Ulrich, B., and J. Pankrath (eds.), Effects of Accumulation of
Air Pollutants in Forest Ecosystems, Reidel, Dordrecht, 47-55.

Mitterhuber, E., Pfanz, H. and W.M. Kaiser (1989) 'Leaching of solutes by the action
of acidic rain: a comparison of efflux from twigs and single needles of Picea abies
(L. Karst.)', Plant, Cell and Environment 12, 93-100.

Sehmel, G.A. (1980) 'Particle and gas dry deposition - a review', Atmos. Environ.14,
983-1011.

Seufert, G. (1988) 'Untersuchungen zum Einfluß von Luftverunreinigungen auf den
wassergebundenen Stofftransport in Modellökosystemen mit jungen Waldbäumen',
Berichte des Forschungszentrums Waldökosysteme, Univ. Göttingen, Reihe A, Bd.
44.

Ulrich, B. (1983) 'Interaction of forest canopies with atmospheric constituents: SO_2,
alkali and earth alkali cations and chloride', in Ulrich, B., and J. Pankrath (eds.),
Effects of Accumulation of Air Pollutants in Forest Ecosystems, Reidel, Dordrecht,
33-45.

OCCURRENCE OF HNO$_3$, NH$_3$ AND NH$_4$NO$_3$ IN TWO FORESTED REGIONS IN AUSTRIA

H. PUXBAUM, CH. ROSENBERG[*] AND M. GREGORI
Institute for Analytical Chemistry
Technical University of Vienna
Getreidemarkt 9/151
A-1060 Vienna, Austria

ABSTRACT. HNO$_3$, NH$_3$ and NH$_4$NO$_3$ concentrations were determined at two sites in eastern and northern Austria using an annular diffusion denuder sampling technique. Measurements were performed simultanousely inside and outside of the forest canopy. Average summer daytime concentrations outside of the canopy were 1.4 ppb HNO$_3$, 1.6 ppb NH$_3$ and 0.8 ppb NH$_4$NO$_3$ in a semirural area in eastern Austria. In a rural area in northern Austria the corresponding levels were 0.5 ppb HNO$_3$, 2.3 ppb NH$_3$ and 0.4 ppb NH$_4$NO$_3$. Diurnal variations showed a different behaviour for the different sites and components. Namely for HNO$_3$ large concentration differences were observed for inside versus outside the forest canopy. Concurrent NH$_4$NO$_3$ measurements show that the decrease of HNO$_3$ inside the forest canopy is not related to NH$_4$NO$_3$ formation and therefore due to deposition.

1. Introduction

In this paper we report about the behavior of the nitrogen-containing components HNO$_3$, NH$_3$ and NH$_4$NO$_3$ in two forested regions in Austria. The behavior of the atmospheric HNO$_3$/NH$_3$ system is well documented for the Los Angeles basin (Russell and Cass 1984, Rusell et al. 1985, Russell and Cass 1986, Hildemann et al. 1984) and for rural areas in North America (Pierson et al. 1989, Anlauf et al. 1985, Cadle et al. 1982). For European sites the data are very sparse. The largest data set for HNO$_3$ and NH$_3$ has been reported from the Netherlands (Erismann et al. 1988). Other data available are from Germany (Höfken et al. 1986), Sweden (Grennfeld 1980) and Austria (Ober 1989, Ober et al. 1987).

The objective of our study was to accumulate data on concentration levels of gaseous and particulate components within the forest atmosphere in one semiurban and one rural area in Austria. The two sampling sites were situated in Wienerwald in the vicinity of the city of Vienna and in Schöneben in upper Austria. The latter being in the region of Böhmerwald. Five independent field studies were carried out in 1986-87, of which three occurred in summer and two in winter. The daytime/nighttime variation was studied in addition to the seasonal variation. Reference samples were collected outside the forest atmosphere concurrently with the sampling under the canopy within the forest. Air was sampled by use of a diffusion denuder technique. The versatile sampling assembly allows simultaneous determination of gaseous and particulate acidic and basic pollutants from one air sample. 146 samples were collected during the course of the work resulting in over 2000 determinations. Possible deposition or production patterns of the investigated compounds were evaluated by comparison of the concentration levels inside and outside the forest atmosphere.

[*] *on leave from Institute of Occupational Health, Helsinki, Finland*

H.-W. Georgii (ed.), *Mechanisms and Effects of Pollutant-Transfer into Forests, 97–107.*
© 1989 by Kluwer Academic Publishers.

98

2. Sampling Sites and Measurement Periods

The semiurban area in Wienerwald is located 10 km northwest of the city center of Vienna (Fig.1). A meteorological station, the Background Station Exelberg, operates since 1982 in the "Rundfunkturm Exelberg" (Puxbaum and Ober, 1987). The tower is situated on a sparsely populated forested hillside about 300 m above the average level of the city. The predominant species in the wooded area, which begins about 50 m from the tower, are beeches with an average height of 5-8 m. Air samples were collected at two different height levels. The lower sampling point was situated at the edge of the forest within the canopy 2 m above ground level. The higher sampling point was placed on a platform on the tower 80 m above the ground level. Standard meteorological parameters, wind speed, -direction, temperature, and relative humidity as well as the trace gases SO_2, NO_x and ozone are registered at the station.

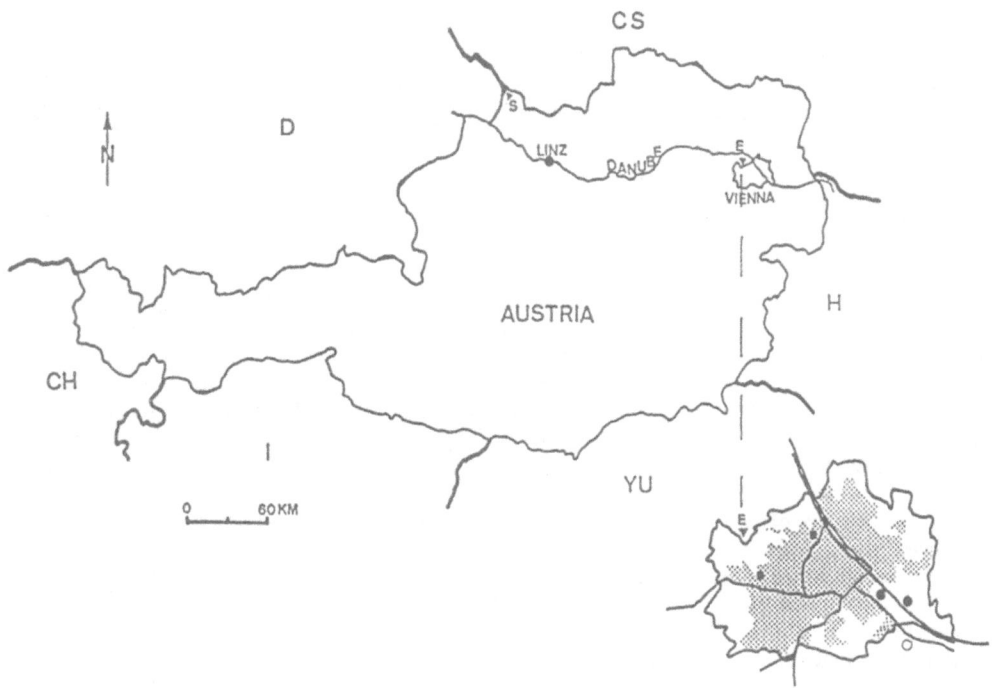

Figure 1. Map showing the sampling sites. E denotes Exelberg and S Schöneben. Insert: filled square = refuse incinerator; filled circle = power plant; open circle = refinery; dotted area = inhabited area

Sampling at the rural site in Schöneben (Fig.1) was carried out in the neighborhood of a meteorological station similar to that on Exelberg. The station is housed in a container placed on an open field. The adjacent forest, about 50 m from the container, was composed of spruce with scattered beech and fir with an average height of 20 m. Air was sampled within the forest and on the open filed close to the container 2 m above ground level. Two sampling runs, 3 hours each, were accomplished daily commencing at noon and midnight middle european time. The diurnal variation was studied by taking sequential samples throughout 24 hours. Days with westerly wind on Exelberg reflect the regional background situation the air masses originating from areas outside the city. Days with easterly winds describe the influence of the city as the winds bring airmasses over the city before entering the sampling point. The wind direction was predominantly westerly during the measuring periods in Schöneben. The time period and the meteorological situation of the five sampling periods, three performed on Exelberg and two in Schöneben, are summarized in Table 1.

Table 1. Sampling periods in 1986-87.

Date	Region	
	Exelberg	Schöneben
1986		
Aug.27- Sept.4.	cloudy, westerly winds, ws = 4.3-8.1 m/s, t =11-21 °C, rh = 54-98 %	
1987		
Feb.2.-5.	clear sky, easterly winds, ws = 0.9-9.2 m/s, t = 0.4- -7.5 °C, rh = 68-90 %	
March 25.- April 3.		cloudy, westerly winds, ws 0.3-5.8 m/s t = -0.7- -9 °C, rh = 62-99 %
July 15.- Aug.22.	clear sky, easterly and westerly winds, ws = 0.6-6.5 m/s, t = 14-28 °C, rh = 41-95 %	
Sept.9-15.		clear sky, westerly winds, ws = 0.1-4.1 m/s, t = 7-22 °C, rh = 56-99 %

ws = wind speed; t = temperature, rh = relative humidity

3. Experimental

An annular diffusion denuder system for differentiation of gaseous and particulate air pollutants was developed (Rosenberg et al., 1988). The denuder assembly consists of five selectively coated denuder tubes and a Teflon filter.

The inorganic acid, hydrochloric and nitric acid are collected on a NaF coated tube whereas the organic acids, formic and acetic acid, are sampled on one coated with KOH. Ammonia is sampled with a H_3PO_4-coated tube. Thermally labile components, sulfuric acid, ammonium chloride and -nitrate are evaporated at elevated temperature (145-155°C). The released anions are deposited on a NaF coated tube while the remaining ammonium as well as ammonium from partially deammoniated neutral sulfate is trapped on a H_3PO_4 tube. Thermally stable ammonium salts are collected on a Teflon filter at the end of the denuder assembly. Ion chromatography with conductivity detection was used for analysis of the ionic components from aqueous extracts of the denuder coatings (Tsitouridou and Puxbaum, 1987; Rosenberg et al.,1988). All reported data are reduced to "normal" conditions (0°C, 1013 hPa) and given in nmol/m^3, the conversion factor for calculating ppb(v) is 44.6 nmol/m3 = 1 ppb(v).

4. Results and Discussion

4.1. NITRIC ACID AND AMMONIUM NITRATE

An average of 43 nmol/m^3 of nitric acid was observed on the Exelberg tower during the late summer period (Aug.27-Sept.4 1986) with low photochemical activity, cloudy sky and westerly winds (Table 2). The ozone and NO$_x$ levels registered on the tower were 48 and 5 ppb, respectively. The measurement period in summer 1987 was characterized by a higher photochemical activity, mostly clear sky and low wind speed. A clear increase in nitric acid concentrations was observed, the average levels being 61 nmol/m^3 during days with westerly winds. This was about 1.4 times higher compared to the fall results in 1986. Furthermore, twice as high levels, 81 nmol/m^3, was found during days with prevailing easterly winds. This finding reflected evidently the photochemical build up of nitric acid originating from anthropogenic precursors. The average ozone levels were 66 ppb during days with westerly and easterly winds as well, whereas the NO$_x$ levels were 3 and 8 ppb, respectively. In Schöneben the nitric acid levels measured in Sept. 1987 averaged about half of those observed on Exelberg in Sept. 1986, being 23 nmol/m^3. On the other hand this value represented about one fourth of the highest levels found on Exelberg downwind the city region. A pronounced trend was found between the daytime levels inside and outside the forest atmosphere. On average 46 % lower values were observed in the forest in the Vienna region, whereas it was 57 % in Schöneben as compared to the corresponding values outside the forest atmosphere. Under normal atmospheric conditions vegetation or soil is not expected to act as a source for nitric acid. Nitric acid is only deposited (Höfken et al. 1987). Our results undoubtedly show the high deposition velocity of nitric acid over a rough surface. This complies with the findings by Meixner et al. (1987) who reported a daytime deposition velocity of approximately 10 cm/s for nitric acid over a humid pine forest.

Additionally, a similar decrease in the nighttime concentrations, as was found for the organic acids (Puxbaum et al. 1988), was also evident for nitric acid. We observed a 53 % nighttime decrease of nitric acid inside the forest, whereas it was 22 % above the forest. The higher decrease in the forest is explained by the more stable atmosphere under the canopy.

Fig. 2 shows the time series diagram of nitric acid concentrations on Exelberg and Schöneben. The concentrations increased during morning hours reaching a maximum at noon on both sites. The levels decreased towards the evening with a minimum around midnight. The variation was more pronounced in Schöneben where the nighttime nitric acid levels were below the detection limit. This reconciles with the fact that during night under humid conditions nitric acid is removed from air isolated from resupply by conversion to ammonium salts (Anlauf et al. 1985). The average relative humidity in the night was over 97 %, whereas it was about 64 % at noon in Schöneben. The corresponding values were around 55 % during day and night on Exelberg. The daytime ammonium nitrate concentrations in summer (Table 2) were generally slightly lower than the corresponding nitric acid values. Although they did exhibit some diurnal variability, the variation was inconsistent and smaller than for nitric acid. This results is explained by the much smaller deposition rate reported for particulate components being over 10 times lower (Cadle et al., 1985) than for gaseous nitric acid. Nonetheless, e.g. on Exelberg during night hours the levels appeared to be about 50 % lower within the forest than above the canopy in summer and about 20 % lower in winter. The observation of a strong diurnal variation of gaseous nitric acid and almost no variation of ammonium nitrate is in good agreement with results reported elsewhere (Shaw et al. 1982; Galasyn et al. 1987).

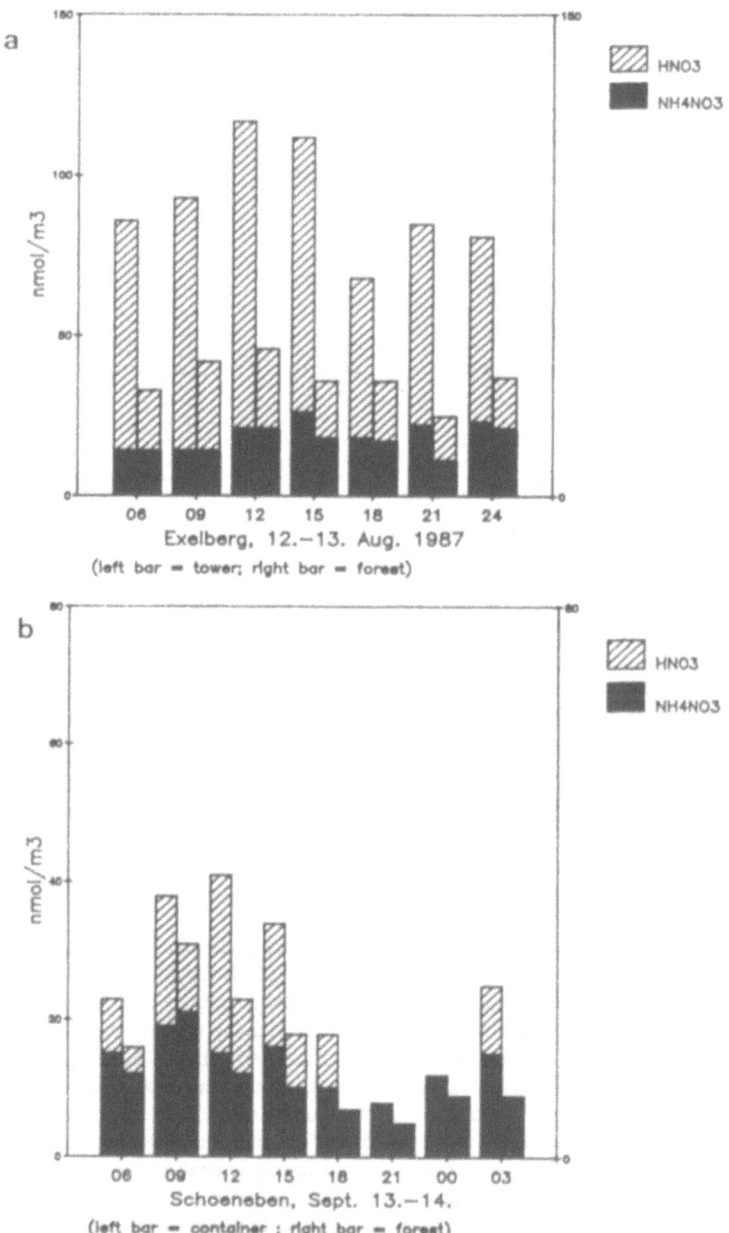

Figure 2. Time series of nitric acid and ammonium nitrate measured on a) Exelberg Aug. 12.-13. 1987 and b) Schöneben Sept. 13.-14. 1987.

Table 2. Ambient concentrations of gaseous nitric acid and particulate nitrate.

Region/date/ component	CONCENTRATION (SD) nmol/m³			
	daytime		nighttime	
	inside[*]	outside[**]	inside	outside
SUMMER				
Exelberg (westerly winds) Aug.27. Sept.4. 1986				
HNO₃	35 (19)	43 (22)	8	36 (2)
NH₄NO₃	44 (35)	30 (13)	36 (24)	60 (36)
(n)	(5)	(5)	(2)	(2)
Exelberg (westerly winds) July 15.–Aug.22. 1987				
HNO₃	26 (13)	61 (31)	15 (7)	50 (24)
NH₄NO₃	21 (10)	29 (20)	18 (3)	45 (19)
(n)	(4)	(5)	(3)	(3)
" (easterly winds) " "				
HNO₃	31 (19)	81 (32)	19 (10)	55 (23)
NH₄NO₃	26 (12)	44 (47)	33 (17)	74 (67)
(n)	(6)	(7)	(5)	(6)
Schöneben (westerly winds) Sept. 9.–15. 1987				
HNO₃	10 (1)	23 (3)	5 (1)	8[***]
NH₄NO₃	17 (9)	18 (6)	34 (26)	43 (29)
(n)	(4)	(4)	(6)	(6)
WINTER				
Exelberg (easterly winds) Feb. 2.–5. 1987				
HNO₃	65 (20)	65 (21)	87 (66)	75 (35)
NH₄NO₃	221 (150)	133 (66)	80 (8)	108 (33)
(n)	(4)	(4)	(3)	(2)
Schöneben (westerly winds) March 25.–April 3. 1987				
HNO₃	21 (9)	35 (11)	26	24 (8)
NH₄NO₃	52 (17)	41 (24)	17	15 (1)
(n)	(3)	(4)	(1)	(2)

SD = standard deviation; n = number of samples
[*] = within forest, 2 m height; [**] = outside forest atmosphere: in Exelberg
      80 m above ground level; in Schöneben open land, 2 m height
[***] = five values below detection limit

The average wintertime nitric acid concentrations were on Exelberg of the same order of magnitude as found in summer during days with westerly winds. No nighttime decrease was observed in winter in accordance with the findings for organic acids. On the contrary, a 20 % increase was registered which could be explained by build up in the inversion layer near ground in the absence of deposition processes. The ammonium nitrate concentrations were about 5 times higher compared with the summer values. Since the Winter measurements on Exelberg were performed during a large scale pollution episode (Ober et al. 1987) it is not possible to derive whether the results reflect a true seasonal variation. On the other hand, the results from the winter measurements in Schöneben showed also an increasing trend in both nitric acid and ammonium nitrate concentrations. Our results contradict the findings of Cadle (1985) and Galasyn (1987) who reported an evident seasonal variation of nitric acid with the highest concentrations in summer and the lowest in fall-winter. The observations referred to were from an urban area (Warren, MI, U.S.A.) and a remote mid tropospheric location (Mauna Loa, Hawaii), as well.

4.2. AMMONIA AND PARTICULATE AMMONIUM

In fall 1986 the average daytime ammonia concentrations were on Exelberg 45 nmol/m^3 while in summer 1987 it was 96 and 71 nmol/m^3 during days with westerly and easterly winds, respectively (Table 3). The corresponding levels observed within the forest were on average 20 % higher. In Schöneben about the same levels were found, but here the values outside the forest were surprisingly 50 % higher compared to the levels in the forest atmosphere. This would reflect an influence from a strong local source. A partly inconsistent nocturnal decrease of the ammonia concentrations, attributed to removal through dry deposition or conversion to ammonium, was observed. This phenomenon was also seen in the time series diagram (Fig. 3). The average nighttime drop was 40 % within the forest and 34 % outside the forest. This result, in turn, reflects that the surface structure, on the study areas, did not significantly affect the removal process.

Table 3. Ambient concentrations of ammonia and particulate ammonium.

Region/date/ component	CONCENTRATION (SD) nmol/m^3			
	daytime		nighttime	
	inside[*]	outside[**]	inside	outside
SUMMER				
Exelberg (westerly winds) Aug.27.-Sept.4. 1986				
NH$_3$	51 (18)	45 (16)	23 (1)	39 (14)
p-NH4$^-$	171 (105)	122 (56)	150 (70)	184 (61)
(n)	(5)	(5)	(2)	(2)
Exelberg (westerly winds) July 15.-Aug.22. 1987				
NH$_3$	102 (47)	96 (48)	59 (26)	68 (54)
p-NH$_4$$^-$	194 (83)	180 (74)	184 (75)	204 (74)
(n)	(4)	(5)	(3)	(3)
" " (easterly winds)				
NH$_3$	98 (43)	71 (48)	73 (40)	100 (70)
p-NH$_4$$^-$	293 (100)	261 (139)	304 (76)	334 (92)
(n)	(6)	(7)	(5)	(6).
Schöneben (westerly winds) Sept. 9.-15. 1987				
NH$_3$	54 (10)	103 (27)	34 (5)	47 (18)
p-NH$_4$$^-$	98 (18)	105 (22)	109 (29)	101 (36)
(n)	(4)	(4)	(6)	(6)
WINTER				
Exelberg (easterly winds) Feb. 2.-5. 1987				
NH$_3$	<dl	<dl	<dl	<dl
p-NH$_4$$^-$	1086 (633)	800 (294)	587 (270)	688 (193)
(n)	(4)	(4)	(3)	(2)
Schöneben (westerly winds) March 25.-April 3. 1987				
NH$_3$	30 (7)[*]	9[b]	19	<dl
p-NH$_4$$^-$	145 (37)	135 (47)	110	89 (3)
(n)	(3)	(4)	(1)	(2)

SD = standard deviation; n = number of samples; <dl = below detection limit
*) = within forest. 2 m height; **) = outside forest atmosphere; in Exelberg 80 m above ground level; in Schöneben open land. 2 m height.
*) = one value below detection limit; b) = three values below detection limit

The wintertime ammonia levels were below detection limit on Exelberg, whereas in Schöneben some samples contained detectable concentrations. It has been calculated that more than 95 % of anthropogenic ammonia emissions derives from animal wastes and fertilizers and that natural emission

104

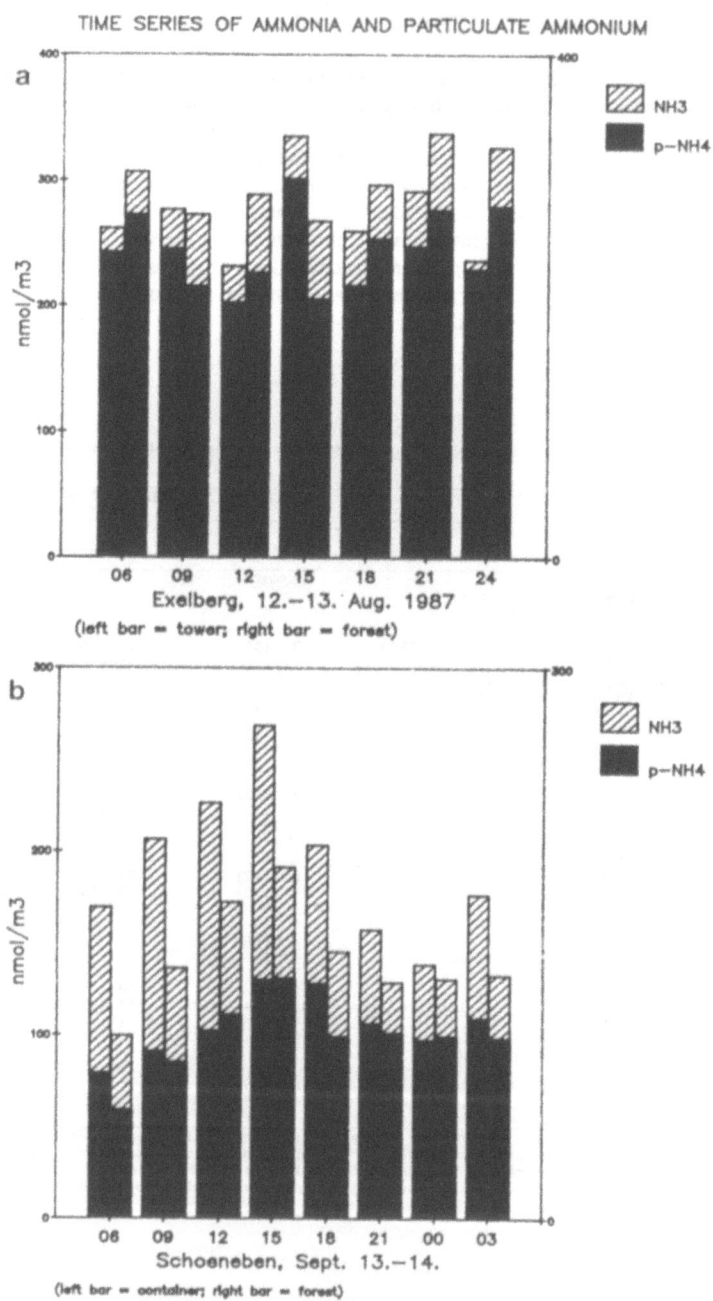

Figure 3. Time series of ammonia and particulate ammonium measured on a) Exelberg Aug. 12.-13. and in Schöneben Sept. 13.-14. 1987.

is only about 10 % of the anthropogenic emission (Buijsman et al. 1987; Nihlgard 1985). A marginal contribution is attributed to traffic. The seasonal variation of the ammonia levels is consistent with the above findings, the emissions being at a minimum during winter. On Exelberg the average daytime concentrations of particulate ammonium varied between 122 and 180/261 nmol/m^3 measured in fall 1986 and summer 1987 during days with westerly/easterly winds. Concurrent measurements within the forest showed equal levels or values which somewhat exceeded the corresponding tower values. No clear daytime/nighttime trend for particulate ammonium was found in our results. The wintertime pollution episode in the Vienna region was again clearly reflected in the levels measured. The particulate ammonium levels amounted to as high as 1086 nmol/m^3 as observed within the the forest. A nighttime decline was seen in the winter samples although a slower deposition rate would be expected caused by the snow cover. The wintertime concentrations of particulate ammonium found in Schöneben were about 40 % lower compared to those observed on Exelberg during days with westerly winds. In Schöneben a nearly 30 % increase of the particulate ammonium levels was observed in winter as compared to summer. Gaseous ammonia and nitric acid concentrations in air are governed by the equilibrium with particulate ammonium nitrate (Fig 4) (Stelson et al. 1979)

$$HNO_3(g) + NH_3(g) \; NH_4NO_3(s)$$

with an equilibrium constant $k = C_{NH3} \times C_{HNO3}$

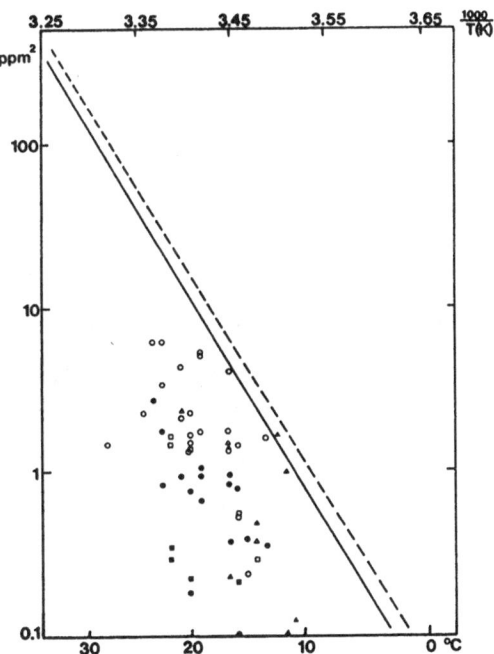

Figure 4. The product $C_{HNO3} \times C_{NH3}$ calculated from data from our work compared with equilibrium theory (Stelson et al. 1979). Exelberg Aug.-Sept. 1986: filled triangle = forest, open triangle = tower: Exelberg July-Aug. 1987: filled circle = forest, open circle = tower; Schöneben Sept. 1987: filled square = forest, open square = container.

Stelson and Seinfeld (1982) derived an expression for the dependence of temperature and humidity. If the relative humidity exceeds the relative humidity of deliquescence (64 % at 25°C and 70 % at 0°C) the dissociation constant is lowered. The product of the ammonia and nitric acid concentrations calculated from our results are depicted in Fig. 4. Over 50 % of the measurements were performed in relative humidities > 60 %. The scatter of the experimental points below the predicted solid NH_4NO_3 dissociation constant agree with the humidity and temperature dependence (Stelson and Seinfeld, 1982; Harrison and Pio, 1983). The lowest values on the diagram refer to measurements within the forest, which is in accordance with the presumably higher humidity in the forest atmosphere.

5. Acknowledgements

This work was performed within the framework of the FIW ("Forschungsinitiative Waldsterben"), financed by the Austrian Ministry for Environment, Family and Youth. Christina Rosenberg acknowledges support from the Finnish Institute of Occupational Health. The activities at the "Exelberg" site were supported by the Austrian Federal Environmental Agency ("Umweltbundesamt") and the Post Administration ("Post- und Telegraphenverwaltung"). Acess to the monitory site in Schöneben was made available by the Local Environmental Groups in Upper Austria ("Amt der Oberösterreichischen Landesregierung").

6. References

Anlauf K. G., Bottenheim J. W., Brice K. A., Fellin P., Wiebe H. A., Schiff H. I., Mackay G. I., Braman R. S. and Golbert R. (1985) Measurement of atmospheric aerosol and photochemical products at a rural site in SW Ontario. Atmos. Environ. 19, 1859-1870.

Buijsman E. D., Maas H. F. M. and Asman W. A. H. (1987) Anthropogenic NH_3 emissions in Europe. Atmos. Environ. 21, 1009-1022.

Cadle S. T. (1985) Seasonal variations in nitric acid, nitrate, strong aerosol acidity, and ammonia in an urban area. Atmos. Environ. 19, 181-188.

Cadle S.H., Countess R.J., Kelly N.A. (1982) Nitric acid and ammonia in urban and rural locations, Atmos. Environ. 16, 2501 - 2506

Cadle S. T., Muhlbaier Dasch J. and Mulawa P. A. (1985) Atmospheric concentrations and the deposition velocity to snow of nitric acid, sulfur dioxide and various particulate species. Atmos. Environ. 19, 1819-1827.

Erisman J.W., Vermetten A.W.M., Asman W.A.H., Waijers-Ijpelaan A., Slanina J. (1988) Vertical distribution of gases and aerosols: the behaviour of ammonia and related components in the lower atmosphere, Atmos. Environ. 22, 1153 - 1160

Galasyn J. F., Tschudy K. L. and Huebert B. J. (1987) Seasonal and Diurnal variability of nitric acid vapor and ionic aerosol species in the remote free troposphere at Mauna Loa, Hawaii J. Geophys. Res. 92, 3105-3113.

Grennfelt P. (1980) Investigation of gaseous nitrates in an urban and a rural area, Atmos. Environ. 14, 311 - 316

Harrison R. M. and Pio C. A. (1983) An investigation of the atmospheric HNO_3-NH_3-NH_4NO_3 equilibrium relationship in a cool, humid climate. Tellus 33B, 155-159.

Hildemann L.M., Russell A.G., Cass G.R. (1984) Ammonia and nitric acid concentrations in equilibrium with atmospheric aerosols: experiments vs theory, Atmos. Environ. 18, 1737 - 1750

Höfken K.D., Meixner F., Müller K.P., Ehalt D.H. (1986) Untersuchungen zur trockenen Deposition und Emission von atmosphärischem NO, NO_2 und HNO_3 an natürlichen

Oberflächen, Kernforschungsanlage Jülich, ISSN 0366-0885

Höfken K. D., Meixner F. and Ehhalt D. H. (1987) Dry deposition of NO, NO_2 and HNO_3. 16th International Technical Meeting on Air Pollution Modeling and its Applications, Lindau, FRG, April 6- 10 1987.

Meixner F. X. and Franken H. H. and Duyzer J. H. and van Aalst R. M. (1987) Dry deposition of gaseous HNO_3 to a pine forest. 16th International Technical meeting on air pollution modeling and its applications, Lindau, FRG, April 6-10, 1987.

Nihlgard B. (1985) The ammonium hypothesis-an additional explanation to the forest dieback in Europe. Ambio 14, 2-8.

Ober E. (1989) Deposition von Ozon und anorganischen Nitraten in Waldgebiete am Exelberg (Wienerwald) und am Jauerling (Waldviertel) unter besonderer berücksichtigung der Entstehung und des Transportes, Dissertation, University of Bodenkultur, Vienna

Ober E., Rosenberg C. and Puxbaum H. (1987) Ionic composition of ammonium salts during an acidic wintertime episode at two sites at different altitude. Eurasap Symposium "Ammonia and Acidification, Bilthoven, The Netherlands, April 13-15.

Pierson W.R., Brachaczek W.W., Gorse R.A.Jr, Japar S.M., Norbeck J.M., Keeler G.J. (1989) Atmospheric acidity measurements on Allegheny Mountain and the origins of ambient acidity in the northeastern United States, Atmos. Environ. 23, 431 - 459

Puxbaum H. und Ober E. (1987) In: Backgroundstation Exelberg. Umweltbundesamnt, Wien, Österreich.

Puxbaum H., Rosenberg C., Gregori M., Lanzerstorfer C., Ober E. and Winiwarter W. (1988) Atmospheric Concentrations of formic and acetic acid and related compounds in eastern and northern Austria. Atmos. Environ. 22, 2841 - 2850.

Rosenberg C., Winiwarter, W., Gregori M., Pech G., Casensky V. and Puxbaum H. (1988) Determination of inorganic and organic volatile acids, NH_3, particulate SO_4^{2-}, NO_3^- and Cl- in ambient air with an annular diffusion denuder system. Fresenius Z. Anal. Chem. 331, 1-7.

Russell A.G., Cass G.R. (1984) Acquisition of regional air quality model validation data for nitrate, sulfate, ammonium ion and their precursors, Atmos. Environ. 18, 1815 - 1827

Russell A.G., Cass G.R. (1986) Verification of a mathematical model for aerosol nitrate and nitric acid formation and its use for control measure evaluation, Atmos. Environ. 20, 2011 - 2025

Russell A.G., McRae G.J., Cass G.R. (1985) The dynamics of nitric acid production and the fate of nitrogen oxides, Atmos. Environ. 19, 893 - 903

Shaw R. W., Stevens Jr R. W., Bowermaster J., Tesch J. W. and Tew E. (1982) Measurements of atmospheric nitrate and nitric acid: the denuder difference experiment. Atmos. Environ. 16, 845-853.

Stelson A. W. and Seinfeld J. H. (1982) Relative humidity and temperature dependence of the ammonium nitrate dissociation constant. Atmos. Environ. 16, 983-992.

Stelson A. W., Friedlander S. K. and Seinfeld J. H. (1979) A note on the equilibrium relationship between ammonia and nitric acid and particulate ammonium nitrate. Atmos. Environ. 13, 367-371.

Tsitouridou R. and Puxbaum H. (1987) Application of a Portable Ion Chromatograph for field Site measurements of ionic composition of fog water and atmospheric aerosols. Intern. J. Environ. Anal. Chem.31, 11-12.

DEPOSITION
OF ORGANIC COMPOUNDS

DEPOSITION OF ORGANIC ANIONS AT A SEMI-RURAL SITE IN CENTRAL EUROPE

W. Elbert and M.O. Andreae,
Max-Planck-Institut für Chemie,
Otto-Hahn-Institut,
Saarstr. 23,
D-6500 Mainz, F.R.G.

ABSTRACT. Wet-only rain samples were collected and analysed for formic, acetic, hydrochloric, nitric and sulfuric acids. The organic acids contributed to free acidity of rain to about 5%, and wet depositional fluxes were lower than 4.0 and 2.0 mmol m^{-2} yr^{-1} for formate and acetate, respectively. We found a strong correlation between formic and acetic acids with a ratio averaged to 1.72, and we observed a slight dependency of the formic-to-acetic acid ratios on air temperature. Although the concentrations of the organic acids showed a distinct maximum during the growing season, highest deposition rates were found later in the year.

1. INTRODUCTION

Organic acids have been reported in the late 70's (and early 80's) as possible contributors to the acidity of atmospheric precipitation, and since then there has been a steadily growing body of data on aqueous phase concentrations of, predominantly, formic and acetic acids from various marine and continental sites of both hemispheres [2,5-14]. As yet however, very little about organic acids in central European rain has been published [15,16].

Formic and acetic acids may contribute up to 60% of the free acidity of rain in remote areas of the world [10,11], whereas in more urban locations, the contribution can be as low as a few per cent [6,17].

In this paper we will report on the measurements made at a semirural site in Central Europe. The concentration and deposition of aqueous formic and acetic acids, the seasonal variability of both concentration and deposition, and the possible contribution to rain acidity from organic acids will be discussed.

2. METHODS

2.1. Sampling Site

Rainwater was collected at a rural site west of Presberg, West Germany

111

H.-W. Georgii (ed.), Mechanisms and Effects of Pollutant-Transfer into Forests, 111–118.
© *1989 by Kluwer Academic Publishers.*

(50°4'N, 7°53'E, elev. 381 m ASC). Presberg is a small village located on the western part of the Rheingau Range which itself is a part of the Taunus Mountains. In most cases, the sampling site was not influenced by the large industrial regions nearby such as the Rhein-Main-Gebiet, which is more than 20 km east of the site, because for 75% of the sampled events the wind came from a westerly (15% northerly and 10% easterly) direction. Between the sampling site and the next industrial center westward of Presberg (e.g. Saarbrücken) are predominantly wooded secondary chains of mountains with minor industries and low population density.

Rainfall amount data were calculated from the volume of the sample taken, and for comparison, the quarterly rainfall amount means were obtained from a nearby station of the Deutsche Wetterdienst (DWD) at Geisenheim [18]. Air temperature and wind direction were recorded at the site, whereas all other meteorological data were taken from local weather maps.

2.2. Collection

Precipitation samples (rain, snow, hail) were collected manually using a vibrating funnel made from very smooth, rolled PTFE foil of 2 mm thickness. The funnel is positioned 1.50 m above the ground and has an area of 0.75 m^2. The angle of inclination of the funnel wall is approx. 40°. A 12 V automobile fan motor equipped with an unbalanced disc is connected to this spring-mounted funnel so that it vibrates vigorously with a frequency of about 25 sec^{-1} thus improving the drainage characteristics of it's rather flat shape. Because of this vibration even small rain drops from drizzle do not stick to the funnel wall, but start to flow down, coalesce, and run into the screw cap glass bottle which serves as a collection vessel. This minimizes the evaporation of small rain drops from the funnel wall and, rainfall events of as little as 0.08 mm h^{-1} have been sampled. As a result of the large opening and the shaking, very short sampling times are necessary: usually 5 to 10 minutes with a minimum of 15 seconds and a maximum of 30 minutes for very heavy to light rains respectively.

With the sampler described above, fractions were sampled during longer events. Usually the beginning of the event was missed, except for a few scattered summer showers which could be sampled from beginning to end. The samples were frozen in the field and kept frozen until analysis. Some of the samples were analyzed for the first time one hour after they were taken, then after subsequent and repeated freezing and defrosting of the samples, no changes in their chemical composition were observed.

2.3. Analysis

The samples were analyzed for inorganic anions by ion exchange chromatography (IC) using a single column system with a Waters IC-PAK A column and potassium hydrogen phtalate eluent. Ions were detected by negative UV detection.

In the course of time, three different analytical methods have been employed for the measurements of the organic anions. First, a labelling technique was used for determination of dicarboxylic acids as well as for low molecular monocarboxylic acids [19]. Second, a system for the analysis of short chain carboxylic acids using ion exclusion chromatography (ICE) and UV detection was developed [20], and third, a commercial ICE system was employed for the routine analysis of formic and acetic acids only, using a Dionex HPICE - AS 1 column with a micro membrane suppressor and octane sulfonic acid as the eluent. The anions were detected with a conductivity detector.

Conductivity and pH of the samples have been determined with commercially available instruments.

3. RESULTS AND DISCUSSION

3.1. Data Base

The data spans the period of Jan. 1984 through Mar. 1989 with the exception of the year 1986. A total number of 26 sampled events gave 132 fractions, from which pH, conductivity, and the concentrations of formic, acetic, hydrochloric, nitric, and sulfuric acids were determined. After initial analysis, monovalent organic acids other than formic and acetic and dicarboxylic acids higher than glutaric were not determined, because concentrations of these species were commonly below 5% of those of formic acid. For the eleven samples from the **Polarstern** cruise (March/April 1987), only the concentration range of the organic acids is included in Table 1.

Table 1. Range of Concentration of Formic and Acetic Acids in Rain at Various Sites of the World.

HFo	HAc	Site	Reference
2-20	2-10	USA	5
<1-56	<1-33	USA	6
<1-23	<1-9	USA	7
2-20	3-102	USA	8
10-100	4-40	USA	9
1-40	5-22	Brazil	10
4-30	2-21	Venezuela	11
4-19	<1-9	Amsterdam/Island	2
3-12	0-2	" " "	13
<1-50	<1-16	Australia	12
<1-58	<1-21	FRG	16
<1-6	<1-3	**Polarstern** cruise (equatorial Atlantic Ocean)	14
<1-28	<1-12	FRG	this work

units are in micromoles per liter.

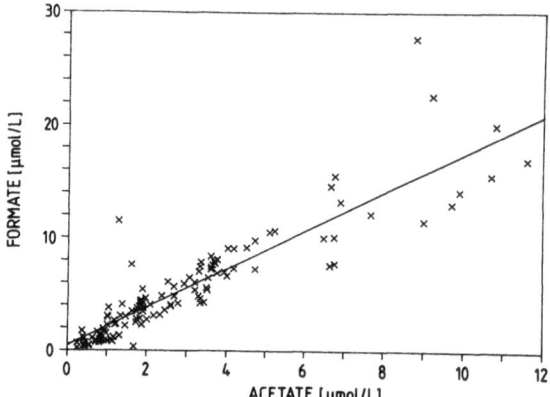

Figure 1. Reduced major axis regression of formate versus acetate in samples of precipitation collected at Presberg, West Germany.

3.2. Ratio and Comparison

Table 1 gives an overview of measured concentration ranges of formic and acetic acids in rain from various remote and urban regions of the world. From this table it can be seen that concentrations of both acids can go down to below 1.0 μmol L^{-1}, and that maximum concentrations of formate normally exceed those of acetate (except Ref. 9). Ratios of formic-to-acetic acid calculated from volume-weighted mean concentrations which were reported in the literature, were in most cases also greater than one [6,7,10,12,13]. Besides this, a plot of the concentration of formic versus acetic acid (Fig. 1) shows high correlations between both with a slope of 1.72, an intercept of 0.39 and a correlation coefficient r^2 of 0.82 (n=132). Similar graphs published by other workers show comparable high correlation, but the slopes differ by a factor of about two (e.g.: 1.51-2.62 in [7], 2.28 in [10] and 3.16 in [12]. Even in the gas phase, good correlation between both acids has recently been found [10,21]. Aside from a possible explanation for the different slopes of the formic-to-acetic acid ratios through different source/sink relationship at the various sample sites already discussed in [6,7,10,21], we suspect the possible additional influence of physicochemical effects other than pH, i.e. temperature. In Figure 2, the ratios of formic-to-acetic acid for all rain samples (fractions) are plotted against temperature, measured at 1.50 m above ground. The two parallel bordering lines are drawn by hand to cover 99% of the data points, and to give an idea of the trend. The slope of these lines is 6.25×10^{-2}. This could mean that a rise in temperature of 15 degrees (K) may double the ratio. If this were true, one could expect a mean ratio of approx. 3.0 at a temperature of 30°C. Such a ratio was found in Venezuelan rain sampled in Oct./Nov. 1988 (36 samples) when daytime temperatures ranged from 25 to 35°C [23].

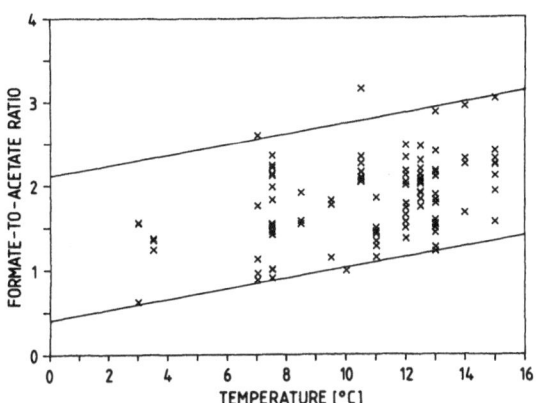

Figure 2. The relationship between the formate-to-acetate ratio and the air temperature.

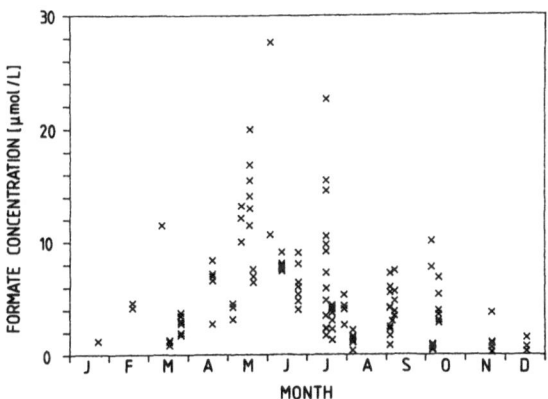

Figure 3. Seasonality of formate concentration in precipitation over the period from Jan. 1984 to Mar. 1989. The data points represent individual fractions.

Because of the correlation mentioned above, results were comparable for both acids. Therefore, the following sections will deal only with formic acid assuming that the same always holds for both formate and acetate.

3.3. Seasonality in Concentration

The seasonality of formate concentration in precipitation is shown in Fig. 3. A distinct maximum appears in late spring which coincides with the peak of the growing season in this region. In other regions of the world, similar seasonalities have been found which were related to production of organic acids by vegetation during the growing and wet season, shown in [7,12], respectively.

116

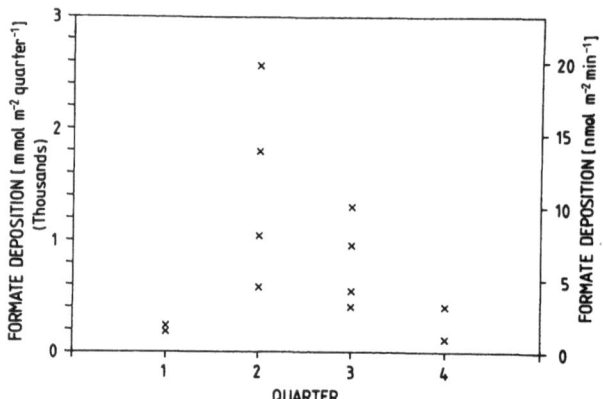

Figure 4. Mean quarterly wet deposition of formate over the period
indicated in Figure 3. The data points represent quarters with sufficient data.

3.4. Seasonality in Deposition

To estimate amounts of organic anions deposited by precipitation, quarterly mean values of the concentrations in rain over the time period
studied were computed and then multiplied by the respective quarterly
mean amount of precipitation. Based on highest concentrations during the
second quarter of the year (see Fig. 3), and moderately constant amounts
of rainfall throughout the year, the highest rates of deposition would
be in spring (Fig. 4). But if one multiplies the concentration of each
fraction by its respective rain rate (measured during sampling), the
resulting deposition rate peak appears later in the year (Fig. 5).
Judging from this, one can estimate, that although the sum of deposited
amounts of organic anions is highest in spring, temporarily very high
(up to 3.09 $\mu mol\ m^{-2}\ min^{-1}$) deposition rates occur in late summer and
early fall, when rain intensities in convective showers are high. If one
assumes the same pattern would occur with strong acids then this may
lead to a very high deposition of free acidity during a short period.

3.5. Contribution to Acidity

Whereas in remote areas of the world, such as Central Amazonia [10],
Bolivar State/Venezuela [11], and Katherine/Australia [22] organic acids
can contribute up to 64% to the total acidity of rain, a distinctly
lower contribution is reported from more rural sites, where mean values
of 19% (Wisconsin/USA [6]) and 16% (Virginia/USA [17]) were found.
     From the data in Table 2, we estimate a contribution to free
acidity in rain from Central Europe of about 5%, when simply the sum of
equivalents of organic acids is divided by the sum of equivalents of the
inorganic acids. This value is still lower than those mentioned above
for rural regions of the United States, hence we conclude that either
the emission rates of inorganic anions (due to anthropogenic activities)

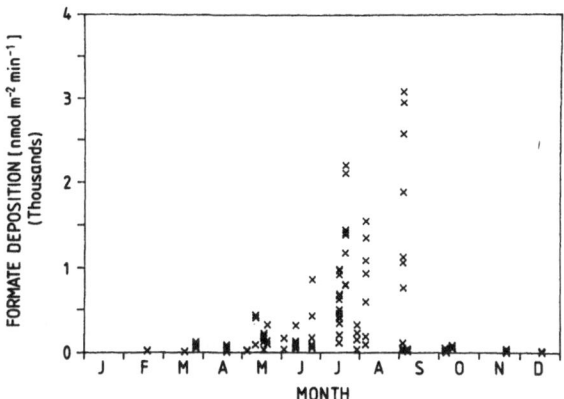

Figure 5. Seasonality of wet deposition of formate over the period indicated in Figure 3. The data points represent individual fractions.

are higher or the sources of organic anions (due to biogenic or photo-chemical activities) are smaller in Central Europe, and it is possible that both could occur simultaneously. In the case of deposition, the same calculation results in a contribution of 7% by weak acids relative to total acidity. Wet depositional fluxes estimated for our central European site (3.66 and 1.76 mmol $m^{-2}yr^{-1}$ for formate and acetate, respectively) are also lower than those reported in [21] for Virginia (4.67 and 4.18 mmol m^{-2} yr^{-1}, respectively).

Table 2. Monthly Mean Values of Concentration and Deposition for Strong and Weak Acid Ions.

Acid	Concentration [μequ L^{-1}]	Deposition [μmol m^{-2}]
Formate	4.9	305
Acetate	2.5	147
Cloride	15.0	930
Nitrate	39.3	2,510
Sulfate	90.0	2,960

4. ACKNOWLEDGEMENTS

This work was partially supported by the Deutsche Forschungsgemeinschaft through its Sonderforschungsbereich 233: Chemie und Dynamik der Hydrometeore.

5. REFERENCES

[1] Galloway, J.N., Likens, G.E., and Edgerton, E.S. (1976), Science 194, 722-742.

[2] Galloway, J.N., Likens, G.E., Keene, W.C., and Müller, J.M. (1982), J. Geophys. Res. 87, 8771-8786.

[3] Lunde, G., Gether, J., Gjös, N. and Lande, M.-B.S. (1977), Atmos. Environ. 11, 1007-1014.

[4] Hoffman, W.A., Jr., Lindberg, S.E., and Turner, R.R. (1980), Environ. Sci. Technol. 14, 999-1002.

[5] Hoffman, W.A., Jr., and Tanner, R.L. (1985), Report No. BNL 51922, Denison Univ., Granville, OH 43023.

[6] Chapman, E.G., Sklaren, D.S., and Flickinger, J.S.(1986), Atmos. Environ. 20, 1717-1725.

[7] Keene, W.C., and Galloway, J.N. (1986), J. Geophys. Res. 91, 14466-14474.

[8] Kawamura, K. and Kaplan, I.R.(1984), Anal. Chem. 56, 1616-1620.

[9] Norton, R.B. (1985), Geophys. Res. Lett. 12, 769-772.

[10] Andreae, M.O., Talbot, R.W., Andreae, T.W. and Harriss, R.C. (1988), J. Geophys. Res. 93, 1616-1624.

[11] Sanhueza, E., Elbert, W., Randon, A., Arias, M.C. and Hermoso, M. (1989), Tellus, in press.

[12] Likens, G.E., Keene, W.C., Miller, J.M. and Galloway J.N. (1987), J. Geophys. Res. 92, 13299-13314.

[13] Galloway, J.N. and Gaudry, A. (1984), Atmos. Environ. 18, 2649-2656.

[14] Elbert, W., Bingemer, H. and Andreae, M.O., unpublished data.

[15] Neftel, A., Breitenbach, S., Elbert, W. and Hahn, J., Proceedings of International Conference on Gas-Liquid Chemistry of Natural Waters, Brookhaven, April 1984, Brookhaven National Report No. BNL 51757.

[16] Schilling, M. (1988), Dissertation, Universität Dortmund, Dortmund.

[17] Keene, W.C. and Galloway, J.N. (1984), Atmos. Environ. 18, 2491-2497.

[18] European Weather Report, Monthly Report of Climatological Data, Deutscher Wetterdienst, Offenbach a.M.

[19] Elbert, W., Breitenbach, S., Neftel, A. and Hahn, J. (1985), J. Chromatogr. 328, 111-120.

[20] Elbert, W., Hahn, J. and Lerch, M. (1989), Intern. J. Environ. Anal. Chem. 35, 149-159.

[21] Talbot, R.W., Beecher, K.M., Harriss, R.C. and Cofer, W.R. (1988), J. Geophys. Res. 93, 1638-1652.

[22] Keene, W.C., Galloway, J.N. and Holden, J.D., Jr. (1983), J. Geophys. Res. 88, 5122-5130.

[23] Hartmann, W.R. et al. (1989), private communication.

Determination of Organic Peroxides

K. Bächmann, J. Hauptmann
Fachbereich Anorganische Chemie und Kernchemie
Technische Hochschule Darmstadt
Hochschulstr. 4
6100 Darmstadt
FRG

ABSTRACT. The development of a method for the determination of peroxycompounds with organic chains in the range C_4 - C_9 is described.

1. Introduction

Most of the theories that have been published on the subject forest decline agree that more than one effect is responsible. In general the hypothesis is forwarded that such an extensive damage can only be explained by the simultanous impact of several damaging influences (multi-causality). The enlarged occurence in regions of clean air, elevated and exposed sites, leads to the conclusion that photochemical reactions play an important role in the synthesis of damaging substances. It is in those regions that the production of organic peroxides can be expected:

1. Hydroperoxides are produced, provided NO_x is present only in low concentrations.

2. Hydroperoxides are formed under conditions when sufficient hydrocarbons are present.

3. Hydroperoxides are produced photochemically, so elevated altitudes (about 1000 m) are of adantage.

A damaging influence of organic peroxides can be traced back to two different effects:

1. Oxidation of SO_2 in cloud water caused by hydro-peroxides.
Lind et al. [1] published investigations on this subject. They came to the conclusion that organic hydroperoxides con-

119

H.-W. Georgii (ed.), *Mechanisms and Effects of Pollutant-Transfer into Forests*, 119–124.
© 1989 by Kluwer Academic Publishers.

tribute to the main oxidants, especially when other oxidants are diminished.

2. The direct influence on leaf and needle surfaces.
The direct influence of fog containing hydrogenperoxide has been investigated by Masuch et al. [2]. They determined different degrees of defects in the cells. Since organic hydroperoxides react in a different way than H_2O_2, the damage on plant surfaces might not be the same.

2. Generation of Organic Peroxides

The generation of the peroxides can be explained according to the following mechanisms. The natural, by vegetation emitted hydrocarbons consist mainly of terpenes like isoprene, cumene or pinene. During daytime these are oxidized photochemically by ozone or hydroxyradicals. Abstraction of one hydrogenradical leads to alcoperoxyradicals (1). The reactions of these radicals depend on the concentrations of NO_x :

1. In the presence of NO_x (that means in urban regions) alcoxyradicals, nitrates and peroxynitrates are formed (2), the former yielding aldehydes and ketones.
2. On the other hand, when hydrocarbons are accumulated in comparison to NO_x (which is the case in regions clean air and at elevated sites), the radicals are reacting with perhydroxyradicals to give primary, secondary and tertiary hydroperoxides (3).

$$(1) \quad RH + OH \longrightarrow ROO + H_2O$$

$$(2) \quad ROO + NO \longrightarrow RO + NO_2$$

$$RO \longrightarrow aldehydes + ketones$$

$$ROO + NO \longrightarrow RONO_2$$

$$RO_2 + NO_2 \longrightarrow RO_2NO_2$$

$$(3) \quad RO_2 + HO_2 \longrightarrow ROOH + O_2$$

Summarizing the stable products of the oxidation, we find aldehydes, ketones, nitrates and organic hydroperoxides.
Aldehydes and ketones can be derivatised with DNSH, separated by HPLC and detected by fluorescence [3], peroxynitrates are determined by gas chromatography. Until now there does not exist a satisfying method for the determination of organic hydroperoxides in the atmosphere.

When developing a convenient method the separation of the peroxides gives rise to considerable problems. Due to the low thermal stability of these compounds a separation by gas chromatography is not possible. Therefore the separation has to be carried out using liquid chromatography. The expected excess of hydrogenperoxide interferes with the separation. Until now, analysis of peroxides was successful only for the determination of the total amount of peroxides or the amount of hydrogenperoxide. The separations were restricted to peroxides of biological interest [4,5].

Usual detection methods for peroxides are : direct UV/Vis absorption [6], polarographic detection [7] and detection of a post-column reaction product. Since UV/Vis detection yields insufficient limits of detection, especially for those peroxides that do not contain chromophoric groups, a post column reaction has to be applied.

Post-column reaction detection of peroxides can be performed via UV/Vis detection (reaction with iodide [8]), luminescence (reaction with p-hydroxyphenylacetic acid [9]) or chemiluminescence (reaction with luminol [10,11]). Because of the low limits of detection the latter seemed to be the most promising method.

One of the mechanisms postulated for the reaction with luminol is shown in Fig. (1). Under catalytical influence of cytochrome c luminol reacts with peroxides to give a chemiluminescent product.

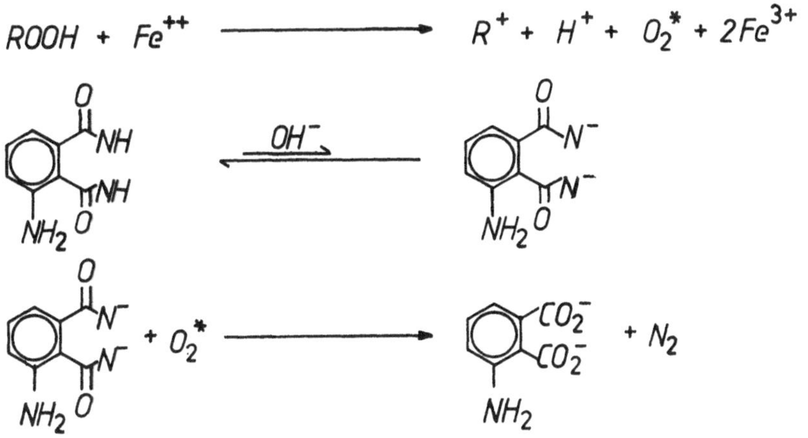

Figure 1. Mechanism of the reaction of luminol with peroxides

122

The reagent solution had to be buffered to pH 9-10. It contained 10 mg luminol and 5 mg cytochrome c per liter borate buffer. The preparation had to be done daily, deareating all solutions thoroughly by bubbling with nitrogen. To optimize the system, a test mixture of three commercially available hydroperoxides (t-butyl-, t-amyl- and cumole hydroperoxide) was prepared. The content of peroxide was determined by iodometry. To achieve a satisfying separation, several column materials had to be applied. Separation on rp-18 material as well as on silicagel proved to be not possible. Therefore mid-polar phases, rp-8 and rp-2 materials were used. On the rp-8 column the separation was found to be incomplete, whereas the resolution of the rp-2 material satisfied the requirements.

Methanol/water mixtures with mixing ratios of about 50/50 were used as eluents. Flow-rates of eluent as well as of reagent should not exceed 0.3 ml/min. Faster flow-rates resulted in incomplete separation and diminished limits of detection (shorter reaction times). The chemiluminescence was detected at 440 nm. Reaction was realized in a knitted open tube reactor with a total volume of 1 ml. A temperature of 52° C was found to be an optimal compromise between a fast reaction and a low degree decomposition of peroxides. A schematic diagramm of the system is shown in Fig. (2). Fig. (3) shows a typical chromatogram for the separation of the three alkyl hydroperoxides together with an excess of hydrogenperoxide (1:50).

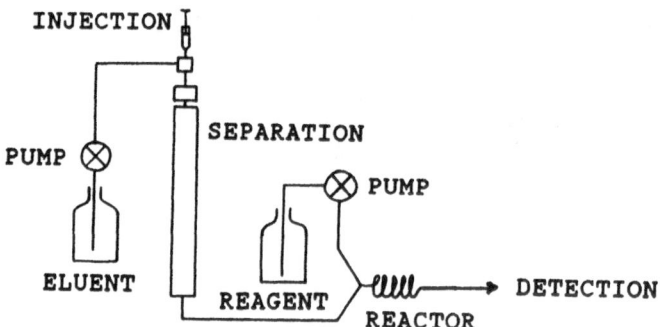

Figure 2. Illustration of the post-column derivatisation system

3. Conclusion

A convenient method for the separation of short-chain hydroperoxides from hydrogenperoxide was developed. The limits of detection are sufficient for measurements in laboratory and the determination of standards. For an adaptation to atmospheric concentrations it will be necessary to apply a different method, for example fluorescence or polarography. Nevertheless, a method for an enrichment procedure prior to analysis is necessary.

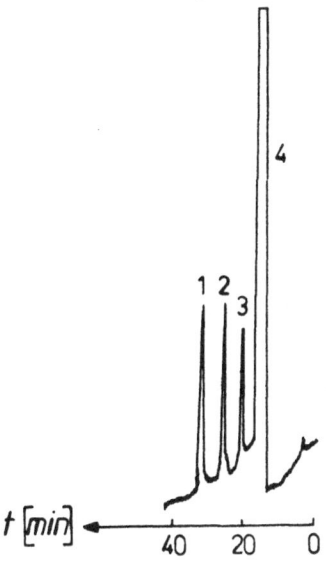

Figure 3. Chromatogram of the peroxide test-mixture. Column: Lichrosorb RP-2 (5µm, 250*4mm), 50% water - 50 % methanole; flow-rate : 0.2ml/min; flow-rate of reagent: 0.2ml/min; temperature: 52°C; detector at 440nm; (1) cumole hydroperoxide; (2) t-amylhydroperoxide; (3) t-butylhydroperoxide; (4) hydrogenperoxide.

Acknowledgements

We are grateful to the BMFT for financial support.

References

[1]   J. A. Lind, A. L. Lazrus and G. L. Kok (1987) "Aqueous phase oxidation of sulfur (IV) by hydrogenperoxide, methylhydroperoxide and peroxyacetic acid", Journal of Geophysical Research 92 (D4), 4171-4177.

[2]   G. Masuch and A. Kettrup (1986) "Effects of hydrogenperoxide - containing fog on young trees", International Journal of Environmental Analytical Chemistry 27, 183-213.

[3]   K. Bächmann and W. Schmied (1988) "Spurenbestimmung
        von Aldehyden und Ketonen in der Troposphäre durch
        Festphasenderivatisierung mit DNSH", submitted to
        Zeitschrift für Analytische Chemie (Fresenius)

[4]   T. Miyazawa, K. Yasudo, K. Fujimoto and T. Kaneda
        (1988) "Determination of phosphatidylcholine hydro-
        peroxide in human plasma by chemiluminescence high
        performance liquid chromatography", Analytical Letters
        21(6), 1033-1044.

[5]   K. Akasaka, H. Ohuri and H. Meguro (1988) "An aromatic
        phosphine reagent for the high performance liquid
        chromatography fluorescence determination of hydroper-
        oxides - determination of phosphatidylcholine hydro-
        peroxides in human plasma", Analytical Letters 21(6),
        965-975.

[6]   R. K. Jensen, M. Zinboand S. Korcek (1983) "High per-
        formance liquid chromatography determination of hydro-
        peroxidic product formed in the autooxidation of
        n-hexadecane at elavated temperatures", Journal of
        Chromatographic Science 21, 394-397.

[7]   M. O. Funk and W. J. Baker (1985) "Determination of
        organic peroxides by high performance liquid chroma-
        tography with electrochemical detection", Journal of
        Liquid Chromatography 8(4), 663-675.

[8]   R. S. Deelder, M. G. F. Kroll and J. H. M. van den
        Berg (1976) "Determination of trace amount of hydrogen-
        peroxide by column liquid chromatography and colori-
        metric detection", Journal of Chromatography 125,
        307-314.

[9]   G. L. Kok, K. Thompson and A. L. Lazrus (1986)
        "Derivatization technique for the determination of
        peroxides in precipitation", Analytical Chemistry
        58(6), 1192-1194.

[10]   Y. Yamamoto, M. H. Brodsky, J. C. Baker and B. N. Ames
        (1987) "Detection and characterization of lipid hydro-
        peroxides at picomole levels by high performance
        liquid chromatography", Analytical Biochemistry
        160(1), 7-13.

[11]   T. Miyazawa, K. Yasudo and K. Fujimoto (1987)
        "Chemiluminescence - high performance liquid chroma-
        tography of phosphatidylcholine hydroperoxide", Analy-
        tical Letters 20(6), 915-925.

The Determination of Phosgene in the Lower Troposphere

K. Bächmann, J. Polzer
Institut für Anorganische Chemie und Kernchemie
TH Darmstadt
Hochschulstraße 4
D-6100 Darmstadt
FRG

ABSTRACT. The importance of decomposition products of chlorinated hydrocarbons in tropospheric chemistry and their effect on plants is discussed. A sensitive analytical method for the determination of phosgene in ambient air as well as on plant particles using thermal desorption preceeding gas-chromatographic separation is described. Results of measuring phosgene in the troposphere and diurnal variation of its concentration are presented.

1. Introduction

Large amounts of halogenated hydrocarbons have been released in the atmosphere in recent years. Most of them are of anthropogenic origin, only few are released from natural sources /1/.
Tab. 1 presents some of these compounds, their tropospheric lifetime and mean concentration. The reactive compounds i.e. the hydrogen containing species and some unsaturated compounds decompose in the troposphere whereas the inert species reach the stratosphere.

TABLE 1. Global distribution of some important chlorinated hydrocarbons in the troposphere /2/

compound	concentration (pptV)	lifetime (a)
CCl_4	145	60 – 100
CH_3CCl_3	130	6 – 10
CH_2Cl_2	32	0,5
$CHCl_3$	30	0,3 – 0,6
CCl_2CCl_2	26	0,4
$CHClCCl_2$	8	0,02

125

H.-W. Georgii (ed.), Mechanisms and Effects of Pollutant-Transfer into Forests, 125–132.
© 1989 by Kluwer Academic Publishers.

In spite of shorter lifetimes of some compounds their ubi-
quitous distribution in the troposphere is ensured. For instance, the trace concentrations measured in urban regions
are only slightly different from those measured in rural
regions. So it cannot be excluded that these compounds or
their decomposition products contribute alone or together
with other compounds to the forest decline.
For example Frank /3/ discussed reactions of natural emitted
terpenes with halocarbons or reactive intermediates to form
toxic decomposition products or the possible accumulation of
these lipophilic species in the wax of the surface of
plants.
One of the most interesting decomposition products of chlorinated hydrocarbons is phosgene.

2. Sources and Sinks

Phosgene has been identified as an atmospheric decomposition
product of several halogenated hydrocarbons as for example
di- und trichloromethane, 1,1,1- trichloroethane, tri- und
tetrachloroethene /4,5/.

$$CHCl_3 + OH \longrightarrow CCl_3 + H_2O$$
$$CCl_3 + O_2 \longrightarrow CCl_3O_2$$
$$CCl_3O_2 + NO \longrightarrow CCl_3O + NO_2$$
$$CCl_3O + O_2 \longrightarrow COCl_2 + ClO_2$$

$$CCl_2{=}CH_2 + O \longrightarrow H_2C\overset{O}{\diagup\diagdown}CCl_2$$
$$H_2C\overset{O}{\diagup\diagdown}CCl_2 \longrightarrow H_2ClC{-}COCl$$
$$H_2ClC{-}COCl + H_2O \longrightarrow H_2ClC{-}COOH + HCl$$

Figure 1. Degradation of selected chlorinated hydrocarbons

Fig. 1a shows the reaction mechanism of the formation of
phosgene from $CHCl_3$ under tropospheric conditions. Hydrogen
containing compounds are degradated by attack of hydroxyl
radicals and hydrogen abstraction, in case of unsaturated
compounds also attack of O_3, O_2 and oxygen radicals are
discussed /4,6/.
In that case the formation of chlorinated acetic acids is
also possible as shown in Fig. 1b.
Beside these gaseous phase reactions also reactions in
liquid phase (hydrometeors) and on solid phases (plants,
soil, aerosols) are important for atmospheric chemistry.
Fig. 2 is an example for correlation of liquid and gaseous
phase during the degradation process of 1,1,1- trichloroethane.

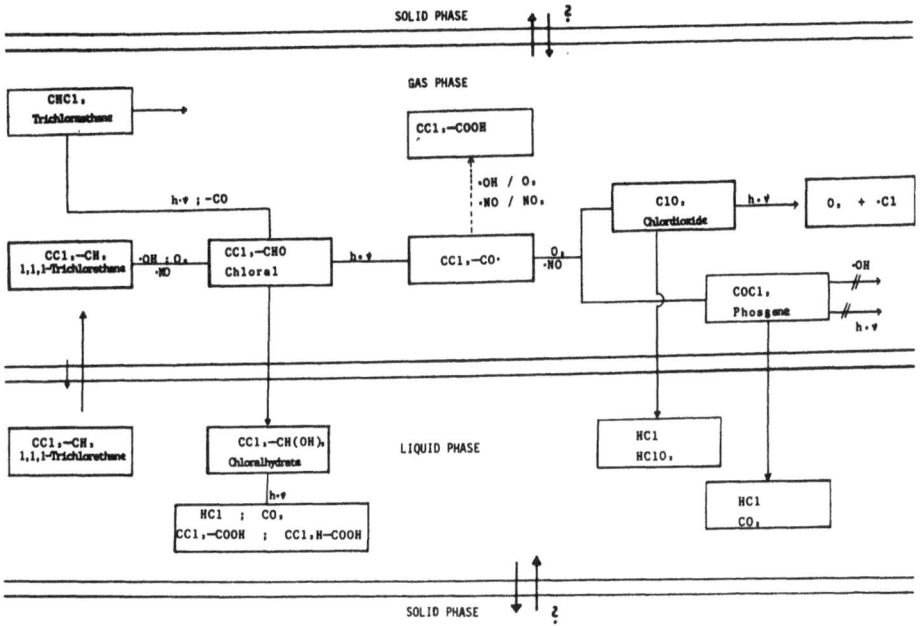

Figure 2. Degradation of 1,1,1-trichloroethene in the troposphere

For solid phase reactions there is a lack of information, but nevertheless these processes are not neglectable as pointed out in Fig. 2.
Among the degradation products are chlorinated acetic acids and phosgene. Sink processes for phosgene are hydrolysis in aqueous particles and heterogenic catalysed decomposition on solid phases. Photolytic processes and gaseous phase hydrolysis are known to be very slow in the lower troposphere /5,7/.
Therefore some authors assume an accumulation of phosgene in the troposphere in future /5,8/.

3. Phosgene Determination in the Troposphere

The main problem of the analysis is the low detection limit which has to be achieved for determination of phosgene under tropospheric conditions and the high reactivity of this compound in contact with solid phases (heterogeneous catalysed decomposition).

Therefore a gaschromatographical detection system has to consist of:
1. a sensitive detector
2. application of a chemical inert column
3. careful enrichment and analytical procedure
4. calibration has to be carried out identical to the sample taking procedure in order to compensate possible losses of phosgene during the analysis.
Until now only Singh /8/ could present a gaschromatographical system able to detect tropospheric phosgene. He used a packed column but due to heterogeneous decomposition of phosgene on active column sites preconditioning of the column with phosgene was necessary.
The system presented here does not suffer from this problem and provides lower detection limits.

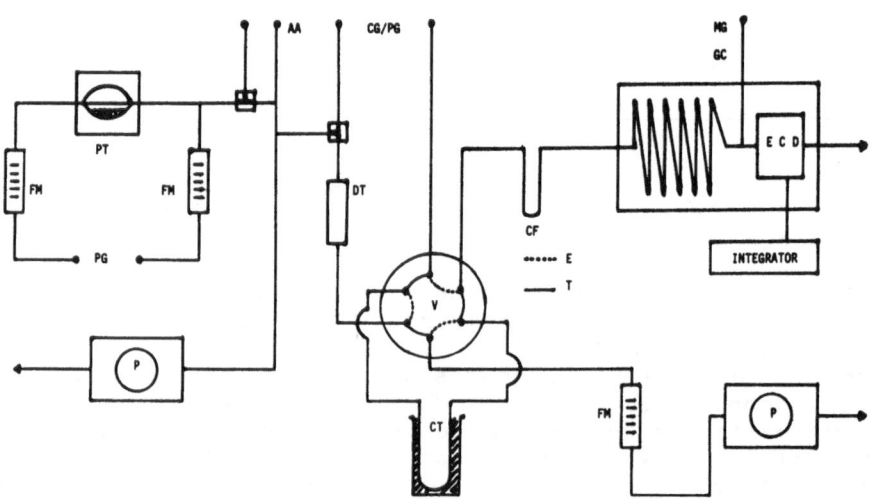

Figure 3. Analytical system: FM: flow-meter; PG: purge gas (N₂); PT: permeation tube; AA: ambient air; CG/PG: carrier gas/purge gas (He); DT: drying tube; V: six port valve; P: pump; CT: cold trap; CF: cold focussing; T: transfer; E: enrichment; GC: gaschromatographical system; MG: make-up gas

Airsamples were collected by a system of two pumps as shown in Fig. 3. Ambient air was sucked through a stainless steel covered teflonloop at the trapping temperature of -196 °C. The humidity of the atmosphere has to be removed to avoid possible interferences of the ECD. This was carried out in a drying tube filled with magnesiumperchlorate.
The analysis was started when conducting carrier gas through the teflonloop by switching the six-port-valve. The analytes

were vaporized by resistant heating of the stainless steel cover, additionally focused in a deactivated capillary column to provide sharper peaks, followed by separation on a non-polar capillary column and ECD.

The calibration was carried out using a dynamic dilution system, phosgene standard was generated with a teflon permeation tube. The content of phosgene in the purge gas was determined with IC as reference method. Phosgene was hydrolysed in water and determined as chloride.

Purge gas was diluted with ambient air which was conducted through a trap filled with NaOH- coated charcoal in order to decompose phosgene. So the phosgene concentration of the whole gas flow was reduced to concentrations as expected in the troposphere.

Enrichment and analysis were carried out analogous to the analysis of airsamples, the amounts of phosgene were varied by changing the sampling volume.

The method described avoids significant losses of phosgene by use of chemical inert tubes and a capillary column, careful enrichment step and analytical procedure. Quantitative detection of concentrations as low as 7 pptV phosgene is possible, a reproducibility of 8 % can be achieved. The high resolution of the capillary column allows simultaneous detection of other ECD sensitive compounds. In addition the method can be adapted easily for collecting samples in the field.

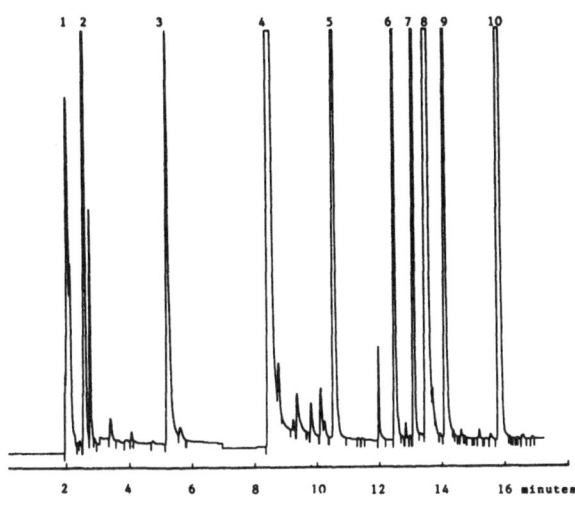

Figure 4. Gaschromatogram of an airsample, 0,6 l, Darmstadt, capillary column, temperature program. 1 O_2/$CHCl_2F$; 2 CCl_2F_2 3 $COCl_2$; 4 CCl_3F; 5 CCl_2F-$CClF_2$; 6 $CHCl_3$; 7 CCl_3-CH_3; 8 CCl_4 9 CCl_2=$CHCl$; 10 CCl_2=CCl_2.

4. Results and Discussion

Fig. 4 presents a chromatogram of an airsample containing
the main ECD sensitive trace compounds in ambient air.
During our investigations concentrations of phosgene which
we determined in airsamples varied between 8 and 87 pptV
during daytime, whereas during nighttime concentrations up
to 143 pptV were present. Resulting from the varying
phosgene concentrations which we measured on several days/
nights a tropospheric lifetime of a few months has to be
assumed. Until now lifetimes of a few years have been stated
/5,8/. Fig. 5 presents the diurnal variation of phosgene.

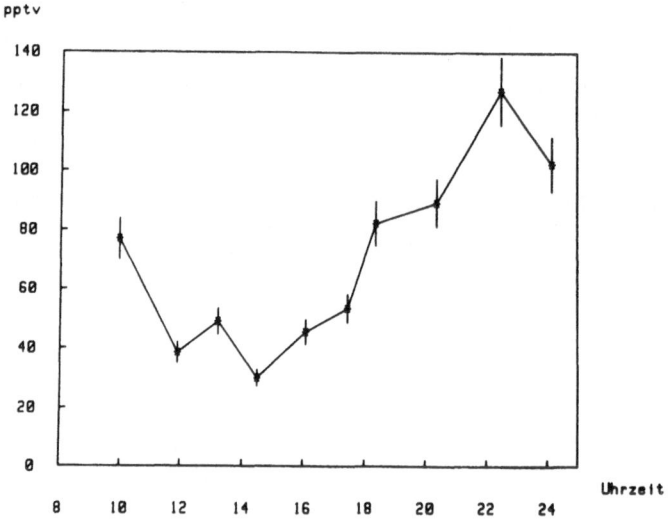

Figure 5. Diurnal variation of phosgene

Measurement in the field proved phosgene to be present in
ambient air even far from urban sites, the concentrations
determined at remote sites were less below those detected in
urban areas.
In order to verify if there is an effect of this compound or
other halocarbons on plants, we analysed needles of pines.
Fig. 6 shows a chromatogram achieved from thermal desorption
of 0,3 g needles in a Helium flow /9/.
Almost all halocarbons significant for ambient air are
present in needles, additional great amounts of phosgene and
other new ECD sensitive compounds which are not identified
yet. Obviously halocarbons and phosgene are enriched in the
wax of the surface of plants and presumably a degradation of
these compounds takes place there.

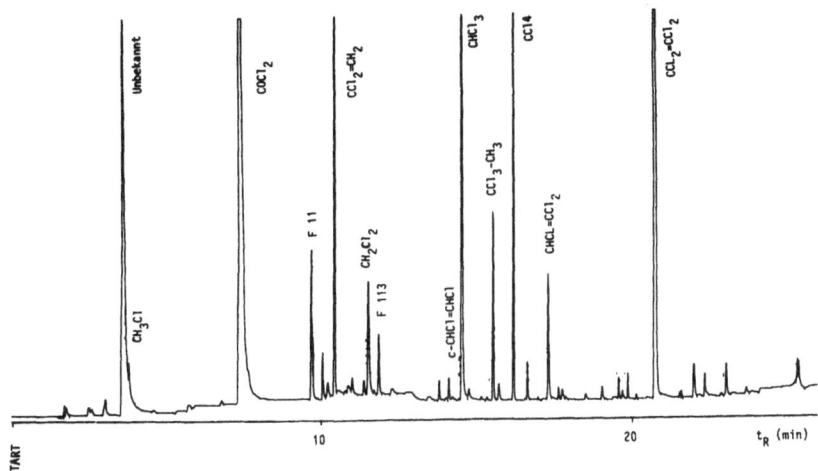

Figure 6. Chromatogram from thermal desorption of pine-needles

5. Conclusion

The analytical procedure described is a sensitive, quantitative method for the determination of phosgene and other halocarbons present in ambient air. Phosgene is present in air up to concentrations of 143 pptV and obviously accumulated in plants. Considerations on the effect of phosgene on plants at those concentrations do not exist yet.

6. Acknowledgement

We are grateful to the BMFT for financial support.

7. References

1 Noy, T. et al. (1987) "Trace analysis of halogenated hydrocarbons in gaseous samples by online enrichment in an adsorption trap, on column cold trapping and capillary gas chromatography", J. Chrom. 393, 343-356.

2 Fabian, P. (1986) Handbook of Environmental Chemistry 4A, Hutzinger, Springerverlag Berlin.

3 Frank, H. (1984), "Waldschäden durch Photooxidantien", Nach. Chem. Tech. Lab. 32, 298-305.

4 Gay, B. et al. (1976) "Atmospheric oxidation of chlorinated ethylenes", Environ. Sci. Technol. 10, 58-67.

5 Helas, G. and Wilson, S.R. (1987) "Considerations on sources and sinks of phosgene in the troposphere", submitted to Tellus.

6 Atkinson, R. (1985) "Kinetics and mechanisms of the gas phase reactions of hydroxyl radical with organic compounds under atmospheric conditions", Chem. Rev. 85, 69-201.

7 Butler, R. and Snelson, A. (1979) "Kinetics of the homogeneous gas phase hydrolysis of CCl_3COCl, CCl_2HCOCl, $CH_2ClCOCl$ and $COCl_2$", J. Air Pollut. Control. Assoc. 29, 833-837.

8 Singh, H.B. (1976) "Phosgene in ambient air", Nature 264, 428-429.

9 Lohleit, M. and Bächmann, K. (1988), unpublished results.

AROMATIC AND CHLORINATED HYDROCARBONS IN FOREST AREAS

JÜRGEN MÜLLER
Umweltbundesamt, Pilotstation Frankfurt
D-6050 Offenbach, Frankfurter Str. 135

ABSTRACT. Aromatic hydrocarbons (AHC) and chlorinated hydrocarbons (CHC) injected into the atmosphere from polluted to remote areas have concentration gradients according to their reaction rates with OH-radicals. Components which have residence times of several days or weeks are enriched and act as photochemical precursors in remote forest areas.

For CHC a slight decrease of the airborne concentrations was found within the last years.

1. INTRODUCTION

The atmospheric burden of aromatic hydrocarbons (AHC) and chlorinated hydrocarbons (CHC) due to man-made activities within the last decades heavily increased (1).

AHC in the atmosphere mostly originate from automobile exhaust gases and CHC mostly by industrial use as solvents and cleaners. Some CHC-compounds like chloroform, 1,1,1-Trichloroethane and carbon tetrachloride have residence times of several months or years and thus reach the stratosphere where they attack the ozone layer (2).

AHC and CHC both are solvents and may dissolve in the wax layer of needles and leaves. Thus, perforating the surfaces acid substances and heavy metals can attack the metabolism of the plants. However, up to date it is not clear if the concentrations of AHC and CHC are sufficiently high to be the inducers of forest decline. Nevertheless, knowledge about the concentrations in polluted and remote areas is indispensable.

2. METHODS OF MEASUREMENT

AHC-components in ambient air are collected on active coal. The components are dissolved with carbon disulphide which is injected into a chromatograph for detection with a FID-detector.

CHC-components are also sampled on active coal but are collected on tenax as well. The analysis is carried out with a headspace-chromatograph or an automated-thermal-desorber chromatograph.

133

H.-W. Georgii (ed.), Mechanisms and Effects of Pollutant-Transfer into Forests, 133–139.
© 1989 by Kluwer Academic Publishers.

In both cases behind the separation columes a FID-detector and an
ECD-detector are installed which detect the AHC- and CHC-components
respectively (3, 4).

The samples were collected at different sites and analyzed in a
central laboratory.

Calibration is carried out by mixing quantitatively the measured
components and handling the solution similar, like the collected
ambient air samples.

3. RESULTS AND DISCUSSION

In Fig. 1 the AHC-concentrations measured at different sites are
represented. The conce n.trations in the flowing traffic are the
highest demonstrating the AHC as contributors to the automobile
exhaust gases. The atmospheric burden in a suburban area is about
ten times higher then at the "Kleiner Feldberg/Taunus station" in
823 m height, a forest area about 30 km away from the center of
the city.

In Fig. 2 the relative portions of the AHC-components are plotted.
The profile of the AHC changes during transportation of the polluted
air from the City to the Taunus-mountains. Due to different residence
times benzene increases and toluene decreases in the relative concentration profile.

By operation with the respective residence times the profile
measured far away from the sources can be retransformed. The
residence times τ_i of the different AHC-components (in days) are
as follows (5):

Benzene	8	d
Toluene	2,5	d
Ethylbenzene	2	d
O-Xylene	1,5	d
m-Xylene	1,1	d
p-Xylene	1,5	d

In connection with the different concentrations c_i of the compounds
between the two sites the following equation exists:

$$\left(\frac{c_i}{100 \cdot \sum c_i} \right)_{Frankfurt} = \left(\frac{c_i / \tau_i}{100 \cdot \sum c_i / \tau_i} \right)_{kl.Feldberg/Ts.}$$

At the remote site the AHC-concentration profile is changed according
to the different residence times.

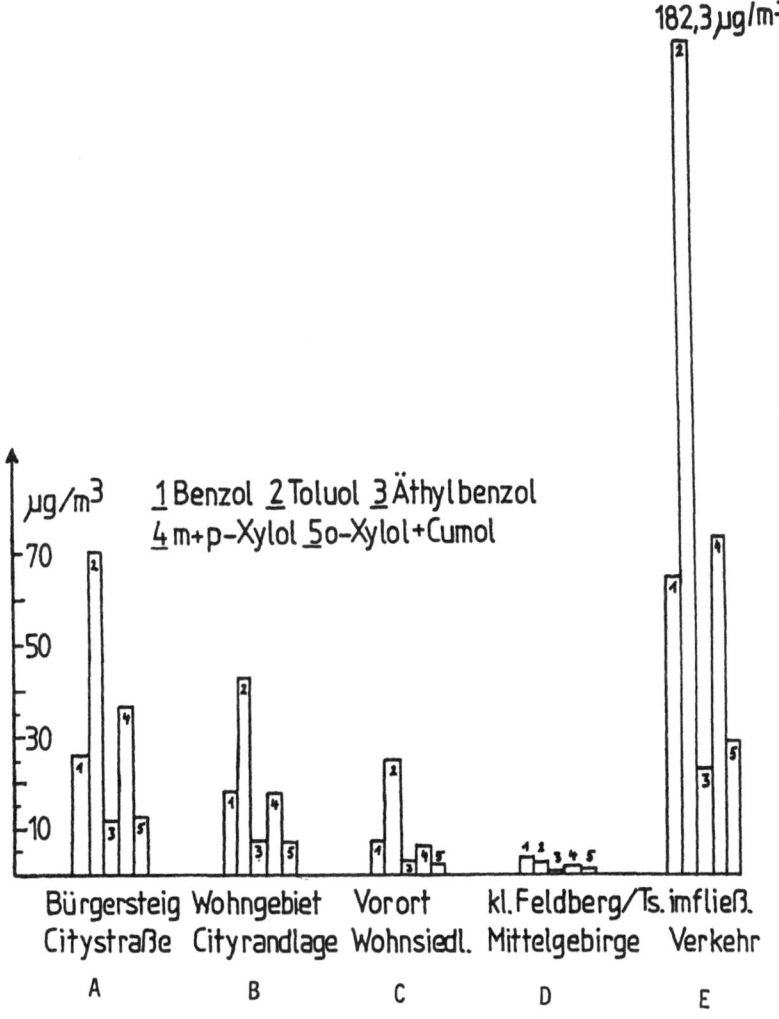

Fig. 1: Benzene (1), toluene (1), ethylbenzene (3), m+p-Xylene (4),
O-Xylene (5)   at different sampling sites:
City-street (A), Residential-City (B), Suburban (C), kl.
Feldberg/Ts. (mountains, 823m height) (D), Flowing traffic (E)

136

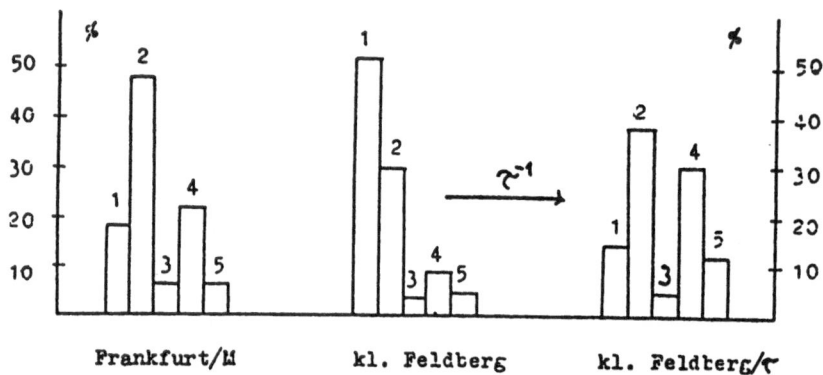

Frankfurt/M   kl. Feldberg   kl. Feldberg/τ

Fig. 2: Transformation of AHC-profile from City to kl. Feldberg/Ts.
30 km away and retransformation by operation with residence
times τ_i.

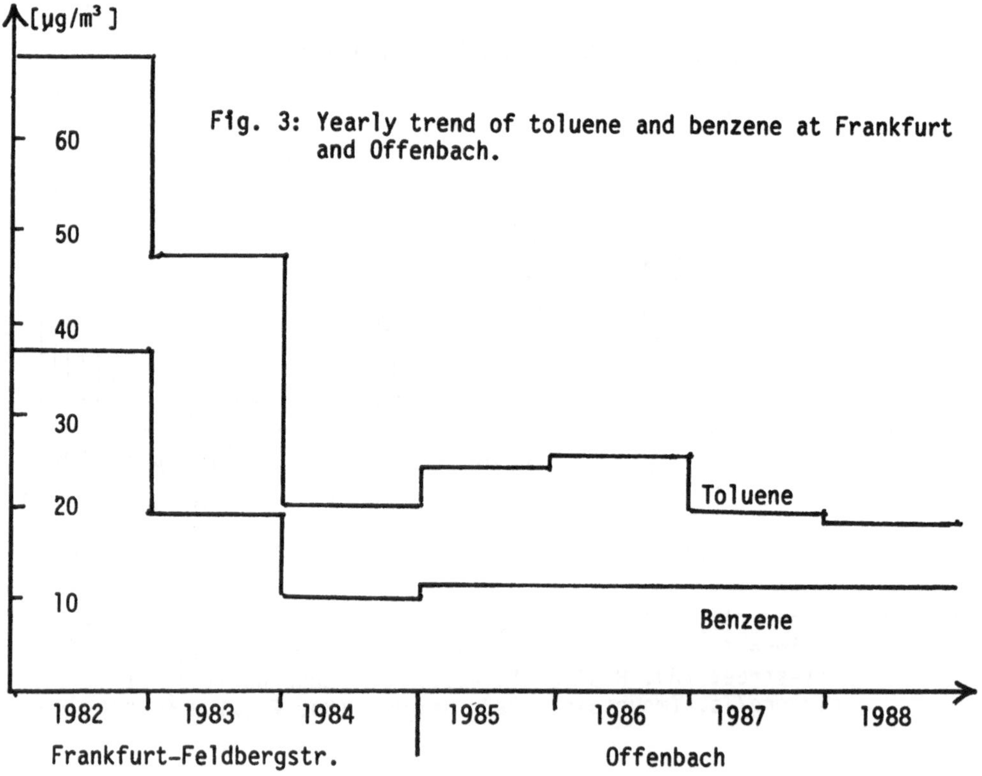

Fig. 3: Yearly trend of toluene and benzene at Frankfurt
and Offenbach.

Toluene

Benzene

1982   1983   1984   1985   1986   1987   1988

Frankfurt-Feldbergstr.   Offenbach

In Fig. 3 the yearly concentrations of toluene and benzene are plotted. The sampling sites in Frankfurt and Offenbach have similar air quality. It can be seen that since 1984 airborne concentration considerably decreased.

In Tab. 1 the concentrations between the sites Offenbach, Schauinsland (Black Forest) and Izana (Teneriffa) are represented. The concentrations are mean values measured in 1987. The values show the concentration gradients between a polluted area in Germany (Offenbach), a remote forest area (Schauinsland), 1200 m height) and a remote site with clean Atlantic air (Izana, 2600 m height):

	Offenbach (µg/m³)	Schauinsland (Schwarzwald) (µg/m³)	Izana (Teneriffa) (µg/m³)	Residence Times
Chloroform	0,22 (100 %)	0,09 (40 %)	0,05 (22 %)	0,4 a
1,1,1 Trichloroethane	2,12 (100 %)	0,89 (41 %)	0,72 (32 %)	3 a
Carbontetrachloride	0,33 (100 %)	0,27 (81 %)	0,21 (63 %)	~10 a
Trichloroethylene	2,50 (100 %)	0,41 (16 %)	0,04 (1 %)	0,03 a
Tetrachloroethylene	2,10 (100 %)	0,55 (26 %)	0,10 (4 %)	0,3 a
Isopentane	12,63 (100 %)	1,49 (11 %)	0,41 (3 %)	3 d
Hexane	3,20 (100 %)	0,44 (13 %)	0,10 (3 %)	3 d
Toluene	16,60 (100 %)	2,30 (13 %)	0,34 (2 %)	2,5 d

Tab 1: Concentrations at sites with different air quality and residence times

138

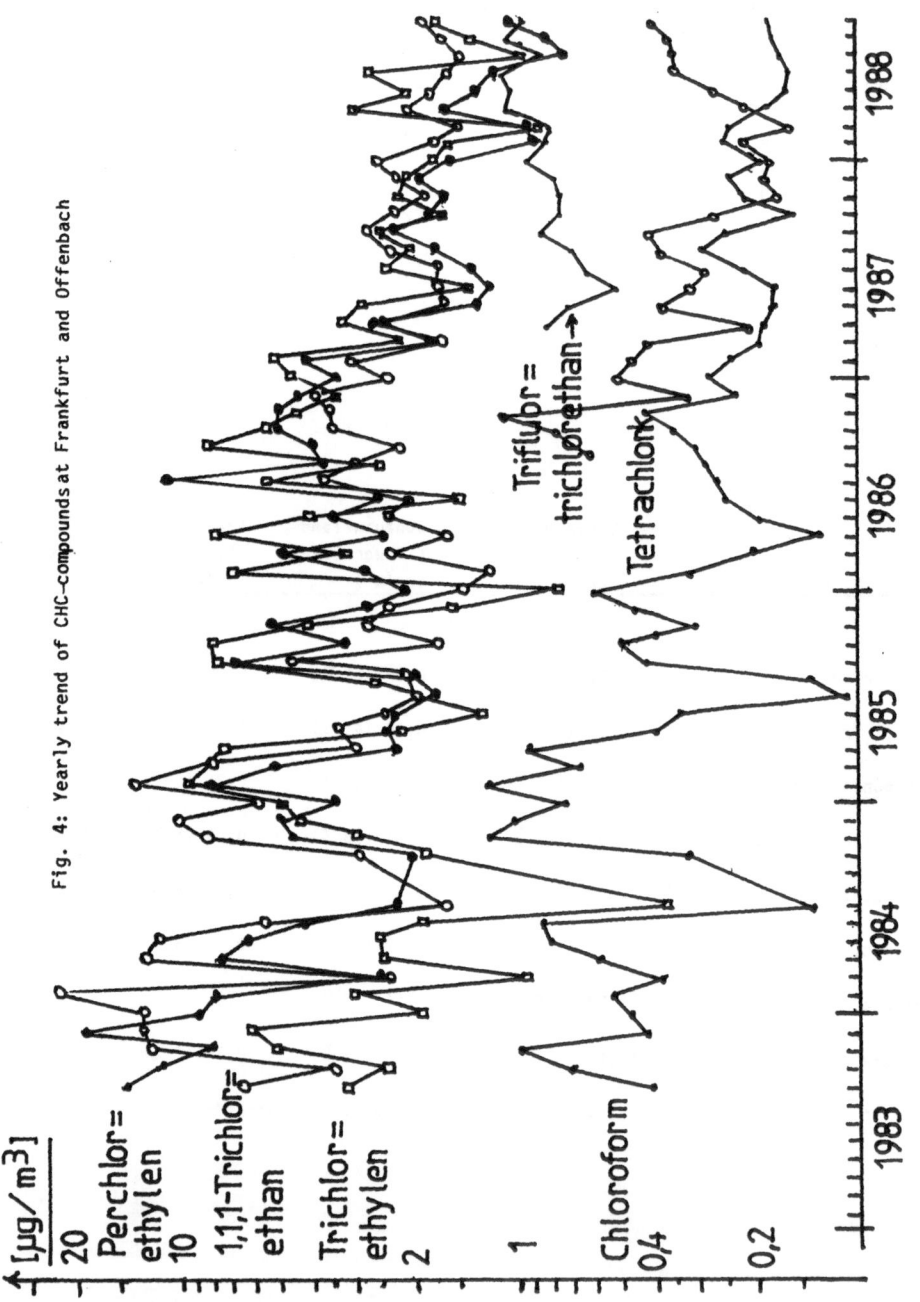

Fig. 4: Yearly trend of CHC-compounds at Frankfurt and Offenbach

From Tab. 1 can be seen that compounds with long residence times have lower concentration gradients. Carbon tetrachloride with a residence time of more than 10 years thus, globally becomes homogeneously distributed.
Compounds with a short life time do not spread far away off their sources. Therefore, the accumulation of a hydrocarbon compound in a remote forest area largely depends on its residence time.

In Fig. 4 the yearly trend of CHC-compounds in the polluted Rhein-Main-area is plotted. It can be noticed that since some years a slight concentration decrease takes place.
This is due to better recycling methods in metal industry and chemical laundries.

REFERENCES

1. Pearson, C. R. (1982) in O. Hutzinger (ed.) Handbook Environmental Chemistry-Antropogenic Compounds, Vol 3/B, Springer-Verlag, Berlin

2. Crutzen, P. J. (1979) Ann. Rev. Earth Planet Science 7: 443

3. Müller, J., F. Riedel (1986), Fourth European "Symposium on Physico-Chemical Behaviour of Atmospheric Pollutants", Stresa (Italy), Proc. c/o Reidel Publ. Comp., Dordrecht/Neth. pp 157-165

4. Ciampa , R. (1983), Diplomarbeit c/o Inst. f. Meteorologie and Geophysik, UNI, Frankfurt/M.

5. Singh, H. W. et al (1981), Atm. Environment, Vol. 15, pp 603-609

CASE STUDIES

THE DEPOSITION OF AEROSOL PARTICLES IN A FOREST USING AN ATMOSPHERIC RESIDENCE TIME MODEL.

R. JAENICKE
Institute for Meteorology
University of Mainz
Saarstraße 21
D-6500 Mainz
Germany

ABSTRACT. Usually forests are expected to act as a "filter" for aerosols. This term and the interpretation of earlier observations triggered the thoughts and calculations of this paper. The results indicate that the forest better should be termed "sink" rather then "filter". The efficiency of a forest to remove particles from the air most probably is caused by the time the aerosol spends in the forest to settle and coagulate. The value of a forest for unpolluted air to the community stems from the fact that on the location of a forest, no polluter is existing. Based on a residence time model of the atmosphere, a model for the forest is developed and compared with published results.

1. Introduction

In the literature about forests and immission control, usually a forest is termed a filter for aerosol particles (Keller et al 1978, Mayer 1984) if dry deposition effects are discussed. Among scientists and in the public the use of this term 'filter' has triggered the idea of polluted air entering a forest and unpolluted air blowing out to the surrounding areas. And indeed as aerosol physics and chemistry understands it, a filter is a medium retaining particles and gases from the passing air and releasing clean air. This hope is supported by the large surface present in forests. So Mayer (1984) states (translated from German):

> *"Because of the high leaf surface index good contacts exist between the tree surface*
> *and the air. This favors positively the dust interception".*

A recent publication (Meister et al 1984) aiming toward the general public in Germany states (translated from German):

> *"Informed agrarians appraise the forest as a filter for poisons of the air. According*
> *to the Bavarian Ministry for Agriculture and Forest, a forest cleans the air of*
> *microparticles of all sizes by combing out twentyfold better than barren land".*

As far as particles are concerned, a filter acts because of several effects: diffusion toward the filter material because of high mechanical mobility, impaction and interception on surfaces because of inertia and high flow velocities, sedimentation because of gravity. A forest might act as a filter, because it offers additional surfaces (diffusion) and obstacles (impaction and interception) as compared to the barren land.

143

H.-W. Georgii (ed.), Mechanisms and Effects of Pollutant-Transfer into Forests, 143–150.

On the other side, observations and calculations show how drastically the wind velocity in a forest is reduced with the greater amount of air being carried over the forest rather then passing through the forest. Thus a forest traps the air instead of supplying air. As a consequence, more honestly the term 'sink' rather than 'filter' should be used, if the removal of aerosol particles in a forest is addressed.

The additional removal mechanisms for a forest - as compared to barren soil - , like impaction and interception, require large wind velocities, which do not exist in dense forests but may be present at the borders and on top of the canopy. Zenker (1954) is often cited for the 'filter effect' of a forest, because he reported in a forest only half the concentration of condensation nuclei as compared to the surrounding air. But he has a different attitude towards the filter effect and claims that the air movement in a forest as rather low and most air is lifted over the forest. Thus only a fraction of the passing air enters the forest. The forest may be seen as a closed container. In that container the air calms and the particles are removed by coagulation. Zenker (1954) finds support for that idea observing that after drastical reduction of the particle concentration outside the forest, the concentration in the forest might remain elevated for some time. He confirms the earlier observations of Rötschke (1937) that in approaching a forest the concentration decline of the condensation nuclei starts in the open land well ahead of the forest. In entering the forest the air contains only one third of the undisturbed particle concentration. The further concentration reduction within the forest is a mere 10 %.

That work also investigated the 'dust particles', particles with radii in the 0.1 μm to 1 μm range. Even in this case, the concentration declines 15 m before the air reaches the forest border. In the forest itself, the dust particle concentration might even be higher than in outside air. This observation might be an indicator for the particle production of a forest. Such particles could be dead or alive biological material, the presence of that hardly has been investigated (Matthias 1987) at all.

Is our understanding of a forest acting with additional surface for diffusion and obstacles for high wind velocities to simple?

2. Applied Models and Results

This work is not concerned with detailed calculations. It rather offers some estimates and compares it with published results. So the effects of diffusion toward external surfaces, coagulation within the aerosol, impaction (interception) with obstacles and sedimentation in forests will be treated.

2.1. IMPACTION AND INTERCEPTION

Aerosol particles moving with the air might eventually intercept with obstacles such as needles or leaves and get deposited. To estimate the collection efficiency of cylindrical objects for aerosol particles, often the evaluation of Prodi and Tampieri (1982) is used. They rely mainly on the work of Ranz and Wong (1952) using a not so well defined calibration aerosol (Jaenicke and Blifford 1972). For jet impactors the collection efficiency curve is well known, but for body collectors like rods, calibrations never have been updated. To estimate the collection efficiency cut-off radius r_{50} (50% collection efficiency) of needles we use the data of Jaenicke and Blifford (1972) which agree with that of Ranz and Wong (1952). This agreement is exclusively for the cut-off radius but not for the shape of the collection efficiency curve.

Fig. 1 shows the calculated cut-off radius r_{50} for various obstacle diameters (D_j) and various wind velocities u. In addition the cumulative rural aerosol volume size distribution (Jaenicke 1988) is shown. It can easily be seen, that for most wind velocities in a forest, only particles larger than 10 µm are collected. A better collection can be expected for high wind velocities as it occurs eventually at forest borders and on top of the canopy. This finding agrees nicely with Mayer's (1984) statement (translated from German):

> *For efficient dust filtering (cement factory) large amounts of dust carrying air should pass a forest with low tree density and a border with minor air flow resistance*.

These low density forests permit high wind velocities and thus good removal. For the center of a dense forest, the effect of impaction seems to be minor.

2.2. DIFFUSION TOWARDS EXTERNAL SURFACES

Thermal diffusion acts as sink for suspended aerosol mass only if aerosol mass is deposited on external surfaces. Such surfaces are the leafs and needles in a forest. With a leaf surface index of 5, a forest can be compared in its surface to volume ratio (0.5 m²/m³) with a hollow sphere with

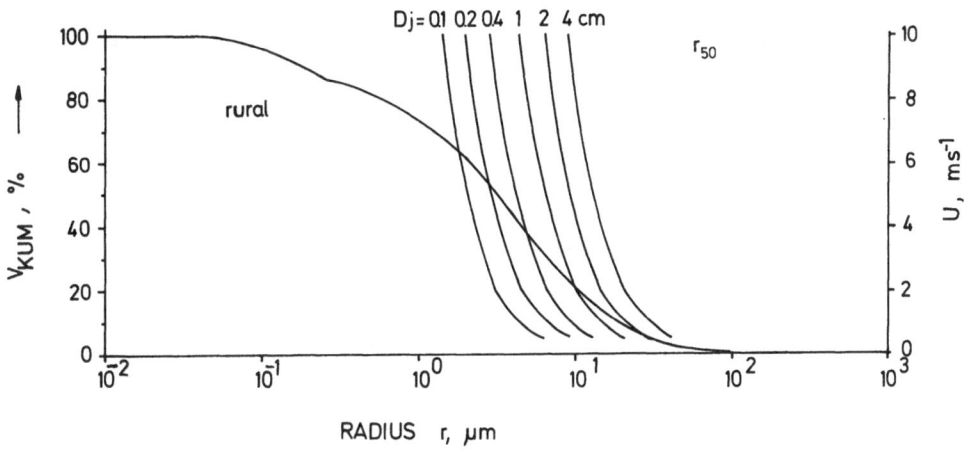

Fig. 1: Collection efficiency cut-off radius r_{50} (50% collection efficiency) calculated for various obstacle diameter (D_j) and various wind velocities u. The rural size distribution shown in its volume cummulative form indicates, that for most obstacles and low wind velocities only particle larger 10 µm are collected with only up to 10% volume of the whole aerosol.

6 m radius. As we will see later, the results will not be effected very much if we use higher leaf surface indexes. However the loss of particles number concentration in such a sphere due to diffusion only is experimentally difficult to observe, because coagulation is acting simultaneously.

The pure diffusion toward external surfaces can be treated theoretically with Fuchs (1964) and the results are shown in Fig. 2 for a period of 24 h. Because of the dependency of the diffusion coefficient from the particle size the overall loss of particle concentration depends on the particle size distribution. For the 'rural' distribution (Jaenicke 1988) used in this paper, it is as low as 3 % by number in 24 h. This reduction is neglegtable compared to the 'loss' by coagulation, as will be discussed next.

2.3. COAGULATION AND SEDIMENTATION

After ruling out impaction and diffusion toward external surfaces as major pathways for particle deposition in a dense forest, the remaining processes can be treated in a single model.

The change of particle concentration as function of time depends on the residence time of the particles in a certain reservoir:

$$N = N_0 \exp(-t/T)$$

with
T = residence time
t = time
N = particle concentration at time t
N_0 = particle concentration at $t = 0$

The residence time itself is particle size dependend. Giant particles have a short residence time because they settle out rather rapidly. For the atmosphere exists a residence time model as function of particles size (Jaenicke, 1978) depending only on sedimentation, coagulation and wet removal. Therefore it is only needed to adjust the atmospheric residence time model to the forest situation.

The coalescence of particles because of thermal diffusion is called coagulation. This way larger particles are formed from smaller ones reducing their concentration. Strictly speaking, this process does not reduce the suspended aerosol mass, but is moving particles mass from a high mobility size range to a less mobile particle size. In other words, aerosol mass is moved into that range (0.1 μm to 1 μm) having the longest residence time. This range is cleaned only through wet removal. So the wet removal size range of the atmospheric aerosol is ironically that with the longest residence time.

The coagulation constant in a monodisperse aerosol is rather small. In a polydisperse aerosol and the presence of immobile larger particles, the coagulation constant is greater by orders of magnitude reducing the Aitken particle concentration quite rapidly. The velocity of this process depends on the particle concentration. In higher concentrations coagulation (occassionally called aging) is much more rapid than in low concentration aerosols.

The residence time model of the atmosphere looks like this (Jaenicke, 1978):

$$\frac{1}{T} = \frac{1}{C_D}(\frac{r}{R})^2 + \frac{1}{C_F}(\frac{r}{R})^{-2} + \frac{1}{T_{wet}}$$

with

C_F = a constant depending on the concentration of Aitken particles
C_D = a constant depending on the sedimentation height
T_{wet} = the residence time depending on the wet removal
r = particle radius
R = reference radius, $R = 1 \mu m$

If we adjust the above values for
  * the sedimentation heights typical in a forest (C_D),
  * the time t the air spends in a forest,
  * an aerosol with a concentration (C_D) of 15000 cm-3 (30000 cm-3) and
  * the wet removal (T_{wet}),

the reduction curves per number in Fig. 2 result. It can be seen how effectively the Aitken particles are removed from their size bin and the giant particles likewise. The range 0.1 μm to 1 μm is less efficiently removed, as was to be expected. It can be seen how minor the diffusion toward external surfaces is compared to the loss of particles by coagulation in their respective size range.

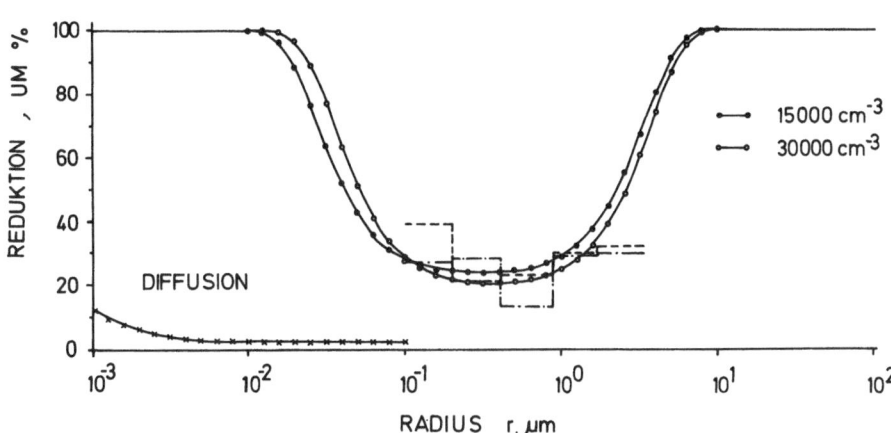

Fig. 2. Theoretical calculations of reduction of particle concentration in a forest using the modified residence time model. In addition the effect of diffusion loss within 24 h is shown and seems to be negligible. The measurements of Höfken et al (1981) are indicated for beech (Buchenwald) and spruce (Fichtenwald).

3. Comparison with Published Data

Data about the deposition of total mass and chemical species in forests are plentiful. Size resolved data are sparse. For the comparison with our estimates, the data of Höfken et al (1981) can be used. The forest they selected was rather dense. The collection of material always covered periods of 3 to 4 days, thus wet removal could not be excluded, but was not reported. The particle size range is 0.1 μm to several micrometer, covering the range of the less effective removal. Fig. 2 includes their results.

Theory and experiment agree in the fraction removed and in the development of a minimum efficiency for removal. As can be seen, the theoretical curves do not vary too much with the particle concentration. This value - which was not reported in Höfken's et al (1981) measurements - seems to be of minor importance.

The agreement shows how good such a simple model can be used to handle the removal processes in a forest.

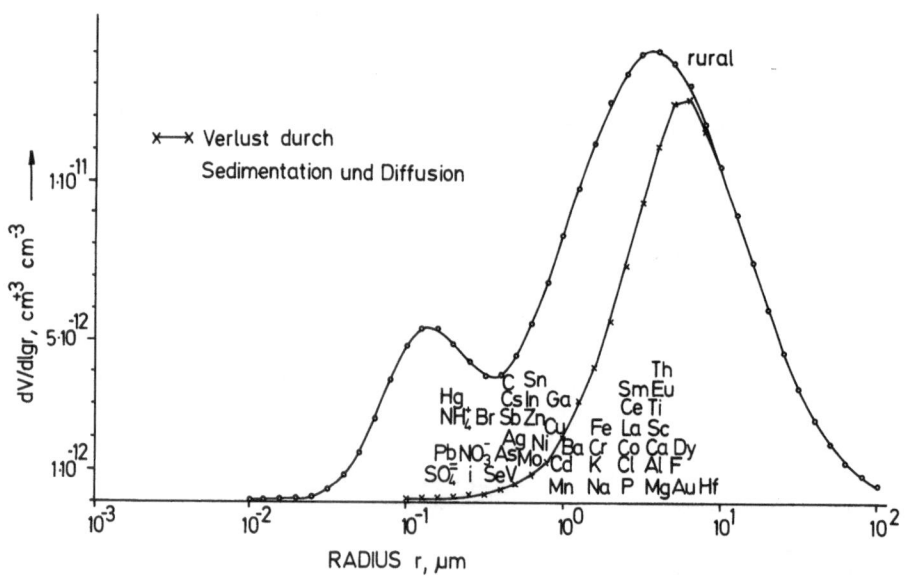

Fig. 3. The loss of volume from a 'rural' aerosol size distribution. The distribution is differential and logarithmic on the ordinate. The cross-curve indicates that portion lost. The loss due to coagulation is not a real loss, because their mass remains suspended. In addition the 'center of gravity' in terms of radius for major elements and ions are indicated according to Rahn (1975).

149

4. Conclusions

In Fig. 3 the above calculated loss of particles in a forest has been applied to a 'rural' aerosol volume size distribution. Effectively removed from the aerosol are the particles in the range greater than some micrometer. In addition Fig. 3 indicates elements and ions of the aerosol in their respective size class according to Rahn (1975). Obviously elements like Pb, Hg, Br, and others together with the ions SO_4 and NO_3 are comparatively less efficiently removed. On the other hand rare earth elements and K, Na, Fe, and P are efficiently removed in a forest.

Could this result indicate that a forest protects itself to a certain extend from particles of the aerosol by capturing particles in border trees or in frontal tree groups, this way 'sacrificing' them? As an inexperienced observer of declining forest symtoms, I have notized tree rows facing the wind perpendicular being more severely damaged than solid forests, in agreement with the above results.

5. References

Fuchs, N.A. (1964) The Mechanics of Aerosols, Pergamon Press, Oxford

Höfken, K.D., Georgii, H.W., Gravenhorst, G. (1981) Untersuchungen über die Deposition atmosphärischer Spurenstoffe an Buchen- und Fichtenwald. Berichte des Instituts für Meteorologie und Geophysik der Universität Frankfurt/Main

Jaenicke, R. and Blifford, I.H. (1974): The Influence of Aerosol Characteristics on the Calibration of Impactors. J. Atmospheric Science 5, 457-464

Jaenicke, R. (1978) 'Über die Dynamik atmosphärischer Aitkenteilchen' Berichte Bunsengesellschaft Physikalische Chemie 82, 1198-1202

Jaenicke, R. (1988) 'Aerosol Physics and Chemistry', in G. Fischer (ed.), Landolt-Börnstein New Series Volume 4 'Meteorology' Subvolume b 'Physical and Chemical Properties of the Air', Springer Verlag Berlin, pp. 391-457

Keller, T. and Flühler, H. (1978) 'Die Bedeutung des Waldes als Immissionsfilter in der Industrielandschaft', Mitteilungen Eidgenössische Anstalt für das Forstliche Versuchswesen 54, 464-475

Matthias, S. (1986) Die biogene Komponente im atmosphärischen Aerosol im Mainzer Raum. Diplomarbeit Universität Mainz

Mayer, H. (1984) Waldbau auf soziologisch-ökologischer Grundlage, Fischer Verlag, Stuttgart

Meister, G., Schütze, C., Sperber, G. (1984) Die Lage des Waldes, Verlag Gruner und Jahr, Hamburg

Prodi, F. and Tampieri, F. (1982) 'The Removal of Particulate Matter From the Atmosphere: The Physical Mechanisms', Pageoph 120, 286-325

Rahn, K.A. (1975) 'Chemical Composition of the Atmospheric Aerosol: A Compilation I.' Extern, Tijdschrift voor Omgevingswetenschappen 4, 286-313

Rahn, K.A. (1975) 'Chemical Composition of the Atmospheric Aerosol: A Compilation II.' Extern, Tijdschrift voor Omgevingswetenschappen 4, 639-667

Ranz, W.E. and Wong, J.B. (1952) 'Impaction of Dust and Smoke Particles on Surface and Body Collectors' Industrial and Engineering Chemistry 44, 1371-1381

Rötschke, M. (1937) 'Untersuchungen über die Meteorologie der Staubatmosphäre' Veröffentlichungen des Geophysikalischen Institutes Leipzig, 2. Serie 11, 1-78
Zenker, H. (1954) 'Waldeinfluß auf Kondensationskerne und Lufthygiene', Zeitschrift für Meteorologie 8, 150-159

INTERCOMPARISON OF WET-ONLY COLLECTORS AND PRECIPITATION SENSORS

P. WINKLER
Deutscher Wetterdienst
Meteorologisches Observatorium Hamburg
Frahmredder 95
D-2000 Hamburg 65

ABSTRACT. An intercomparison of 11 various types of wet only collectors has shown that large differences in the chemical composition occur, especially for trace metals for which it exceeds those of the main ions. Washing and rinsing experiments have shown that large amounts of trace metals are absorbed or adsorbed in the sampling device before entering the sampling bottle. If all collectors were steered by a single precipitation sensor, the instruments agreed within +/- 10 % for the main ions, but not for the trace metals. The investigation of sensors under outside conditions and in a laboratory testing chamber gave comparable results at the onset of precipitation. The behaviour of sensors at the end of rain events was also investigated and large deficiencies have been established.

1. Introduction

In a field experiment of two years 20 wet only collectors of 11 various types have been investigated on their ability to collect wet deposition of anions, cations and the trace metals Fe, Pb, Cu, Cd. It has been found that differences in the chemical composition between the samples of the different instruments exceed 20 %. This large differences originate from (a) the insufficient response of the precipitation sensors at the onset of precipitation, (b) the wetting loss of rain water in the sampling device, and (c) interaction of precipitation water with materials of the collection device.

In this present paper we will address especially the less pronounced agreement in trace metal concentrations as compared to the main ions. Since the results are described in detail in Winkler et al. (1989) we will present here only a short summary of our findings of a more general interest.

2. Intercomparison of wet only collectors

It was found that the most powerful tool for characterizing the individual samplers was the definition of a suitable reference value for each trace constituent. This reference value was created by taking the average of the individual values of those collectors which had
- a sensor with sufficiently low threshold intensity,
- not interaction between precipitation and material of the sampling device, and
- no pollution of the sample by insects, leafs and so on.

It could be shown, that the individual samplers characteristically deviated from this reference value. The concentration of nearly all trace constituents, for example, was above the reference value if a

151

H.-W. Georgii (ed.), Mechanisms and Effects of Pollutant-Transfer into Forests, 151–155.
© *1989 by Kluwer Academic Publishers.*

sampler had a very high wetting loss causing a memory effect which rises the concentration. Or, in other cases, the concentration regularly remained below the reference value, if the precipitation sensor of that particular collector had a poor sensitivity.

Now, while the correlations for those collectors free of contamination problems remained within certain limits for the main ions, the scatter became large for the trace metals. Therefore, some experiments were made to find out the reasons.

In a first experiment the funnels of the collectors were rinsed with a certain amount of distilled water which corresponded to a precipitation height of about 1.4 mm. The rinsing solution was analyzed for ions and trace metals and we found the trace metal content to be much above the content of the main ions. When the rinsing solution was filtered in order to distinguish between the dissolved and particulate form of metals it was found that more than 50 %, for Cu and Pb more than 80 %, of the metals was bound to particles. Such particles may be adsorbed to the funnels until they eventually are washed off during events with sufficiently high intensity. In other cases some metals from adsorbed particles may become redissolved by precipitation with sufficiently low pH-value, because their solubility is pH-dependent. This hypothesis was confirmed by experiments in which the funnels were washed with diluted HNO_3 (1%), again with an amount corresponding to about 1.4 mm precipitation height. Appreciable concentrations of trace metals were obtained, exceeding the average concentrations in precipitation. Even in a subsequent rinsing experiment with double distilled water high concentrations of metals were found.

It is concluded that
a) the larger scatter between the collectors for trace metals as compared to main ions can be understood in terms of absorption of metal containing particles on the funnel walls;
b) the wet deposition is measured incorrectly unless both, the dissolved and the undissolved fractions of trace metals are determined;
c) the interpretation of daily precipitation samples with respect to trace metals must be difficult, since a more or less high fraction of metals may remain in the funnel or is redissolved by an event with low pH-value;
d) daily cleaning of the funnels by washing with diluted HNO_3 is not an appropriate tool to overcome the problems because the rinsing solution must be analyzed as well which means that the analytical work is doubled.

In another experiment in which improved collectors have been run by controlling them by only a single sensor, it was found that the concentrations of the main ions agreed within 10 %. Again such a good result was not achieved for the metals. Thus, while it can be concluded that the present state of the art allows to consider the determination of the wet deposition of the main ions standardized as long as certain guidelines with respect to funnel shape, material, and sensitivity of the sensor are fulfilled, this aim is still not reached for the determination of the wet deposition of trace metals. Here fundamental experiments are to be accomplished before this goal can considered to be reached.

Another important aspect should be mentioned from which differences in the determination of the wet deposition may originate although the material of our intercomparison does not allow any conclusion whether this point is of practical importance. Sevruck (1988) has shown that, because a collector is an obstacle in the wind field, the wind velocity is increased above the funnel opening by about 20 %. Consequently the small droplets at the lower end of the size distribution may be blown away and be hindered from entering the funnel. Since these small droplets may carry higher solute concentrations (Georgii and Wötzel, 1970), a shift in the composition of the collected sample may be induced by such an effect.

3. Intercomparison of precipitation sensors

A precipitation chamber has been developed, in which the response of sensors to low precipitation intensities can be investigated (fig. 1). Water of an electrical conductivity comparable to precipitation is pumped at a controlled flow rate to a nozzle where it is sprayed by pressurized air. The rain drops produced are measured by a distrometer (Joss and Waldvogel, 1967) from the output of which the precipitation intensity is calculated. The sensor to be tested is standing aside of the distrometer being there exposed to the same precipitation intensity. A computer program controls the essential parameters like water flow, air flow and selects the conductivity out of three possible stocks. The signal of the distrometer and the sensor under test are registered and evaluated by the computer. A testing program can be run automatically.

An example of the response curves of two sensors to increasing intensity are depicted in fig. 2. These curves obtained in the laboratory are steeper than those obtained under outside conditions (see Winkler, 1988; and Winkler and Jobst, 1988), however, a sensor which was found to be poor under outside conditions proved to be poor in the testing chamber.

By means of the results of these laboratory tests it was tried to find, which of the adjustable parameters is of most importance for a good response: Threshold resistance between the electrodes, heating power and surface temperature, shape of the sensor and so on. Since these investigations were not yet finished we can only give some preliminary information while the complete details of the intercomparison will be presented in the final report of the respective research project (to be published in 1989).

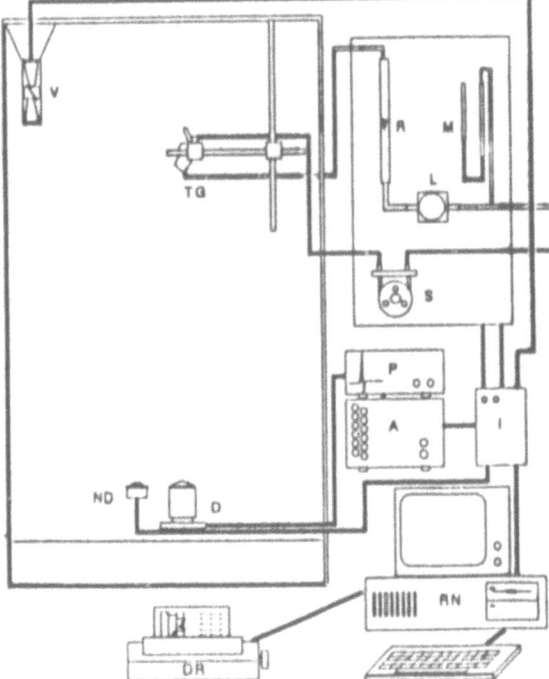

Figure 1: Testing chamber for precipitation sensors. TG = drop generator, D = distrometer, ND = precipitation sensor, V = fan, R = rotameter, M = manometer, L = needle valve, S = pump, I = interface, P = processor, A = analyzer, RN = computer, PR = printer.

154

When the surface temperature of sensors is measured as function of the air temperature two types of behaviour were distinguished:

a) Sensors, the heating of which is thermostated, show no stable but slowly decreasing surface temperature when the airtemperature is lowering. However, the temperature difference to air temperature increases as the air temperature decreases.

b) Sensors heated with constant energy show different behaviour. Their temperature difference to air temperature increases as the air temperature increases but at low temperatures near the freezing point the surface temperature is only slightly above the air temperature and becomes zero only a few degrees below the freezing point.

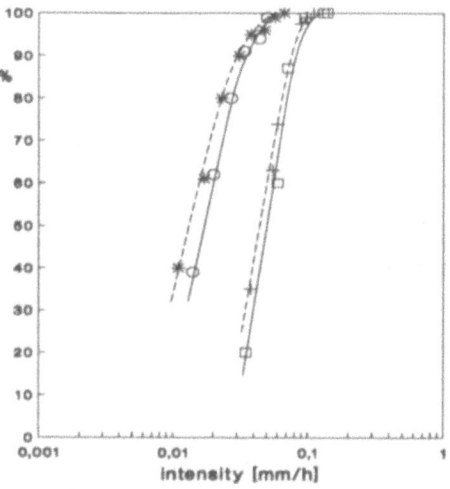

Figure 2: Response of two precipiation sensors as function of precipitation intensity. In case a the electrical conductivity of rain water was 50 µS/cm, in case b 5 µS/cm.

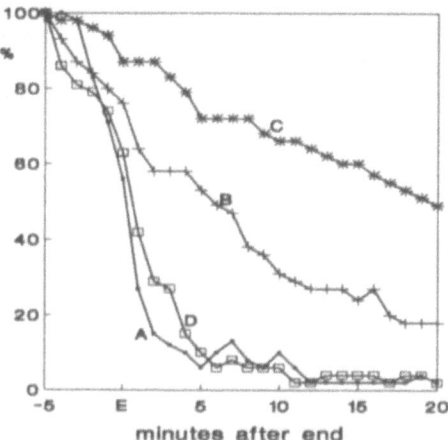

Figure 3: Behaviour of precipitation sensors 5 minutes before and 20 minutes after the end E of rain events. Average over 60 events of more than 5 minutes duration and a productivity above 0.06 mm.

Sensors of type b become poor at high surrounding temperature because they get to hot and evaporate precipitation of slow intensity immediately after contact. And they become poor at temperatures near or below the freezing point, because the surface does not get warm enough in order to melt snow which then is detected only with large delay.

Finally, the behaviour of sensors at the end of precipitation events is of importance. If the end of events is not detected within a few minutes dry deposition is collected. An example of the average behaviour (60 events) of some selected sensors is depicted in fig. 3. The end of the events is marked by E and has been defined by the fact that no precipitation was seen by the distrometer and during the subsequent 10 minutes no further drops were detected. We realized some sensors to switch off already several minutes before the nominal end. Another sensor needed more than 20 minutes in 50 % of all events to recognize to precipitation end.

The registrated opening time of a wet only collector thus cannot considered to be a representative measure of the precipitation duration. If a sensor shows some deficiencies in detecting both the beginning and the end of a precipitation event the duration may apparently be given correct but the amount of precipitation is not collected completely.

Acknowledgement: The investigations on the precipitation collectors has been conducted in cooperation with the Institute of Inorganic and Applied Chemistry of the University of Hamburg. The work was financially supported by the Bundesministerium für Forschungs- und Technologie under grants 07431073 and 07431018.

References

Georgii, H.W., Wötzel, D (1970) "On the relation between drop size and concentration of trace elements in rain water". J. Geoph. Res. 75, 1727-1731.

Joss, J., Waldvogel, A. (1967) "Ein Spectrograph für Niederschlagstropfen mit automatischer Auswertung" PAGEOPH 68, 240-246.

Sevruk, B., (1988) "Towards the universal precipitation gauge of the future". Vaisala News 113, 12-14.

Winkler, P., (1988) "Response of precipitation sensors to precipitation". In: WMO/TP No. 222: Instruments and observing methods, Report No. 33: Papers presented at the WMO Technical Conference on Instruments and Methods of Observation, Leipzig 16-20 May 1988, pp. 269-274.

Winkler, P., Jobst, S. "Comparison of various precipitation gauges and sensors for the determination of wet deposition". In: K. Grefen, J. Löbel (eds.) Environmental Meteorology Kluver Acad. Publ. Dordrecht, pp. 193-200.

Winkler, P., Jobst, S., Harder, C. "Meteorologische Prüfung und Beurteilung von Sammelgeräten für die nasse Deposition". Final Report. GSF-Berichte (in press).

TRANSMISSION OF AUTOMOBIL EXHAUST WITHIN A FOREST

THEO SATTLER and WOLFGANG JAESCHKE
Johann Wolfgang Goethe-Universität
Zentrum für Umweltforschung
Robert Mayer-Str. 7-9
D - 6000 Frankfurt am Main

ABSTRACT: In a forest lane with a highway in the "Frankfurter Stadtwald" field experiments dealing with the emission and transport of automobil exhaust have been conducted over a period of several years. Attention was also paid to the immission of pollutants in the surrounding forest and its dependency on meteorological parameters. Simultaneously a numerical model of the dispersion and transport of pollutants from a street into a forest was developed. Based on a close interaction between measurements and model development the calculations could be optimized. Finally, the model was validated by the aid of the measured data with respect to several boundary conditions like meteorological parameters, seasonal changes, traffic situations etc..

1. Introduction

It is commonly accepted, that the exhaust of the automobiles is one of the main sources of atmospheric pollution. The question to what extend the automobil gases are responsible for the increasing damages of forests should be given greater attention. Therefore, at the Zentrum für Umweltforschung of the University of Frankfurt/Main a field measuring program was designed. One of the topics of that program was the determination of the source strength of a highway in a forest lane and the investigation of the transport of the emitted pollutants into the forest.

As suitable measuring site for this endeavour the Babenhäuser Landstraße was chosen. It is a highway with four roadways leading through the Frankfurter Stadtwald. Since this street is highly frequented by local and long distance traffic, a high source strength could be expected, which enables measurements of the traffic exhaust in the forest in distances up to 100 m from the street. The direction of the street is situated from the southeast to northwest (307°), which means ortogonal to the main wind direction. Therefore, it could be assumed, that the main part of the measured data could be used in the foreseen model calculations. They should treat the transport of the pollutants into the forest downwind to the street. The considered forest consists mainly out of beeches. Measurements of NO_x, O_3, CO and hydrocarbons were performed over more than two years in order to observe the transport and the behaviour of the pollutants during all seasons with a wide range of climatic situations. For this purpose an automatic measuring system was developed amd constructed. A detailled description of the measuring technique was given in Sattler and Jaeschke (1987) and Sattler (1989).

Parallel to the measurement activities a diagnostic model of the transport of pollutants downwind to a street was developed. This model, which was based on the theories of turbulent diffusion and chemical reaction kinetics, was designed for a comparison of the measured data and the calculated predictions. As prerequisit for the desired interaction between measurements and model calculations

157

H.-W. Georgii (ed.), Mechanisms and Effects of Pollutant-Transfer into Forests, 157–167.
© *1989 by Kluwer Academic Publishers.*

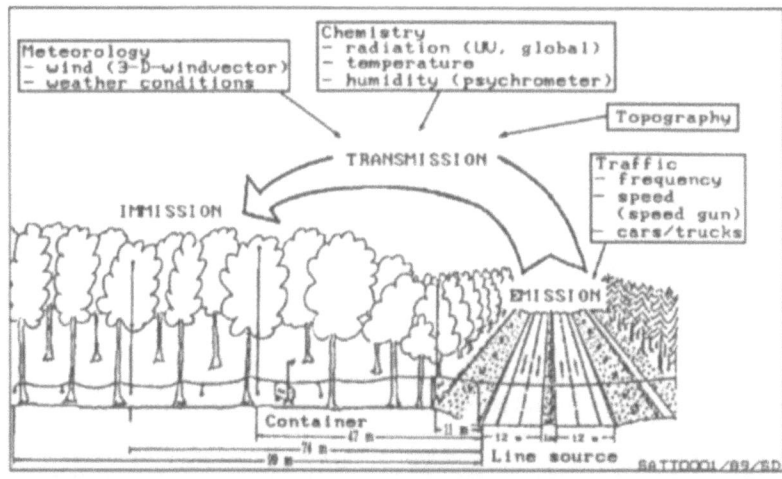

Fig. 1: Causality-chain emission-transmission-immission.

meteorological parameters like radiation, wind speed and wind directions and other quantities like traffic speed and frequency had to be measured. With these informations the whole chain of causality between emission, transmission and immission could be considered (Fig. 1).

2. NO_x-Source strength

Common combustion motors are sources for NO and NO_2. Compared to the NO-exhaust the direct NO_2-emission is only a very small fraction. However, after NO has left the exhaust pipe, it is partially oxidized to NO_2 via various pathways. This is mainly occuring in the presence of Ozone. Because of this interaction the NO_x-emission was considered as the sum of NO and NO_2 and NO_x was treated like a chemical inert gas.

In order to study the automobil exhaust in the considered Babenhäuser Landstraße in more detail, the frequency and the speed of the traffic had to be ascertained. The speed of the cars was continuously monitored by a RADAR speed gun which was similar to those used by the traffic police. From the data measured over a period of several months a mean velocity of the cars in the order of 90 km/h could be calculated with a standard deviation of 16 km/h. The frequency was determined by the aid of automatically counting induction slopes. The results gained by these instruments were randomly tested by students, who counted the cars in the street over a period of 24 hours. From these data a mean frequency of 25.000 vehicles per working day was obtained.

For the measurement of the automobil exhaust at several points a rope was tightend between two masts which had been erected on both sides of the road. By the aid of this rope teflon tubes and cables could be led to 15 measuring points situated in the space of the forest lane in a plane perpendicular to the direction of the street. The exact positions of the measuring points can be seen in Fig. 2. The air to be sampled from all these points was sucked by a central pump to a measuring station which was positioned on one side of the road. Beside the concentration of the air pollutants the temperature, humidity and UV-radiation were also measured in the forest lane of the street. All data were registered as half hour mean values. Since the concentrations were only measured in one x-z

Fig. 2: Streampath in the street layer.

area perpendicular to the y-direction of the street axis, the determination of the source strength could be calculated by the aid of an only two dimensional consideration. Therefore it had to be assumed that the concentration in the forest lane in y-direction is more or less homogeneous. The transport of pollutants in with concentration C can now be defined as the flux Q through a line element dl in the considered x-z area:

$$\vec{Q_t} = \vec{w_t} \cdot C \cdot \vec{dl} \tag{1}$$

Index t means transposed. C is supposed to be constant on the line element dl. By integration one gets for the mass flux

$$f = \int_{x,z} \vec{w_t} \cdot C \cdot \vec{dl} = \int_{x,z} (w_x, w_z) \cdot \begin{pmatrix} C_x & 0 \\ 0 & C_z \end{pmatrix} \cdot \begin{pmatrix} dl_x \\ dl_z \end{pmatrix} \qquad [g/m \cdot s] \tag{2}$$

By the assumption of constant wind velocity w and constant concentration C on a line element the mass flux for lines perpendicular to the x- or z-direction is :

$$f = \vec{w_t} \cdot C \cdot \vec{l} = w_x \cdot C_x \cdot l_x + w_z \cdot C_z \cdot l_z \qquad [g/m \cdot s] \tag{3}$$

The vector of the wind velocity was registered as half hour mean value. Its standard deviation in x,y and z direction was separately registered. This was important for the estimation of turbulent transport, which enhances the whole mass transport. The turbulent diffusion plays a significant role for the whole consideration of pollutant transport in the forest lane with its high traffic frequency. Its influence is firstly caused by the high roughness of the complex forest terrain with its road lane and secondly by vortexes caused by the high speed of passing cars.

For the calculation of the source strength of the street the following balance was considered: Source strength = conv. transport (output - input) + turb. transport (output).

Each calculation of the NO_x-emission was based on the data of one week. These were more than 300 half hour mean values and standard deviations of 3-D-windvectors, traffic frequency, NO and NO_2-concentrations. The NO_x immission data were measured at 9 different measuring points. In Fig. 3 the calculated NO_x-emissions are plotted against the frequency of the traffic.

The correlation coefficient between the calculated NO_x-emission and the traffic frequency is 0.7. The critical value of the student-test for a comparison with the correlation coefficient is 0,188 (for 300 values and significant-level 99.9%). Thus the dependency between the source strength and the traffic frequence could be demonstrated with a high significance.

160

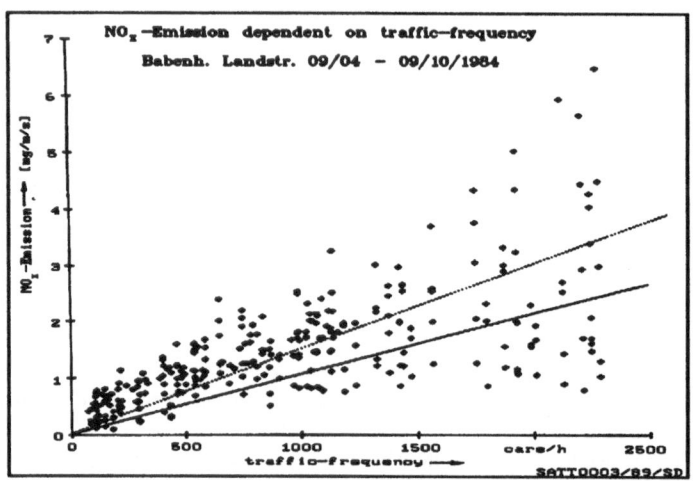

Fig. 3: Balanced NO_x-emission from Babenhäuser Landstraße. Strait line: 3.82 g/km (100 km/h, Waldeyer, 1978). Dashed Line: The calculated emission factor 5.61 g/km (90 km/h, ZUF, 1987).

Deviations of the emission values mainly occured at higher traffic frequencies. Some reasons can be considered. The cars, which are passing the measuring points during a half hour period, may be of different ages and conditions. The part of heavy trucks may also differ significantly up to the extrem situation of no trucks at the weekend because of truck driving prohibition. Also variations in the acceleration behaviour of the cars must be taken into account. In the afternoon, when the cars leave the city of Frankfurt with higher acceleration, the emissions are higher then in the morning when the cars are passing the measuring station on their way downtown with relative low speed. Finally an influence of the relative humidity could be observed. At higher humidities higher concentrations of NO_x were registered, which means, that under these conditions an exchange of air masses in the forest lane is inhibited.

The source strength was calculated in the dimension mg/m · s and is dependent on the traffic frequency. In order to get an independent value, the data of the calculated NO_x emissions can be devided by the respective traffic frequency. The mean value of these ratios yields the so-called emission factor for NO_x:

$$e_{NO_x} = E\left(\frac{f}{Kfz/h}\right) \cdot 3600 = 5.61 \qquad [g/km] \qquad (4)$$

This is higher than a NO_x emissionfactor given by the Technische Überwachungsverein (TÜV) Rheinland (Waldeyer, 1978). Their value of 3.82 g/km, which was determined in 1975 by tests in the laboratory, is lower than our factor, which is based on measurements in the real world. This can easily be seen from Fig. 3, where the TÜV NO_x-emissionfactor is also plotted as straight line. The deviation is probably caused by an increase of the NO_x-emission of the cars in recent years, because of the higher compression of modern combustion motors. During the time of our experiments in the year 1986 and 1987 the influence of the catalysator technique was obviously not yet detectible. The reason may be that in 1986/87 only a minor part of the cars was equipped with this technique.

3. Transmission of the pollutants in the forest

3.1 EXPERIMENTAL

After determining the NO_x-sourcestrength of the considered street the transport of NO_x into the forest was investigated in more detail. For that purpose both a numerical transport model was developed and data for the calibration and validation of that model were measured in a long lasting field experiment for more than one year. Similar to the method already described for the street lane 15 measuring points were installed now in the forest northeast to the street. These points were positioned between the ground and the canopy of the beeches in distances up to 100 m from the road. They all were in an area perpendicular to the axis of the street. During this part of experiments the measuring station was nearly in the middle of the measuring area (Fig. 1).

Beside the air pollutants the 3D-windvector (8 m height), the dry and moist temperature (3 m height), the UV- and global radiation was measured. This was possible, because the beeches had no foliage up to a height of 15 m. The measurements of radiation and wind velocity below the canopy were performed in order to get informations about this parameters in a forest below the canopy. As expectable, large deviations between summer and winter could be observed, because of the seasonal changes of the foliage in the canopy. During winter time the differences of the measured radiation and air movement below and above the canopy were not as large as during summer, where the foliage absorbs radiation and disturbs the air movement. Wind velocity and radiation above the canopy were measured up to a hight of 50 m at a meteorological tower which is situated 2,5 km southwest of the Babenhäuser Landstraße. A detailed description of the measurements and the obtained data as well as several case studies were given by Sattler (1989).

3.2 MASSTRANSPORT IN THE FOREST

In order to describe the diffusion of pollutants in the turbulent atmosphere the data of wind velocity and concentration are splitted into half hour mean values and turbulent deviations around this mean values. This allows the transition of the diffusion-equation valid for a short time period to the turbulent diffusion-equation for mean values. The result is the well known differential equation for the concentration field of a pollutant i in the space of its dispersion

$$\frac{D\,C_i}{D\,t} = \vec{\nabla}\,(K \cdot \vec{\nabla}\,C_i) \; + \; R_i \; + \; D_i \tag{5}$$

Where $\vec{\nabla}$ is the Nabla-operator, while D represents the operator of the temporal differentiation. C_i is the mean value of the concentration of the component i and K is a tensor which is used for the description of the turbulent diffusion. R_i stands for the mass flux density $(g/s \; m^3)$ per volume which is caused by conversion processes such as homogeneous chemical reactions. Finally, D_i describes the rate of deposition or absorption of the compound i.

The knowledge of boundary conditions enables a simplification of the diffusion- equation such that it can be solved analytically or, if this is not possible, that it can be solved at least numerically.

The goal of each model is the simulation of effects that several input-parameters have on central quantities. In the context here the central quantity is the concentration $C_i\,(x,z)$ of a compound i at a certain point (x,z) in the forest. It is influenced by meteorological parameters like wind field and the tensor of exchange, by orographic parameters like the ruoghness of the surface, by parameters which stear chemical conversion processes, like temperature, humidity and UV-radiation and finally by parameters which trigger the emission like frequency and velocity of the cars in the traffic.

By a number of simplifications and assumptions for the boundary conditions equation (5) can be reduced in order to get the problem more handy:

1. In a turbulent stream the exchange coefficient K is commonly a second order tensor. Assumed, that the axis of the considered coordination system coincides with the main axis of the tensor, its mixed terms can be cancelled.

2. The parameters used in the model calculation were registered during the experiments as half hour mean values. It is therefore assumed, that the immission and the source strength remain constant in the considered time period of half an hour. This means $dC/dt = 0$.

3. The y-axis of the system corresponds to the axis of the roadway in street direction. Assumed that the source strength and the influence of the terrain are homogeneous in y-direction, all y-terms are cancelled in the diffusion- equation.

4. The experiments have shown, that vertical movements of the air are very low compared to the air streams in horizontal directions. Therefore vertical movements can be neglected.

5. The considerations do not include total calm. Therefore, the turbulent diffusion in x-direction is neglected, because in this direction diffusion is much lower than convection.

6. Compared to the turbulent diffusion, Fick's diffusion plays only a minor role and can be neglected. The same is true for pressure- and termo-diffusion as well as diffusion caused by other external forces.

7. It is assumed that wind field and exchange coefficient in the forest are homogeneous in x-direction and that they are only dependent on the height z.

With these assumptions the equation of diffusion can be written more familiar:

$$ w_x \frac{\partial}{\partial x} C_i = \frac{\partial}{\partial z} K_{zz} \frac{\partial C_i}{\partial z} + R_i \tag{6} $$

Wind field w, exchange coefficient K and the term of conversion R are still unknown. These quantities will be calculated or defined in the following section.

Height z	$w_x(z) =$	$K_{zz}(z) =$
$35m = h \leq z$	$\frac{w_*}{X} \ln\left(\frac{z-d}{z_o}\right)$	$X \cdot w_* \cdot (z - d)$
$15m = b \leq z \leq h = 35m$	$w_x(h) \cdot \exp\left[-\frac{h\, w_*^2}{K_o w_x(h)}\left(1-\frac{z}{h}\right)\right]$	$K_o = X \cdot w_* \cdot (h-d)$
$z \leq b = 15m$	$A_2 z^2 + A_1 z$	$K_o = X \cdot w_* \cdot (h-d)$

SATAR001/89/SD

Tab. 1: Wind- and diffusion-profile for the different canopy-layers.

3.3   WIND FIELD AND EXCHANGE COEFFICIENT

The three dimensional wind data measured below the canopy were compared with the wind data measured above the canopy at the nearby meteorological tower and at the Frankfurt Rhein Main Airport. It turned out that the wind speed below the canopy normalized with the wind speed above the forest was lower by a factor of 2 when the canopy had leaves compared to winterly conditions with no foliage. As expected, the resistance of the foliage plays thus an important role on the wind field in the forest. Therefore, the difference between foliage and no foliage had to be taken into account when the roughness length and the fraction velocity were calculated. According to Slinn (1982) wind- and diffusion profiles can be described by:

$$\frac{\partial}{\partial z}\left(K_{zz}\,\frac{\partial w_x}{\partial z}\right) = c\,\alpha\,w_x^2 - \frac{1}{\rho}\left(-\frac{\partial p}{\partial x}\right) \tag{7}$$

c	$=$	drag coefficient of vegetation dependet on z
α	$=$	surface of foliage per volume dependet on z (1/m)
ρ	$=$	air density ($= 1290$ g/m^3)
p	$=$	pressure (g/m $\cdot$ s^2)
w_x, K_{zz}	$=$	dependent on z.

The connected resistance term expresses, that the change of the vertical diffusion with height is dependent on a resistance which is normalized on the square of the wind velocity. Now a horizontal stratification consisting out of three layers was introduced. The first layer is the air space above the forest which means z>h = 35 m, the second layer is the canopy (b = 15 m<z<35 m = h) and the third layer is the space between canopy and ground z<15 m.

Placing the respective boundary conditions and some simplifications (Sattler, 1989) one yields a layer-specific wind- and diffusion profile. The used quantities are: h = 35 m (height of the trees), b = 15 m (height of the stems), d = 28 m (zero-plane displacement height), z_{OB} = 0,7 m (roughness length with foliage), z_{OU} = 0,14 m (roughness length without foliage).

A wind profile defined in that way allows the simulation of a secondary maximum of the wind speed in the layer of the stems, as it was observed by measurements of the wind speed inside the forest (Fig. 4).

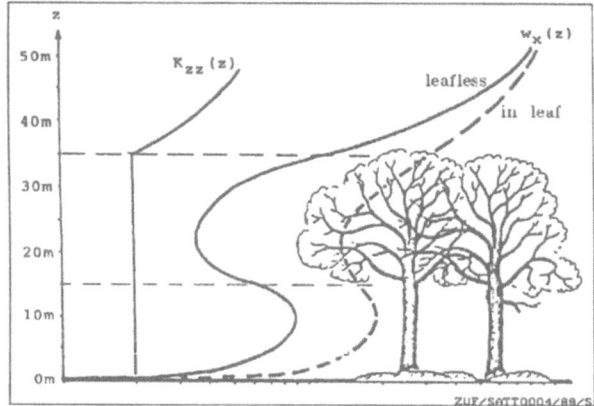

Fig. 4: Wind- and diffusion-profile in canopy.

Reaction	Reaction const.(*) $(ppm^{-n} \ min^{-1})$ (1 atm, 297°K)	According to
$O_3+NO \longrightarrow NO_2+O_2$	$k_1=2.8 \ E1$ $=2.8E3 \cdot exp(-\frac{1370}{T})$	Kasting and Ackerman (1985)
$HO_2+NO \longrightarrow HO+NO_2$	$k_2= 1.2 \ E4$	Calvert and Stockwell (1983)
$NO+CO+O_2 \longrightarrow HO_2+CO_2$	$k_3= 3.93 \ E2$	— " —
$HO+Alkene \longrightarrow RO_2$	$k_4= 3.8 \ E4$	— " —
$RO_2+NO \longrightarrow RO+NO_2$	$k_5= 1.2 \ E4$	— " —
$NO_2+h\nu \longrightarrow O+NO$	$j_{1-max}=4.7 \ E-1$ (**)	Becker et al.(83)
$O+O_2+M \longrightarrow O_3+M$	$k_6= 2.16 \ E1$	Calvert and Stockwell (1983)
$O_3+h\nu \longrightarrow O(^1D)+O_2$	$j_{2-max}=2 \ E-3$ (**)	Becker et al.(83)
$O(^1D)+H_2O \longrightarrow 2 \ HO$	$k_7= 2.4 \ E5$	Calvert and Stockwell (1983)
$HO+NO+M \longrightarrow HONO+M$	$k_8= 7.6 \ E3$	— " —
$NO+NO_2+M \longrightarrow HONO_2+M$	$k_9= 1.48 \ E4$	— " —

Tab. 2: Reaction velocities used in the developed model. (*): n = order of reaction minus 1.
(**): Value for July, high noon, clear sky, 51°N.

3.4   REACTION KINETICS

Though the distance of the considered transport of pollutants within the forest is only in the order of 100 m chemical reactions are of importance. Especially the reaction between nitrogene oxides, ozone and hydrocarbons are of great influence on the dispersion of the compounds because some of these reactions are very fast. Such reactions, especially of the NO_x - O_3 cycle, which were assumed to be relevant in the rather short residence time of the pollutants in the considered space, are compiled in table 2. They were essentially chosen from an article of Calvert and Stockwell (1983).

The most important oxidant for NO is Ozone, which either reaches the forest lane by transport from higher layers of the atmosphere or is generated in the airspace of the lane by photodissociation of NO_2. Other oxidants in this pathway are HO, HO_2 and peroxiradicals. Their formation is accelerated by the presence of CO and Ethen. Whereas NO is oxidized, NO_2 is reduced mainly by photodynamic processes. The later reaction, which generates O_3, is the main cause of radical chain reactions in the lower atmosphere, because in atmospheric layers close to the earth surface NO_2 is the strongest absorber of sunlight. Compared to this reaction the absorption by O_3 and its photodynamic degradation is only of minor efficiency. The generated Singulett D Oxigen reacts rapidly with water vapour and OH radicals are formed. These radicals remove NO_x by forming nitrious acid. The decomposition of HNO_2 into OH and NO is rather slow and is not considered in our model. In the literature sometimes large deviations of the rate constants of the relevant reactions can be found (Kolar, 1981). Thus the model is only as good as the certainty of these constants.

The reaction rate of a trace gas is formulated by the temporal differentiation of its concentration. It is proportional to the product of the educt concentrations. Thus differential equations for the temporal removal of NO, NO_2, O_3, O, O^1D, RO_2, HO_2 and HO can be formulated. Assuming a steady state of the O, O^1D, RO_2, HO_2 and HO concentrations three equations are remaining. They describe the most important chemical conversion processes during the transport distance considered in our dispersion model (Sattler, 1989).

$$\frac{d(NO_2)}{dt} = k_1(O_3)(NO)-j_1(NO_2)+j_2(O_3) \cdot \frac{k_3(CO)-k_9(NO_2)+k_5(Alk)}{k_8(NO)+k_9(NO_2)+k_5(Alk)} \tag{8}$$

$$\frac{d(NO)}{dt} = -k_1(O_3)(NO)+j_1(NO_2)-j_2(O_3) \cdot \frac{k_3(CO)+k_8(NO)+k_5(Alk)}{k_8(NO)+k_9(NO_2)+k_5(Alk)} \tag{9}$$

$$\frac{d(O_3)}{dt} = -k_1(O_3)(NO) + j_1(NO_2) - j_2(O_3) \tag{10}$$

3.5   MODEL CALCULATION

With the arguments of the previous sections the model desrcibed by equation (6) is completely defined and in general it can be solved by numerical operations. However, for the final solution some boundary conditions are missing. The line source of the street is assumed to be situated in the middle of the forest lane. Therefore the transport from the middle to the border of the lane must be distinguished from the transport inside the forest. In the airspace of the lane the turbulence is governed by completely other influences then in the forest. On the one hand the restistances of the stems and leaves are missing and on the other hand fast running cars cause strong vortexes and turbulences.

The dispersion of the pollutants between street axis and forest was therefore described by a Gaussian model. It provides a vertical concentration profile at the border of the wood which can be used as boundary condition for the model of the further transport into the forest. A consideration of chemical reactions during the transport in the space of the forest lane is crucial, because of the strong turbulences which are caused by the cars . This problem was solved by including concentration data, which were measured at the border of the street, in the model calculations. With respect to NO_x the source strength f (NO_x) in a fictive middle line of the street can be calculated by the emission factor e $(NO_x) = 5.61$ (g/km) and the frequency of the traffic. The knowledge of the source strength of NO_x and the ratio of NO and NO_2 which has been measured at the border of the street, enables the separate calculation of individual pseudo-source strengths of NO and NO_2. This leads via equation (11) to vertical concentration profiles of NO and NO_2.

$$C_i(0,z) = \frac{\sqrt{2} \cdot f_i}{\sqrt{\pi} \cdot w_x \cdot \sigma_z} \exp\left(-\frac{z^2}{2\sigma_z^2}\right) + C_{Hi} \quad ; \qquad (i=1,2) \tag{11}$$

The background concentration C_H was measured at the meteorological tower nearby. σ_z can be calculated with the NO_x data, measured at the border of the street. A vertical concentration profile of O_3 at the border was also calculated with measured data by the aid of a exponent function (Sattler, 1989).

The vertical profiles were used as boundary values for a grid model. A finite difference scheme was used and the differential quotients were approximated by difference quotients. The x- and z-coordinates were devided in n-1 and m-1 steps of the length Δx and Δz, such that a grid was formed. At the points of the grid the difference quotients were calculated. This led to a system of n times m algebraic equations. The differential quotients could easily be approximated by Taylor series.

After all these operations the final model equation can be given

$$C(x + \Delta x, z) = (M+R)\frac{\Delta x}{w(z) \cdot \Delta z^2} + C(x,z) \tag{12}$$

where R is a term for the chemical reactions of the considered air pollutant and M represents the following term:

$$M = C(x,z+\Delta z) \cdot K(z+\Delta z) - C(x,z) \cdot [K(z) + K(z+\Delta z)] + C(x,z - \Delta z) \cdot K(z) \tag{13}$$

According to the measurements the model was also calculated with half hour time periods. However, some restrictions were made. Time periods with wind speeds below 0.02 m/s were discriminated because horizontal diffusion was neglected in the model and therefore calm conditions could not be treated. A wind direction between 187° and 247° was another restrictive condition. It should guarantee, that during the considered time period the occuring transport led from the street into the forest.

In order to test the sensitivity of the model special half hour data sets were chosen. Firstly they should represent situations in different seasons and secondly they should include extreme values (minima or maxima) with respect to the source strength of the street f, the wind speed w, the O_3 concentration and the UV-radiation. For reasons of demonstration the measured and calculated concentration data under comparison were plotted as bars in a diagramm. As can be seen in Fig. 5 the single bars are plotted in such a way that they are approximately located at the places where the concentration was measured during the experiments or calculated with the model. By the aid of three planes the concentration of two pollutants are displayed with four lines of bars. Two lines belong to each pollutant. The bars in the front line represent the measured and the bars behind the calculated concentrations.

From various situations one case will be shown here. Beside the source strength of the street transport and chemical conversion of the pollutants are strongly influenced by the ozone-concentration of the air in the environment of the forest lane. Ozone is mostly generated during the day by photodissociation. In that case it is correlated with the UV-radiation. However it may also be transported from higher layers of the atmosphere down to the ground. This was sometimes indicated by maxima of the ozone-concentration during the night. A situation with a high ozone-concentration is reflected in Fig. 5.

While the measured ozone-concentrations are increasing with increasing distance from the street and the measured NO_2-concentrations are decreasing, the calculated values show nearly the contrary. Only at higher altitudes the agreement between measured and calculated data is rather satisfying. At higher altitudes and farer distances from the street general a better agreement between model and reality could be observed. This may be explainable with a longer integration time of the model.

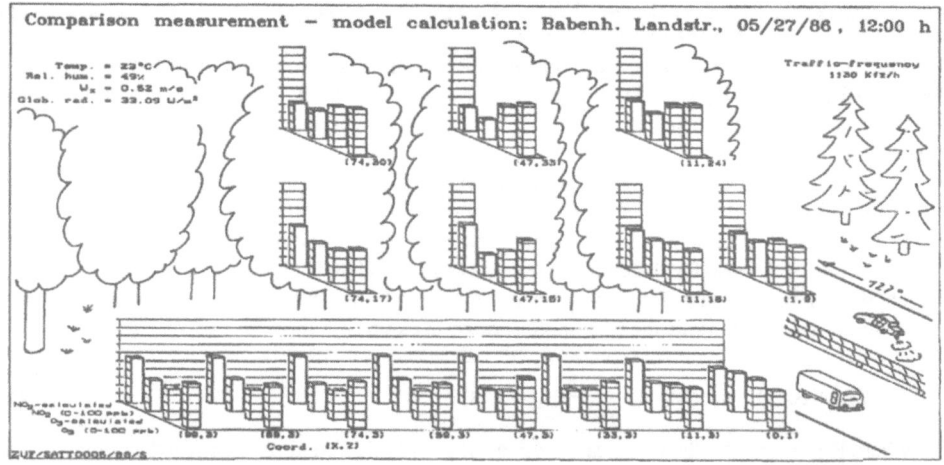

Fig. 5: NO_2- and O_3-concentrations (high O_3-level, May).

4. Discussion

Dispersion and reaction of air pollutants were simulated with a two-dimensional model. In the model calculations a dynamic equilibrium between the reactants was assumed. In most cases an included Gaussian-submodel provided meaningful boundary conditions. This could be demonstrated by the rather good agreement of the calculated Gaussian concentration values and the data which were measured at the border of the forest. The model showed sensible reactions on the variation of certain parameters. It turned out, that the chemistry played an important role on the dispersion of nitrogene oxides. In contrary to open areas in a forest the UV-radiation is only of minor importance. Especially in our case, where short transport times were considered, photochemical reactions can be neglected. Ozone from farer sources is the main oxidant for the oxidation of NO.

During the transport into the forest the oxidation of NO could be described realistically by the model especially in the presence of high ozone-concentrations. The agreement between model and measurement was better at lower and medium wind speeds. Also in cases of longer integration times the agreement was generally very good. Altogether the model could be validated by the measured data in a very satisfying manner. Therefore, in future it can be used for assessments of the immissions of pollutants in forests close to streets also in situations where no measurements are possible.

5. Acknowledgements

This work is a result of a project "Untersuchung von Immissionsschäden im Frankfurter Stadtwald unter besonderer Berücksichtigung der Emission von Verbrennungsmotoren - Immissionsbelastung in der Umgebung einer Straßenschneise in Abhängigkeit vom Kfz-Aufkommen." sponsored by the BMFT (037396 7).

6. References

Becker, K.H., Löbel, J. und Schurath, U. (1983): Bildung, Transport und Kontrolle von Photooxidatien. In: Umweltbundesamt (Hrsg.): Luftqualitätskriterien für photochemische Oxidatien. Berichte 5/83, E. Schmidt, Berlin.

Calvert, J.G. and Stockwell, W.R. (1983): Deviations from the O_3-NO-NO_2 photostationary state in tropospheric chemistry. Can. J. Chem. 61, 983-992.

Kasting, J.F. and Ackermann, T.P. (1985): High atmospheric NO_x levels and multiple photochemical steady states. J. Atmos. Chem. 3, 321-340.

Kolar, J. (1981): Anteil der Stickstoffdioxid-Immission an der gesamten Stickstoff-Immission in Städten. Staub - Reinhalt. Luft 41, 85-91.

Slinn, W.G.N. (1982): Predictions for particle deposition to vegetative canopies. Atm. Environ. 16, 1785-1794.

Sattler, Th. (1989): Ausbreitung Kfz-bedingter Spurenstoffe im Baumbestand. Mit Programmlistings zu Experiment und Modellrechnung in Pascal. Maraun, Frankfurt.

Sattler, Th. und Jaeschke, W. (1987): Automatisierte Bestimmung von Immissionsprofilen anorganischer Luftschadstoffe in verkehrsreichen Waldschneisen. Staub - Reinhalt. Luft 47, 261-266.

Waldeyer, H. (1978): Die Darstellung des Kraftfahrzeugverkehrs als Linienquelle - Bereitstellung der erforderlichen Basisdaten. In: Abgasimmissionsbelastungen durch den Kraftfahrzeugverkehr. TÜV Rheinland, Köln.

DETERMINATION OF THE GASEOUS AIR POLLUTANT UPTAKE OF A SPRUCE BRANCH BY MEANS OF THE ENCLOSURE TECHNIQUE

K. BAUMANN, G. BAUMBACH
Institut für Verfahrenstechnik und Dampfkesselwesen (IVD)
Abt. Reinhaltung der Luft, Universität Stuttgart
Pfaffenwaldring 23, D-7000 Stuttgart 80

ABSTRACT. With respect to generally known problems and conditions by applying the vegetation enclosure technique to determine the CO_2 gas exchange and, with this the air pollutant uptake by the plant, a chamber system which holds a spruce branch in the tree canopy has been developed and tested in a field experiment at a measuring site in the northern Black Forest. During an initial test phase chamber housing absorptions of air pollutants NO, NO_2, O_3, SO_2 as well as CO_2 were continuously measured under real variational conditions. The absorbed fractions were taken into account for the determination of the net gaseous air pollutant uptake of an enclosed branch of a highly damaged 80 years old Norway spruce during a continuous measuring period of 19 summer days.

1. MOTIVATION

Already in 1905 it was Wieler /1/ who first dicovered the photosynthesis and the aspiration as central physiological processes with which directly damaging effects on trees can be detected. By means of initial gas exchange measurements he determined an impairment of photosynthesis of beech and spruce by sulphurous acid. This subject was not taken up till 1958, when H. Keller /2/ proved distinct influences of pollutants on the photosynthetic activity of trees.

By the discussion of "neuartige Waldschäden" the CO_2 gas exchange measurements became internationally accepted as a method to determine possible effects of the environment, including atmospheric pollution, on growth and development of plants. Further investigations aimed at a constructive optimum of the gas exchange chamber to meet the ambient climatic conditions like humidity, temperature, solar radiation, etc., so that the net photosynthesis, the respiration and transpiration of the enclosed vegetation could be measured most accurately in order to get a quantitative status of the overall plant vitality /3 to 11/.

However, the measures which are necessary to control the chamber climate compete with the exact measurement of the photosynthetic active or passive air pollutant uptake by the plant, because indefinite absorption processes may take place at internal chamber installations. With respect to the problems described in the quoted literature, and as a result of a fruitful contact to Prof. Dr. W. Koch, Institut für Fortsbotanik, Ludwig-Maximilians-Universität München /12/ a special chamber system has been developed and tested in a field experiment at a heavily damaged forest area of the northern Black Forest.

H.-W. Georgii (ed.), Mechanisms and Effects of Pollutant-Transfer into Forests, 169–176.
© *1989 by Kluwer Academic Publishers.*

2. SET UP OF THE CHAMBER SYSTEM IN THE FIELD

The field experiments are carried out at a site where the dept. Reinhaltung der Luft (IVD) is conducting air pollution measurements for several years /13, 14/. The measuring station - in the following context named "Schöllkopf" is situated on a smooth hill near the town of Freudenstadt in the northern Black Forest at an elevation of 840 m asl. The forest stand consists of 80 years old, heavily damaged Norway spruces (60 %) and white firs (40 %), and averages to a height of 25 m.

A measuring tower which was erected with friendly support by the Energie-Versorgung Schwaben AG (EVS) rises 17 m above the average tree canopy. In the lower canopy range of a damaged Norway spruce at an elevation of 19 m above ground a tower platform allows the installation of the enclosure chamber. The applied system of emission measurements by means of gas sampling through long, preheated teflon tubes and computer controlled commutation of three different gas flows is depicted by the flowchart of Figure 1. The continuously measured NO-, NO_2-, O_3-, SO_2- and CO_2-concentrations as well as the dewpoint temperature are reduced to half-hour mean values and recorded on floppy disks. Because of the magnetic valve control system, each gas stream (three in total) is analysed only for a fraction of each 30 minutes, so that this principle actually represents a quasi-continuous measurement for air pollutants.

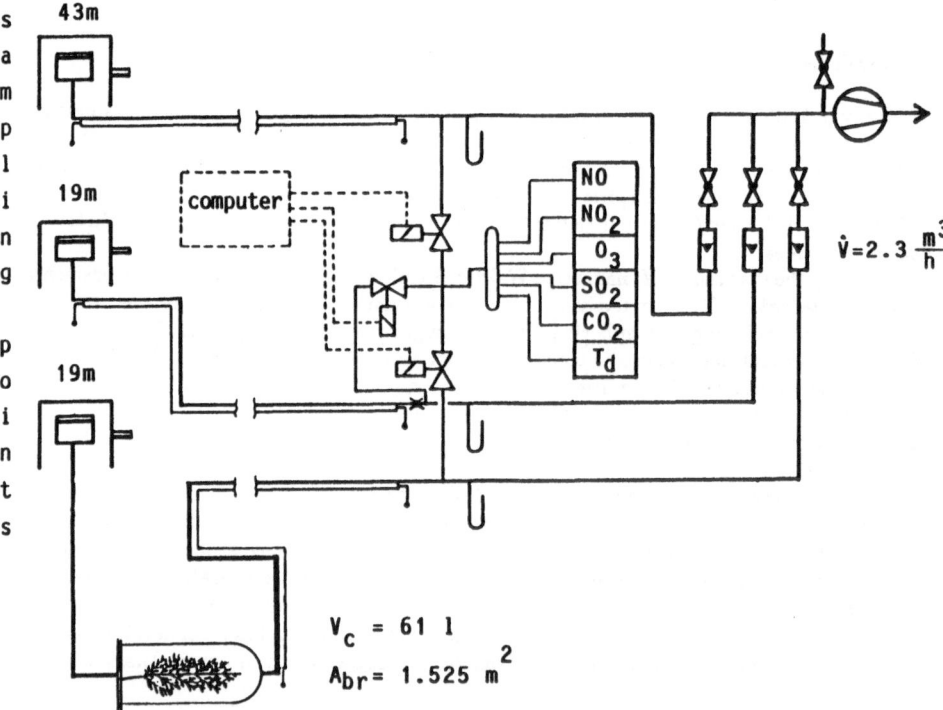

Figure 1: System flowchart of gas sampling and computer controlled alternating concentration measurement covering two reference locations and an enclosed spruce branch. A_{br}... surface area of branch, V_c... net volume of chamber, $\dot{V}$... sample gas flow, T_d... dewpoint temperature

The ambient air at the receptor point in 19 m above ground is first analysed directly and after the magnetic valve shift secondly analysed after passing through the gas exchange chamber with the enclosed spruce branch. In a third step the air at an elevation of 43 m is measured as reference.

The sample gas flow $\dot{V}$ is kept constant at 2.3 m^3/h. The net chamber volume is 61 l and the surface area of all photosynthetic active needles amounts to 1.525 m^2. The total volume of the enclosed branch is approximately 1 l. This results in an air change rate of 0.64/min within the chamber. A very resistant, UV transparent, high optical quality plexiglass /15/ is used as chamber housing material. The chamber form is built to meet a optimum volume-to-inner-surface-area-ratio as well as an optimum streamline configuration.

The measurements described here, are taken in August 1988. The branch enclosure chamber is covered by a gauze to avoid possible inner overheating. Thus, the microclimate of the enclosed branch approximately corresponds to that of a naturally half shaded branch in the lower tree canopy.

3. DETERMINATION OF CHAMBER SPECIFIC ABSORPTIONS

During the field experiment the chamber absorptions of all gaseous air pollutants (and CO_2) are continuously measured at the same time. For this, a second, identical empty chamber is set bypass controlled in series to the branch enclosure chamber /16/, so that these specific microclimatic conditions due to daytime dependent transpiration and respiration can be transmitted into the empty control chamber. Hence, instant chamber absorptions can be measured and expressed as half hour mean values dependent on the momentary climatic conditions inside the branch enclosure chamber. This comprises the initial assumption, that the mass of air pollutants absorbed by the inner chamber surface is proportional to the amount of air pollutants at the chamber inlet. This has been confirmed in good approximation for all air pollutants under investigation by the following correlations.

Figure 2 represents correlations between the simultaneously measured NO-, NO_2-, O_3-, and SO_2- concentrations at the empty control chamber inlet (abscissa) and outlet (ordinate). For correlation coefficients r between 0.964 for SO_2 and 0.998 for O_3 the average chamber absorption rates amount to 7 % for hydrophylic gases like SO_2 and NO_2, and 4 % for O_3. There is no absorption detected for NO and for CO_2 (not depicted in Fig. 2) at good correlations of 0.988 and 0.979 respectively.

The relatively high humidity produced by the enclosed branch has a major impact on hydrophilic gases like SO_2 and NO_2. All relative absorption losses found this way are taken into account for further investigations concerning the calculation of net gaseous air pollutant deposition onto the enclosed branch.

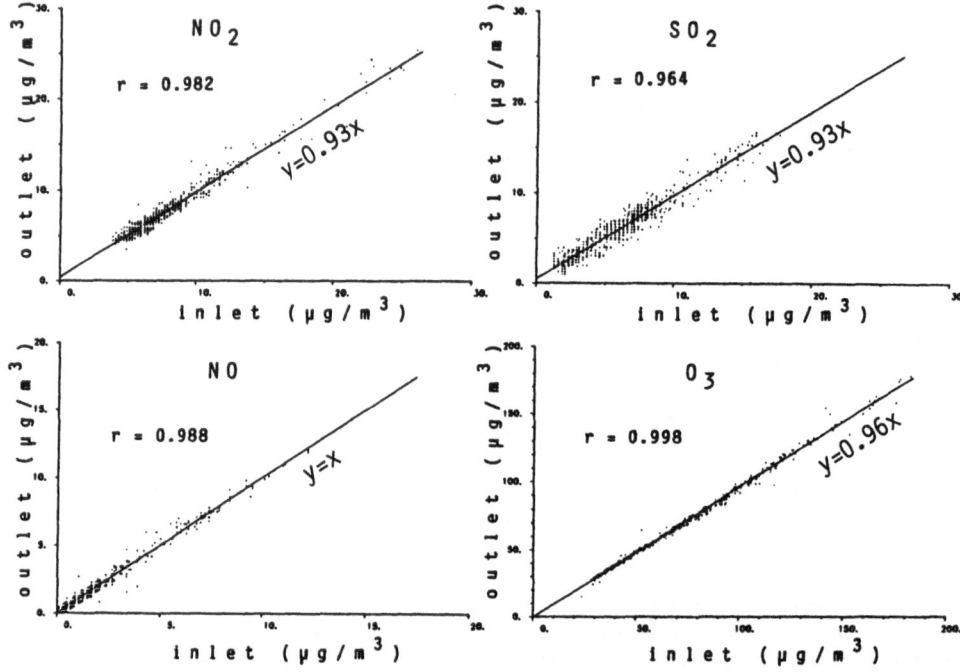

Figure 2. Correlations of air pollutants uptake by the inner chamber surface under variational climatic conditions

4. NET AIR POLLUTANT UPTAKE OF A SPRUCE BRANCH

Figure 3 depicts a possible presentation of ozone deposition onto a spruce branch determined by means of such enclosure technique: the net ozone uptake by the branch is plotted versus the ambient ozone concentration. For linear regression the ozone uptake correlates with increasing ozone offer at $r = 0.864$, which initially can be explained by the occurrence of highest ozone concentrations during midday hours with intensive solar radiation and also most intensive assimilation. But relatively high ozone values can be registered at the Schöllkopf site also during night hours /17/. It is important to note, that certain instantaneous ozone levels do not always meet the same physiological conditions of the plant, so that with this presentation it is hard to tell whether the branch takes up ozone actively or passively. But for a logarithmic regression (dashed line in Fig. 3) a lowest ozone concentration can be assumed below which no further absorption takes place by assimilation, and plant growth may not be influenced directly.

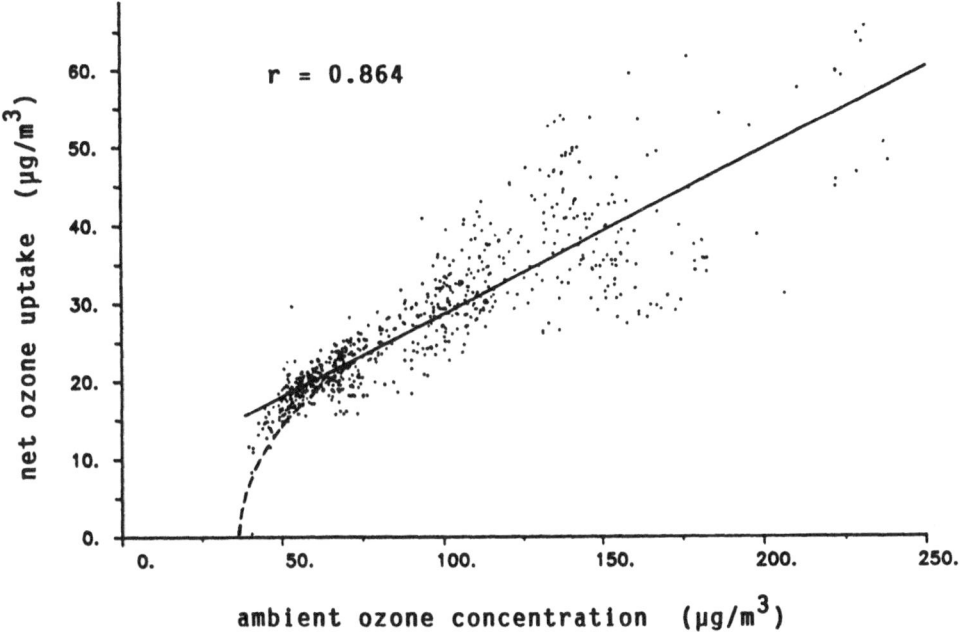

Figure 3. Correlation between ambient ozone concentration and net ozone uptake by the enclosed spruce branch. For logarithmic regression a lowest concentration can be determined below which no uptake occurs.

The opening of the stomata aperture depends mainly on solar radiation and temperature, and therfore, it is subjected to the respective daily variations. Hence, it is of interest to focus on the mass flow of air pollutants onto photosynthetic active parts of the branch by evaluating average daily cycles separated by intensive sunny and less sunny days. This is done by Figure 4: The mass flows of NO_2, O_3 and CO_2 onto the needles are plotted for a sunny period (left) and for the directly following cloudy period (right) in August 1988.

The massflow F of each quantity to be measured can be obtained by

$$F = \dot{V}/A_{br} \, (C_{in} - C_{out} - C_{ch})$$

where the net concentration balance is assumed to be steady state during the averaged measuring time, with the measured concentrations at the chamber inlet C_{in}, and at the chamber outlet C_{out} and the above determined relative concentration loss by chamber absorption C_{ch}. Thus a complete turbulent mixing inside the chamber is furthermore assumed to exist all the time.

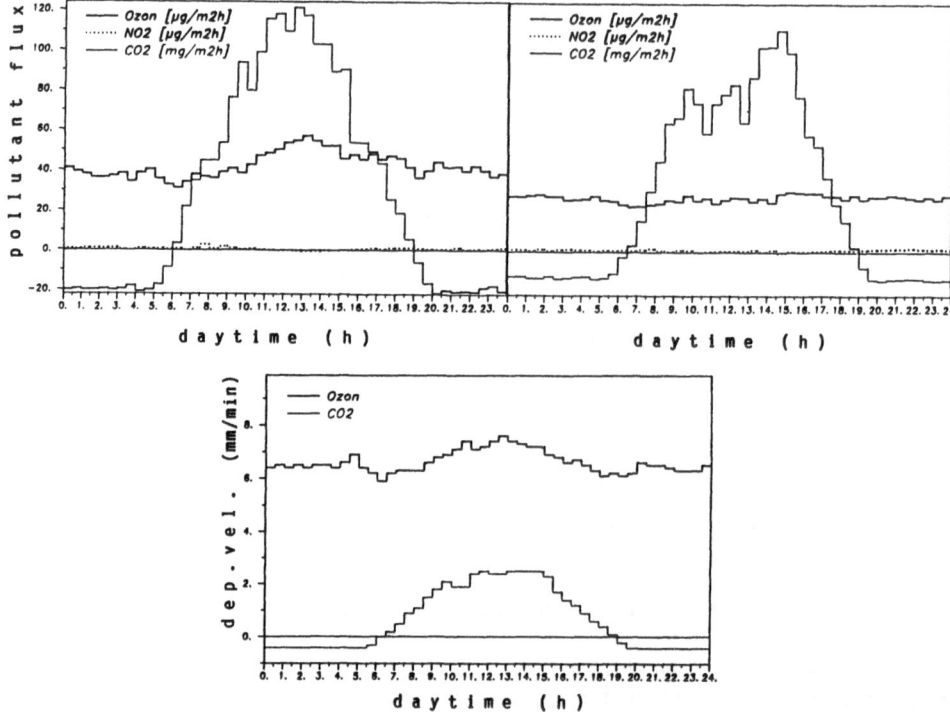

Figure 4. Average diurnal variations of NO_2, O_3, and CO_2 mass flows onto the photosynthetic active needle surface area of an enclosed spruce branch for a sunny (left) and for a cloudy (right) weather period. Calculated deposition velocities for CO_2 and O_3 (bottom) for the total measuring period.

Because of the very low concentration levels of NO and SO_2 during the measuring phase, no mass flows could be determined for this air pollutants. There is no uptake of NO_2 detectable, independent of the existing weather conditions. The branch respires CO_2 at night, and aspires CO_2 during daylight hours when assimilation takes place. The CO_2 uptake by aspiration is more intensive under conditions of clear sky, high solar radiation and temperature. That is, when the ozone uptake is also at maximum (see Fig. 4 left). At night and at cloudy days an evenly leveled ozone uptake can be observed - there is indeed a certain fraction of "passive uptake" of this highly reactive air pollutant in a sense of superficial reaction on the branch needles.

Deposition velocities can be expressed as mass flows (F) divided by absolutely measured ambient concentrations (C_{in}), which for this special case here can be interpreted as reciprocal needle surface resistance (see Fig. 4 bottom). The resistance against ozone is almost always relatively low, whereas the highest resistance against CO_2 occurs naturally at night during respiration.

5. DISCUSSION AND CONCLUSION

The deposition velocity (v_D) in general can be defined as reciprocal of the total resistance acting "naturally" against the mass transport from the free atmosphere into the mesophyl. This overall resistance (r_o) is composed of fractional resistances which describe the aerodynamic turbulent exchange in the atmosphere (r_a), the molecular turbulent diffusion at the surface boundary layer (r_b), as well as the actual absorption by the needle (r_c), where distinctions in chemical reactions, solutions and adsorptions, and stomatal diffusion processes must be made

$$v_D = r_o^{-1} = (r_a + r_b + r_c)^{-1}$$

With the branch enclosure technique presented here, r_a and r_b cannot be determined. This method is good only for measuring the absorption of less hydrophilic gases like ozone; the absorption of hydrophilic SO_2 and NO_2 can only be determined for their gas phase, since wetting and dew do not occur inside the chamber. Considering this it is possible to examine a net air pollutant balance while measuring qualitatively the momentary physiological status of the plant.

This method shall not compete the aerodynamic gradient profile method nor the eddy correlation approach, but it shall contribute to a better understanding of different deposition processes. In this respect a major advantage of the enclosure technique can be seen in its possibility to perform measurements under inversional atmospheric conditions, when no turbulent air exchange exists and relatively high pollution loads occur.

ACKNOWLEDGEMENTS

This report is generally based on a research program sponsored by the German government through the Bundesminister für Forschung und Technologie (BMFT), record no. 0339112. The ambitious help by Andreas Rochowiak, who maintained the equipment of the measuring station was greatfully appreciated.

REFERENCES

/ 1/   Wieler, A.: Untersuchungen über die Einwirkung schwefliger Säure auf die Pflanzen. Gebr. Borntraeger, Berlin, 1905

/ 2/   Keller, H.: Beiträge zur Erfassung der durch schweflige Säure hervorgerufenen Rauchschäden an Nadelhölzern. Beihefte z. forstw. Zentralblatt Heft 10, 1958

/ 3/   Heath, R.L.: Initial events in injury to plants by air pollutants. Ann. Rev. Pl. Physiol. 31, pp 395 - 431, 1980

/ 4/   Schulze, E. D. and W. Koch: Measurements of primary production with cuvettes. Unesco, 1971. Productivity of forest ecosystems. Proc. Brussels Symp. pp 141 - 157, 1969

/ 5/   Koch, W. and H. Roth: Eine große Präzisions-Gaswechselanlage mit getrennter Grün- und Bodenzone und ihre Leistungsfähigkeit dargestellt am Beispiel der Fichte. Photosynthetica 10 (1), pp 71 - 82, 1976

/ 6/   Eller, B.U. and W. Koch: Globalstrahlung innerhalb und außerhalb von Gaswechselkammern. Photosynthetica 11 (3), pp 268 - 275, 1977

/ 7/   Bogenrieder, A. and R. Klein: Does the exclusion of UV influence photosynthetic gas exchange measurements? Problems of climatic conditions in gas exchange chambers. Flora 169, pp 510 - 523, 1980

/ 8/   Lange, O. L., J. Gebel, E. D. Schulze and H. Walz: Eine Methode zur raschen Charakterisierung der photosynthetischen Leistungsfähigkeit von Bäumen unter Freilandbedingungen - Anwendung zur Analyse "neuartiger Waldschäden" bei der Fichte. Forstw. Cbl. 104, pp 186 - 198, 1985

/ 9/   Adaros, G. and H.-J. Daunicht: A movable, dewpoint-controlled daylight growth chamber, equipped for gas exchange measurement at high ventilation rates. Angewandte Botanik 59, pp 415 - 424, 1985

/10/   Atkinson, C. J., W. E. Winner and H. A. Mooney: A field portable gas-exchange system for measuring carbon dioxide and water vapour exchange rates of leaves during fumigation with SO_2. Plant, cell and environment 9, pp 711 - 719, 1986

/11/   Bames, G.: CO_2-Gaswechselmessungen immissionsbelasteter Bäume. MS-Thesis at the University of Hohenheim, 1986

/12/   BMFT-research program 0339157: Kontinuierliche Schadstoffmessungen am Meßturm von Herrn Prof. Koch im Bayerischen Wald, Aschenbrennermarter. 1 April 1987 to 30 June 1988

/13/   Baumbach, G., K. Baumann, F. Dröscher: Luftverunreinigungen in Wäldern. Institut für Verfahrenstechnik und Dampfkesselwesen (IVD), University of Stuttgart, Dept. Reinhaltung der Luft, Report No. 5, 1987

/14/   Baumbach, G., K. Baumann, F. Dröscher: Luftqualität in Freudenstadt. IVD, University of Stuttgart, Dept. Reinhaltung der Luft, Report No. 6, 1987

/15/   Plexiglas GS 2458 (1986), 3 mm, ID-No. 225 - 15. UV transmission guaranteed for 20.000 exposure hours. F. Röhm GmbH D-6100 Darmstadt 1

/16/   Bührle, W.: Konstruktion einer Gaswechselanlage zur Ermittlung der Schadgasaufnahme von Fichten im Freiland. MS-Thesis at the University of Stuttgart, IVD, Dept. Reinhaltung der Luft, 1988

/17/   Baumbach, G., K. Baumann: Ozone in forest stands - examinations to its occurence and degradation. In H.-W. Georgii (ed.): Mechanism and effect of pollutant-transfer into forests, Kluwer Academic Publ., Dordrecht 1989

MODELLING THE UPTAKE OF SO$_2$ INTO LEAVES OF FOREST CANOPIES – Dynamic and steady state considerations.

Badeck, F.-W., Kohlmaier, G.H. and Plöchl,M.
Institut für Physikalische und Theoretische Chemie
der J.W.Goethe Universität
Niederurseler Hang
6000 Frankfurt 50

ABSTRACT: Using a network for diffusion and reaction of SO$_2$, uptake via the stomates and metabolization in various compartments of the leaves of coniferous and deciduous forest canopies is taken into consideration. On the basis of the uptake rates computed for leaves we predict within the validity of our dynamic model a mean annual uptake of 0.35 g S·m^{-2}·yr^{-1} for beech canopies and a corresponding uptake of 0.79 g S·m^{-2}·yr^{-1} for spruce canopies. Taking into consideration the most important tree species growing in the FRG we compute a weighted mean average of 0.55 g S·m^{-2}·yr^{-1} for stomatal uptake. On the basis that the dry deposition of S within the FRG amounts to 0.8·10^6 t·yr^{-1}, corresponding to an average of 3.2 g S·m^{-2}·yr^{-1} it can be concluded from our calculations that between 6 and 16% of the total dry deposition can be ascribed to stomatal uptake. We can show explicitly for a reduced leaf/canopy model that SO$_2$ uptake is controlled by the relative proportions of diffusive transport and biochemical reactions within the leaf.

1. INTRODUCTION

Calculating the deposition of atmospheric trace gases applying the flux balance approach for tree crown layers [1] there always remains some uncertainty in the calculation of the dry deposition inside the crowns. The difference between precipitation in open terrain and throughfall plus stem flow cannot be ascribed to interception alone because a correction term for the sink/source function of the living biomass needs to be included. The term comprises stomatal and cuticular uptake, exhalation, and leaching. With the experimental methods currently available, the components of this correction term cannot be fully resolved. Following Elston & Monteith [2] one of the basic functions of models is to link our knowledge about adjacent levels of organization in order to provide additional information about the system behaviour.

In this sense the model presented here predicts the share of stomatal uptake in total dry deposition. The core model describing the

177

H.-W. Georgii (ed.), Mechanisms and Effects of Pollutant-Transfer into Forests, 177–184.
© 1989 by Kluwer Academic Publishers.

uptake of SO_2 into the leaves and its subsequent metabolization has already been presented elsewhere in greater detail [3,4] and in the present context will only be characterized briefly.

2. MODEL STRUCTURE

The model describes the diffusive fluxes and reactions of several sulfur species with a system of coupled ordinary differential equations.

SO_2 diffuses from the air space surrounding the leaves into the intercellular air space thereby passing through the leaf boundary layer and the stomates. Gaseous SO_2 dissolves in the liquid phase of the cell walls and dissociates into HSO_3^- (hydrogen sulfite), SO_3^{2-} (sulfite), and H^+ (protons). Inside the cell wall as well as in the subsequent compartments the concentrations of these species are determined by a pH-dependent equilibrium.

SO_2, HSO_3^- and SO_3^{2-}, i.e. the sulfur species of the oxidation state +IV, are diffusing through the plasmalemma into the cytosol and furtheron into the chloroplasts and the vacuole. Elimination reactions considered in the simulation are the oxidation of S(IV) to S(VI), i.e. sulfate, in cytosol and chloroplasts, and the reduction of S(IV) to S(-II) in the chloroplasts alone. These processes are taken into consideration in the model in substitution for further elimination reactions (e.g. oxidation inside the cell walls) and transport processes which remove S(IV) and its reaction products from the leaves.

Concentration changes in every single compartment of the cell are simulated by differential equations comprising terms which describe the exchange processes with other compartments and terms decribing elimination reactions.

Parameters varying as a function of light intensity or temperature are calculated in seperate modules using hourly values taken from the data set of the test reference year of the DWD (Deutscher Wetterdienst).

3. RESULTS

The fundamentals in the interpretation of the simulation runs of the complex dynamic numerical model shall be discussed now by developing a reduced steady state system comprising only two compartments. One compartment represents the atmospheric gas phase, the second compartment depicts all liquid compartments of the leaf in one single volume which is set equal to the total cellular volume. In the same way all the different rate constants for the elimination reactions in different cellular compartments are substituted by one single first order rate constant. Therefore this reduced model cannot claim to have the same quantitative validity as the numerical simulation model described before, but shall be regarded as an explorative tool to describe some system characteristics which are still valid for the dynamical model simulation of greater complexity.

3.1 REDUCED MODEL

Diffusive transport from the surrounding air into the leaf is described by Fick's law of diffusion. The change of the total S(IV) concentration with the leaf $c_i(t)$ can than be written:

$$\frac{dc_i}{dt} = \frac{c_a - c_{i,gas}}{d \cdot R} - k \cdot c_i \qquad (1)$$

where the effective thickness of the liquid phase is denoted by d [cm], the total resistance to diffusion by R [$s \cdot cm^{-1}$], the gas phase SO_2 concentration by c_a, the total S(IV) concentration in the liquid phase by c_i, the partial pressure of SO_2 over the liquid phase, respectively its concentration equivalent by $c_{i,gas}$, and the reaction constant of elimination reactions by k [s^{-1}].

Calculating $c_{i,gas}$ from c_i both the process of dissolution and dissociation must be considered. The fraction of dissolved SO_2 in solution is given by $c_i \cdot X_{SO_2}$ with: X_{SO_2} = mole fraction of the gaseous species SO_2 in the liquid phase. The concentration of dissolved SO_2 is related to the corresponding partial pressure of SO_2 above the solution, p_{SO_2}, by:

$$c_i \cdot X_{SO_2} = H \cdot p_{SO_2} \qquad (2a)$$

in which p_{SO_2} can be converted to concentration units:

$$c_{i,gas} = p_{SO_2} / (p_{tot} \cdot V_M) \qquad (2b)$$

with: H = Henrys constant [$M \cdot atm^{-1}$]; V_M = molar volume at a given temperature and p_{tot} = standard total pressure in the gas phase. Finally $c_{i,gas}$ of eq. (1) can be expressed as funtion of the total S(IV) concentration c_i with eq. (2a) and (2b) to yield:

$$c_{i,gas} = \frac{c_i \cdot X_{SO_2}}{H \cdot p_{tot} \cdot V_M} \qquad (2c)$$

or vice versa the enrichment of S(IV) due to dissolution and dissociation of SO_2 can therefore be written $c_i = A \cdot c_{i,gas}$ where the parameter A is defined by:

$$A = (H \cdot V_M \cdot p_{tot}) / X_{SO_2} \qquad (2d)$$

In the steady state the solutions for the total S(IV) equilibrium concentration (c_i) and the uptake rate (F_{in}) can be written:

$$c_i^{\text{steady state}} = \frac{c_a}{d \cdot R \cdot k + A^{-1}} \qquad (3)$$

$$F_{in}^{\text{steady state}} = d \cdot k \cdot c_i^{\text{steady state}} = \frac{c_a}{R + A^{-1}/(k \cdot d)} \quad (4)$$

where an uptake or deposition velocity for the leaf can be defined as:

$$v_{in}^{\text{leaf}} = \frac{1}{R + A^{-1}/(k \cdot d)} \quad (5)$$

The influence of the metabolization rates onto the uptake rate of SO_2 can be investigated by considering the sensitivity of the system for changes in the reaction constant of elimination reactions. For small rates of elimination, $k \to 0$, the product $d \cdot R \cdot k$ will be small compared with A^{-1} such that equation (3) is reduced to:

$$\lim_{k \to 0} c_i^{\text{steady state}} = c_a \cdot A \quad (3a)$$

Under these circumstances the system is in the state of thermodynamic equilibrium between the gas phase and the dissolved phase of SO_2. In the case that no elimination reactions have to be considered the equilibrium concentration of S(IV) is maximal and depends only on the external concentration of SO_2 and the factor A describing the enrichment because of dissolution and dissociation. As soon as the system is equilibrated, the influx of SO_2 comes to a standstill, $F_{in} \to 0$, (eq. 4) as k goes to zero.

For rates of elimination different from zero a steady state equilibrium is established. Depending on the parameters controlling the influx (resistances for diffusion, rates of elimination) a steady state concentration for S(IV) now is established inside the plant. It is lower than the equilibrium concentration for the thermodynamic equilibrium.

If the velocities of the elimination reactions are that high ($k \to \infty$) that $d \cdot R \cdot k \gg A^{-1}$, the S(IV) concentration in the liquid phase will be nearly zero, $c_i \to 0$. The uptake rate reaches a maximum which can be denoted by:

$$\lim_{k \to \infty} F_{in}^{\text{steady state}} = c_a/R = v_{in, \text{max}}^{\text{leaf}} \cdot c_a \quad (4a)$$

$$\text{with:} \quad v_{in, \text{max}}^{\text{leaf}} = 1/R$$

For this case the uptake velocity is given by the reciprocal of the resistance. Multiplied with the leaf area index (area of leaves per unit area of ground), LAI, of the canopy under consideration an uptake velocity for the canopy, v_{in}^{can}, can be calculated. For photosynthetically active plants with fully open stomates it adopts values between 0.25 and 10 $cm \cdot s^{-1}$.

Comparing the calculated uptake velocity range derived from the above considerations with the mean deposition velocities of about 0.8 $cm \cdot s^{-1}$ it becomes evident that in the limiting case of very high elimination velocities the total dry gaseous deposition could be ascribed to the stomatal uptake of gaseous SO_2. Simulation runs with the more detailed numerical model however compute uptake rates which are up to a factor 4 smaller than the dry deposition rates which were given e.g. by Georgii and Perseke [5] who reported values between 0.6 and 2.3 g $S \cdot m^{-2} \cdot yr^{-1}$ for regions with low pollutant concentrations.

3.2 SIMULATION RUNS WITH THE NUMERICAL MODEL

Based on the uptake of SO_2 computed by the leaf model the pollutant uptake into the overall tree crown space for all forests in the FRG was computed. In order to get a first approximation the LAIs for the main tree species j, their respective seasonal changes (eq. 6), and the total area of the stands by species j, area (j), have been taken into consideration.

$$F_{in}^{can}(j,t) = F_{in}^{leaf}(j,t) \cdot LAI(j,t) \qquad (6)$$

with: $F_{in}^{can}(j,t)$ = flux to the canopy of species j per unit area. Integration over the period of one year (eq. 7) and summation over all forest stands (eq. 8) yields an approximation of the mean annual uptake of gaseous pollutant species into the leaves of the forests of the Federal Republic of Germany.

$$\overline{F}_{in}^{can}(j) = \frac{1}{a} \int_0^{1 a} F_{in}^{can}(j,t) \ dt \qquad (7)$$

$$D_{tot} = \sum_{i=1}^{j} \overline{F}_{in}^{can}(j) \cdot area(j) \qquad (8)$$

where $\overline{F}_{in}^{can}(j)$ denotes the mean annual uptake in a canopy of the species j per square meter, D_{tot} the total yearly deposition via the stomatal path in all forests.

The model is to be developed furthermore in order to reflect the changes of stomatal resistances and pollutant concentration as a function of the position of a single leaf layer inside the crown and water availability. This implies the necessity of coupling the leaf model to a model describing gaseous transport in the atmospheric boundary layer and inside the crown space in order to provide a realistic concentration profile.

A simulation run using the monthly mean SO_2-concentration for the years 1978-1982 (monitoring network of the Umweltbundesamt) yields a total uptake of about 40,000 t $S \cdot yr^{-1}$ by the forests of West Germany or 0.55 g $S \cdot m^{-2} \cdot yr^{-1}$ respectively (fig 1). Compared with an estimated mean dry deposition of 3.2 g $S \cdot m^{-2} \cdot yr^{-1}$ the sorption of SO_2 via the stomatal path as computed by the model stands for about 1/6 of the total dry deposition per unit area.

The effect of a variation in the velocity of the elimination reactions is demonstrated in fig. 2. The results presented there have

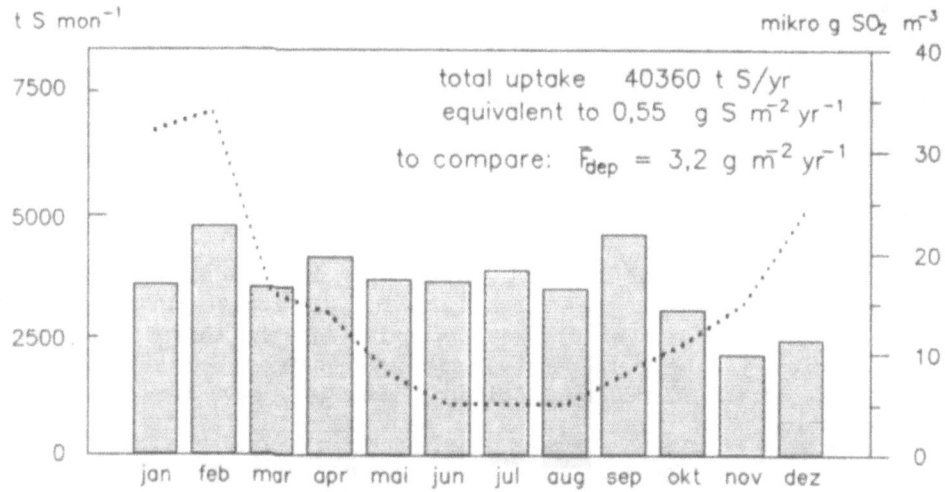

Fig 1.   Total monthly uptake of gaseous SO_2 via the stomates in West German forests. The dotted curve shows the mean monthly SO_2-concentration according to the data of the UBA (ordinate at the right hand side).

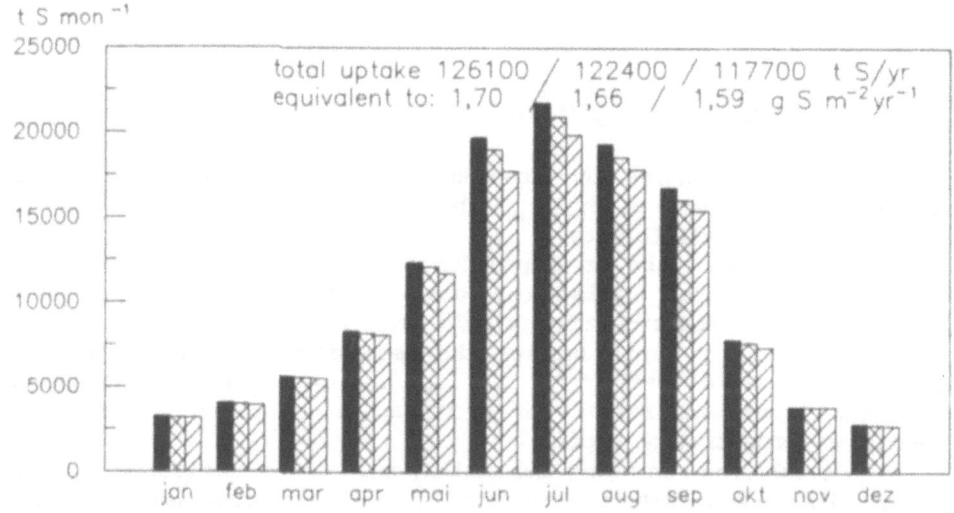

Fig 2.   Comparison of the monthly uptake of West German forests at different oxidation rates for S(IV). Filled column for $k = 5 \cdot 10^{-2}$ ms^{-1}, cross hatched column for $k = 5 \cdot 10^{-4}$ ms^{-1}, hatched column for $k = 5 \cdot 10^{-5}$ ms^{-1}. The input for this simulation was a constant atmospheric SO_2 concentration of 10 μg m^{-3}.

been computed by model runs using oxidation coefficients of 10^{-2} s^{-1}, 10^{-4} s^{-1} and 10^{-5} s^{-1} respectively. Apparently the dynamics of the system is relatively insensible within the suggested range of k; a reduction of the elimination rate by a factor of 1,000 leads to a reduction in the uptake of sulfur dioxide only by about 7%.

4. DISCUSSION

Summarizing, the influence of metabolic processes on the uptake of gaseous SO_2 can be characterized as follows:

Metabolic processes decisively influence the establishment of the steady state equilibrium of S(IV) inside the plant, thereby maintaining a permanent gradient between the atmospheric SO_2 concentration and the concentration of SO_2 solubilized in the mesophyll cell walls. Computer simulations based on rate constants for elimination reactions which were derived from literature data on physiological experiments (references in [3]) lead to the conclusion that a steady state equilibrium is established which is far away from the thermodynamic equilibrium for dissolution and dissociation. The function $F_{in}(k)$ exhibits a slope $\ll 1$ in the k-values derived from experimental data. F_{in} decreases only by 7% when k is decreased by a factor of 1000, which proves that F_{in} is already close to the asymptotic maximum given by c_a/R in the reduced model. Since the atmospheric pollutant concentration and the stomatal resistance change in time the potential maximal sorption can be calculated in the numerical model by:

$$F_{in,max} = \frac{1}{T} \int_{t=0}^{T} \frac{c_a(t)}{R(t)} \, dt \qquad (9)$$

The ratio $F_{in}/F_{in,max}$ can be taken as an expression of the proximity of the uptake rate corresponding to the established equilibrium and the maximal rate. In a simulation run using the parameter set for beech leaves, with $k = 5 \cdot 10^{-3}$ s^{-1} for sulfite oxidation, this ratio came out with: $F_{in}/F_{in,max} = 0.94$.

From these results it can be concluded that the system is close to the asymptotic maximum for the uptake rate and therefore only weakly sensible to changes in k. For the internal concentration of dissolved SO_2 in turn this means that it is already close to but not yet zero. This result is in accordance with the conclusions several authors have drawn from chamber experiments. They based their calculations on the assumption [6] that mesophyll can be regarded as a perfect sink for SO_2 with the concentration of SO_2 in mesophyll set equal to zero.

Although this is true for dissolved SO_2 alone in the range of accuracy which can be realized in chamber experiments, this does not hold for the total S(IV)-concentration, which may be considerably high because of the enrichment caused by dissociation. The simulation runs yield total S(IV) concentrations in the range of some μM to mM in cytosol and chloroplasts while $c_{1,gas}$ is computed to amount to 0.001 through 0.1 of the atmospheric SO_2 concentration for oxidation rates of $k = 10^{-2}$ and $k = 10^{-5}$ respectively.

We conclude that a small but not neglegable part of the SO_2 dry deposition occurs via stomatal uptake Even if in the light of these results only a smaller part of dry deposition can be ascribed to the uptake of gaseous SO_2 via the stomates it may lead to an accumulation of potentially toxic species in cellular liquid phases.

5. ACKNOWLEDGEMENT

We thank the Bundesministerium für Forschung und Technologie for financial support. K. Siebke, E.-O. Siré, and C. Wientzek partizipated in earlier stages of model development which is kindly acknowledged.

6. REFERENCES

[1] Ulrich, B.; Mayer, R.; Matzner, E. (1986) Flüssebilanz des Kronenraums. In: Ökosystemforschung (Ellenberg, H.; Mayer, R. Schauermann, J., eds.) Ulmer, Stuttgart: 405-411.

[2] Elston, J.; Monteith, J.L. (1975) Micrometeorology and Ecology. In: Vegetation and the Atmosphere. Volume 1: Principles. (Monteith, J.L., Hrsg.) Academic Press, London: 1-12.

[3] Siebke, K.; Badeck, F.-W.; Kohlmaier, G.H.; Plöchl, M.; Wientzek, C. (1989) Modelling Pollutant Exchange between Plant and Environment: Uptake and Metabolization of Sulphur Dioxide by different Leaf Cell Compartments. In: Modelling in Ecotoxicology (S.E. Jörgensen, ed.). Elsevier, Amsterdam, in press.

[4] Kohlmaier, G.H.; Janecek, A.; Lüdeke, M.; Kindermann, J.; Siebke, K.; Badeck, F. (1989) Modelling Pollutant Exchange between Plant and Environment. ISPRA Lectures from ISPRA Courses Oct. 12 - 16, 1987, in press.

[5] Georgii, H.-W.; Perseke, C. (1979) Some Results on Wet and Dry Deposition of Sulphur Compounds. In: European Symposium, Physico-chemical Behaviour of Atmospheric Pollutants, Ispra.

[6] Unsworth, M.H. (1981) The Exchange of Carbon Dioxide and Air Pollutants between Vegetation and the Atmosphere. In: Plants and their Atmospheric Environment (Grace, J.; Ford, E.D.; Jarvis, P.G.; eds.), Blackwell, Oxford: 111-138.

MODELLING THE EFFECT OF SULFUR DIOXIDE, HYDROGEN SULFITE AND SULFITE ON
THE METABOLISM OF PLANTS

M. PLÖCHL, F.W. BADECK, G.H. KOHLMAIER
Institute for Physical and Theoretical Chemistry
Johann Wolfgang Goethe University
Niederurseler Hang
6000 Frankfurt 50
Federal Republik of Germany

ABSTRACT. Decrease of gross photosynthesis and increase of dark respiration are discussed as main effects of sulfur dioxide and its derivatives on vegetation. The nature of inhibition of the ribulose bisphosphate carboxylase as the main enzyme of the Calvin cycle is examined in model simulations. First simulation runs are compared with measurements under controlled conditions. Further simulation runs combine effects of climate and air pollution under ambient conditions. The effect of SO_2 on net photosynthesis is measurable at relative high SO_2 concentrations (more than 100 ppb) but seems to be neglectable at ambient conditions.

1. INTRODUCTION

The effect of sulfur dioxide on the metabolism of plants has been determined experimentally at different levels of aggregation: 1. determination of the inhibition of the lead enzyme of the Calvin cycle ribulose bisphosphate carboxylase (RuBP carboxylase) at isolated chloroplasts by addition of a range of buffered sulfite solutions, 2. by short term SO_2 exposures in climate chambers at relatively high levels of SO_2, 3. by photosynthesis and respiration measurements in open top chambers at ambient and slightly elevated SO_2 concentrations.

It has been stated that under stress caused by sulfur dioxide the apparent photosynthesis of plants is decreased (Black & Unsworth, 1979; Keller, 1978). This effect could be the result of a decreased gross photosynthesis and/or an increase in dark respiration. We shall attempt here to present a model for the interaction of SO_2 with the carbon metabolism which integrates the different levels of information. The main focus of our attemption rests on a model for carbon dioxide

185

H.-W. Georgii (ed.), Mechanisms and Effects of Pollutant-Transfer into Forests, 185–192.
© 1989 by Kluwer Academic Publishers.

assimilation, originally developed by Lommen et al. (1971), which is modified to describe the effect of competitive and noncompetitive binding of SO_2 to the RuBP carboxylase. As will be shown in the appendix the model of carbon assimilation depends not only on the relative atmospheric concentrations of CO_2 and SO_2, but also on the ambient temperature and light intensity.

2. DETERMINATION OF THE NATURE OF INHIBITION

In a series of investigations several groups have tried to determine the extend and the nature of inhibition of the RuBP carboxylase by HSO_3^-/SO_3^{2-}. Ziegler (1972) found with isolated chloroplasts of *Spinacea oleracea* a competitive inhibition with an inhibition constant K_i= 3.0 mM. Gezelius & Hällgren (1980) investigated extracted chloroplasts of *Pinus sylvestris* and *Spinacea oleracea*. They determined a K_i of 13.0 mM (7.0 mM for spinach) of an noncompetitive or perhaps mixed type.

Like Ziegler Khan & Malhotra (1982) found a competitive inhibition with K_i= 2.2 mM for extracted chloroplasts of *Pinus banksiana*. Parry & Gutteridge (1984) concluded for a mixed type of inhibition with K_i= 1.5 mM for the Dixon plot and K_i= 12.5 for the Cornish-Bowden plot. They experimented with extracted chloroplasts of *Triticum aestivum* and *Spinacea oleracea*, too. Comparing these differing results the differences become very small for the relevant hydrogencarbonate concentrations of about 8 to 12 mM within chloroplasts (fig. 1).

Figure 1. Comparison of several in vitro experiments determing the type of inhibition of RuBP carboxylase.

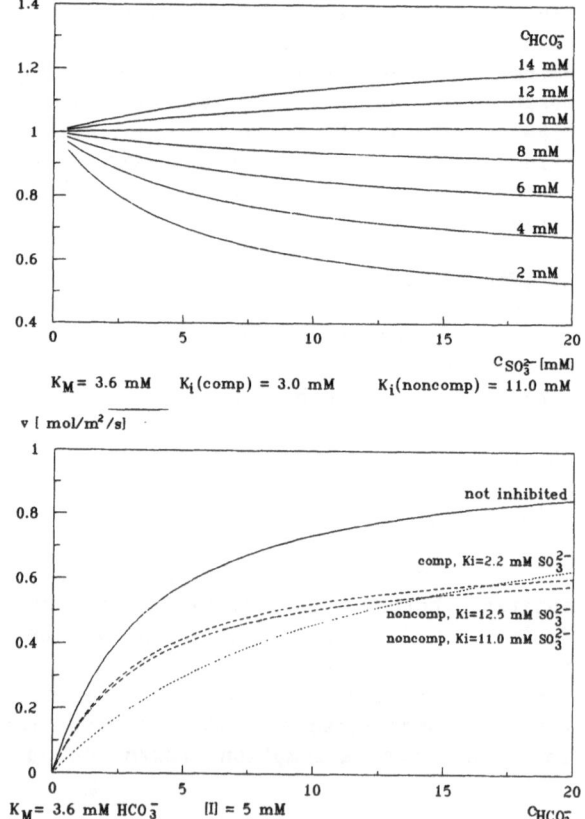

As a central conclusion one can state that for modelling the decrease of photosynthesis under SO_2 stress it is not essential whether the inhibition of RuBP carboxylase is competitive or noncompetitive, the effect in reduction will be nearly the same extend if one uses the corresponding set of data. If we assume competitive inhibition of RuBP carboxylase by SO_2 we follow the literature in setting

$$K_M' = K_M (1 + [I]/K_i) \tag{1}$$

in which K_i is a corresponding inhibition constant for the enzyme inhibitor complex:

$$K_i = [E] \cdot [I] / [EI] \tag{2}$$

In the case of noncompetitive inhibition it is assumed that the rate of carbon turnover is decreased by the binding of SO_2 to an affector site not identical with the substrate binding site. Again we adopt from the conventional literature that the photosynthetic capacity is reduced with a

$$P_{ML}' = P_{ML} (1-1/(1+K_i/[I])) \tag{3}$$

in which P_{ML} is the photosynthesis rate at saturating HCO_3^- in the chloroplast and at saturating light.

In the follow up discussion we examine photosynthesis measurements in climate chambers and on experiments in open top chambers with corresponding relevant SO_2/SO_3^{2-} concentrations.

3. MODELLING THE EFFECT OF SULFUR DIOXIDE ON PHOTOSYNTHESIS UNDER CONTROLLED CONDITIONS

The simulation model of sulfur dioxide uptake - described elsewhere in great detail (Siebke, 1988; Siebke et al., 1989) - was extended such that atmospheric SO_2 concentration, temperature and light can be varied like in chamber experiments under controlled conditions or measurements under field conditions. The formulations of dissolution of SO_2 and the metabolic rates have been extended to temperature dependence. The model for photosynthesis as described in the appendix is incorporated.

Darrall (1986) did a series of investigations with short time fumigations of a couple of plant species under controlled conditions and measured the response of net photosynthesis. Model runs will be compared with experiments with *Hordeum vulgare* fumigated with 100, 300 and 500 ppb SO_2 for 2 hours. Light intensity (PAR) was 450 $\mu E \cdot m^{-2} \cdot s^{-1}$ and air temperature was 291 K. The maximum carbon fixation rate is extrapolated to be 12.2 $\mu mol \cdot m^{-2} \cdot s^{-1}$, the minimum stomatal resistance with 250 $s \cdot m^{-1}$. As shown in fig. 2a and b the effect on the carbon uptake rate (net photosynthesis) in the experiment is to some extend higher than the effect on carbon fixation rate (gross photosynthesis)

in computer simulation run. One can suppose that this differences are due to an increase in "dark respiration". Uritani & Asahi (1980) described that wounded or infecteded tissues have an increased mitochondrial respiration even under daylight conditions. Like these tissues leaves under air pollution stress have to enhance their metabolic pathways for degradation of cell toxins, for repair reactions (e.g. at the cell membranes) and especially for buffering the protons of dissolution of SO_2. As the leaves will react with an enhanced metabolism they should have a mitochondrial respiration under daylight and an increased respiration in the dark, too. That means that under air pollution stress up to 10 % (at 500 ppb SO_2) of the assimilated CO_2 will be lost by additional "dark respiration".

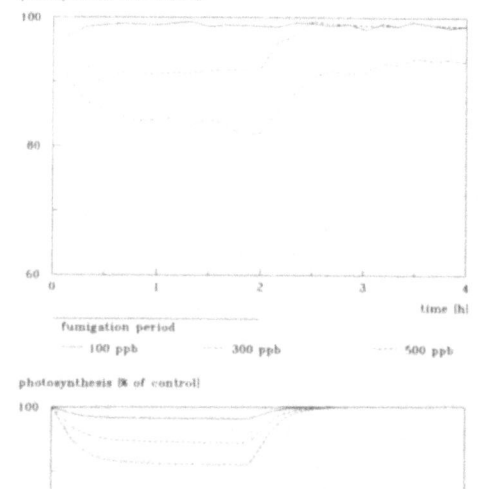

Figure 2. a) Net photosynthesis of barley, fumigated with different concentrations of SO_2 (after Darrall, 1986). b) Gross photosynthesis of barley in computer simulation experiment.

4. MODELLING THE EFFECT OF SULFUR DIOXIDE ON PHOTOSYNTHESIS UNDER FIELD CONDITIONS

As measurable effects on net photosynthesis are obtained only at high concentrations of SO_2 (more than 300 ppb), one wonders if there could be any effect of SO_2 under ambient conditions (2 to 10 ppb). As at low temperature and high light intensities the uptake of SO_2 is enhanced (low stomatal resistance, high solubility) and the enzymatic reactions at all may be decreased, one can suppose that sulfur species will accumalate within the chloroplasts to a greater extend. This higher concentration may lead to an increased inhibition of photosynthesis.

Keller (1978) kept a couple of three year old clonal spruce seedlings, *Picea abies*, within open top chambers under ambient conditions

(5-10 ppb SO_2), and fumigated others with 50, 100 and 200 ppb SO_2 during the whole vegetation period. Trees treated with elevated SO_2 concentrations showed significant lower uptake rates of CO_2. Keller supposed that this is due to disfunctions of the stomatal pores.

We are using Keller's data to examine whether it is possible to get a combined effect of SO_2 and climatic stress (temperature and light are considered here) or not. Climate data are obtained from the test reference year data set of the Deutscher Wetterdienst. In this set extreme climatic conditions prevail about the 35th day (cold and high light intensities) and during the 190th to the 200th day (nearly 30°C).

Decreases in carbon fixation rates are very small (fig. 3a, b) both for normal ambient SO_2 concentrations (1-10 ppb) and elevated levels of SO_2 (40 ppb). Comparing the results shown in fig. 3a and b indicates that everytime the climate is favourable for an enhanced SO_2 uptake the photosynthetic rate is low and vice versa, i.e. the relative reduction in photosynthesis is higher at low temperature and high light intensities than at conditions normal for the vegetation period but in their absolute amounts they are nearly equal. At the 35th day the reductions in maximum gross photosynthesis rate are 1.76% at ambient SO_2 concentrations and 6.23% at 40 ppb SO_2. At the 190th day these reductions amount to 0.02% at ambient SO_2 concentrations and to 0.49% at 40 ppb SO_2.

a)

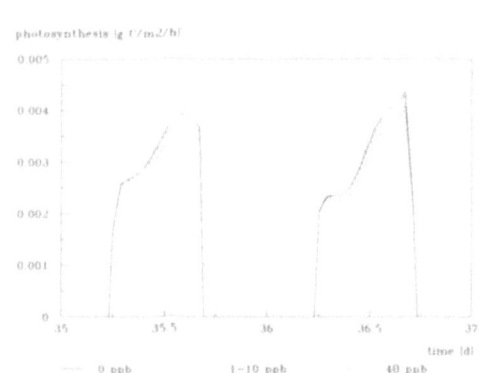

b)

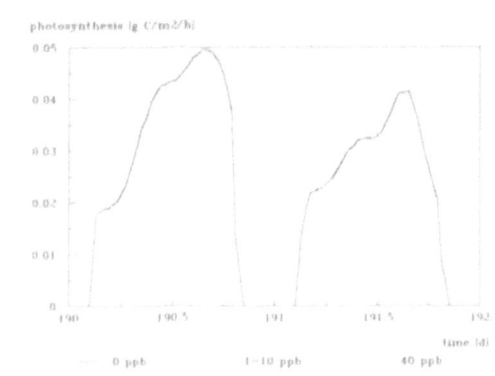

Figure 3. Gross photosynthesis of spruce at several SO_2 concentrations (computer simulation experiments). a) 35th and 36th day of year. b) 190th and 191st day of year.

The proceeding conclusions are based on the inhibition of carbon fixation by the ribulose bisphosphate carboxylase, while no further effects are considered, like e.g. dark respiration and stomatal conductance. In tab. 1 results are shown for computer simulations with the assumptions that the minimum stomatal resistance is increased by 25 %., i.e stomatal pores are partly paralized. The simulation is based on experiments done by Keller & Häsler (1986), who showed an increase in stomatal resistance of spruce fumigated with 25 ppb SO_2. The results compared with those of simulations without stomatal disfunction and with measurements of Keller (1978) show that the reduction in net photosynthesis can not be fully explained by a paralization of stomates.

TABLE 1. Relative reduction of photosynthesis for several computer simulations and measured by Keller (1978)

	1-10 ppb	40ppb
35th day	1.76	6.23
190th day	0.02	0.49
190th with stomatal disfunction		2.07
Keller, July		3.00

5. DISCUSSION

Experiments with the chlorophyll flouresence method (Schmidt et al., 1988) showed that the inhibition of photosynthesis will mainly occur in the Calvin cycle and not in the electron transport chain. But inhibition of ribulose bisphosphate carboxylase cannot be the central effect of SO_2 on plant metabolism under ambient conditions. Keller & Häsler (1986) prefer an increase in stomatal resistance and Black & Unsworth (1978) an increase in dark respiration as main effects of SO_2. Stomatal conductance is increased in the darkness and slightly decreased in the light this leads to an enhanced transpiration and otherwise to a decreased CO_2 uptake and therefor to an decrease in net photosynthesis. The net carbon uptake rate is furthermore decreased by the enhanced dark respiration. This enhancement persists up to 24 hours longer than fumigation.

Modelling these effects presume detailed models of stomatal function and therefor water exchange of the whole plant, as well as models considering the main processes which consume energy from respiration.

6. ACKNOWLEDGEMENT

We thank the Bundesministerium für Forschung und Technologie for financial support. K. Siebke, E.-O. Siré and C. Wientzek partizipated in earlier stages of model development which is kindly acknowledged.

7. REFERENCES

Black, V. J.; Unsworth, M. H. (1979). Effects of low concentrations of Sulphur dioxide on net photosynthesis and dark respiration of Vicia faba. J. Exp. Bot. 30: 473-483.

Darrall, N.M. (1988). The sensitivity of net photosynthesis in several plant species to short-term fumigation with Sulphur dioxide. J. Exp. Bot. 37: 1313-1322.

Gezelius, K.; Hällgren, J.-E. (1980). Effect of SO$_3^{2-}$ on the activity of ribulose bisphosphate carboxylase from seedlings of Pinus sivestris. Physiol. Plant. 49: 354-358.

Keller, T. (1978). Einfluss niedriger SO$_2$-Konzentrationen auf die CO$_2$-Aufnahme von Fichte und Tanne. Photosynthetica 12(3): 316-322.

Keller, T. & Häsler, R. (1986). The influence of a prolonged SO$_2$ fumigation on the stomatal reaction of spruce. Eur. J. For. Path. 16: 110-115.

Khan, A.A.; Malhotra, S.S. (1982). Ribulose bisphosphate carboxylase and glycollate oxidase from Jack Pine: Effects of sulphur dioxide fumigation. Phytochemistry 21(11): 2607-2612.

Lommen, P.W.; Schwintzer, C.R.; Yoccum, C.S. & Gates, D.M. (1971). A Model Describing Photosynthesis in Terms of Gas Diffusion and Enzyme Kinetics. Planta 98, 195-220.

Parry, M.A.J.; Gutteridge, S. (1984). The effect of SO$_3^{2-}$ and SO$_4^{2-}$ ions on the reactions of ribulose bisphosphate carboxylase. Journal of experimental botany 35(151): 157-168.

Schmidt, W.; Schreiber, U. Urbach, W. (1987). SO$_2$ injury in intact leaves, as detected by chlorophyll flourescence. Z. Naturforsch. 43c: 269-274.

Siebke, K. (1988). Modellierung des Kurzstreckentransports von SO$_2$ im System Blatt. Diplomarbeit Fachbereich Biologie d. J. W. Goethe-Universität Frankfurt: 146 p.

Siebke, K.; Badeck, F.-W.; Kohlmaier, G.H.; Plöchl, M.; Wientzek, C. (1989). Modelling Pollutant Exchange between Plant and Environment: Uptake and Metabolization of Sulphur Dioxide by different Leaf Cell Compartments. In: Modelling in Ecotoxicology (S.E. Jörgensen, ed.). Elsevier, in press.

Uritani, I. & Asahi, T. (1980). Respiration and related Metabolic Activity in Wounded and Infected Tissues. In: The Biochemistry of Plants Vol. 2 (D.D. Davies, ed.). Academic Press: 463-485.

Ziegler, I. (1972). The effect of SO$_3^{2-}$ on the activity of Ribulose-1,5-disphosphate carboxylase in isolated spinach chloroplasts. Planta 103: 155-163.

APPENDIX

Lommen et al. (1971) described the carboxylation of ribulose bisphosphate as a Michaelis Menten kinetic:

$$P = \frac{P_M}{1 + \frac{K}{\bar{C}_c}} \tag{1}$$

with P_M being light and temperature dependent

$$P_M = \frac{\alpha \ L}{(\ 1 + \alpha^2 \ P_{ML}^2 \ L^2 \ )^{\frac{1}{2}}} \tag{2}$$

$$P_{ML} = G \ (T_K) \ P_{MLT} \tag{3}$$

$$G \ (T_K) = \frac{c_2 \ T_K \ e^{-\frac{\Delta H^{\ddagger}}{RT_K}}}{1 + \ e^{-\frac{\Delta H_1}{RT_K}} \ e^{\frac{\Delta S}{R}}} \tag{4}$$

To eliminate C_c they compare the production rate with the uptake rate, discribed by Fick's law:

$$P = \frac{C_a - C_c}{R} \tag{5}$$

Solving eq. 1 and 5 for C_c and setting them equal one obtains after solving the quadratic term:

$$P = \frac{(C_a + K + R \ P_M) - [(C_a + K + R \ P_M)^2 - 4 \ C_a R \ P_M]^{\frac{1}{2}}}{2 \ R} \tag{6}$$

with

$$P_M = \frac{\alpha \ L \ P_{MLT} \ G \ (T_K)}{\left[\ P_{MLT}^2 \ G^2 \ (T_K) \ + \alpha^2 \ L^2 \ \right]^{\frac{1}{2}}} \tag{7}$$

α initial slope of light saturation curve
c_2 empirical constant
C_a atmospheric CO_2 concentration
C_c CO_2 concentration within the chloroplast
$G(T_K)$ temperature function of carboxylation
$\Delta H^{\ddagger}$ energy of activation for the enzyme catalized carboxylation
ΔH_1 energy of activation for denaturation equilibrium
K Michaelis-Menten constant for carboxylation
L photosynthetically active radiant flux
P_M rate of photosynthesis at saturating C_c
P_{ML} rate of photosynthesis at saturating C_c and light
P_{MLT} rate of photosynthesis at saturating C_c, light and optimum T_K
R sum of resistances of diffusive pathway for CO_2
T_K absolute temperature

ACCUMULATION OF HEAVY METALS IN OAK WOOD FROM POLLUTED REGIONS

F. QUEIROLO*, P. VALENTA AND S. STEGEN
Institute of Applied Physical Chemistry
Nuclear Research Center (KFA) Jülich
P.O. Box 1913
D-5170 Jülich
S.-W. BRECKLE
Department of Ecology
University of Bielefeld
D-4800 Bielefeld
Federal Republic of Germany

ABSTRACT. Concentrations of ecotoxic heavy metals Cd and Pb and of essential heavy metals Cu and Zn were determined in annual growth rings from a polluted region of Königstein (F.R.G.) and an unpolluted region of Valdivia (Rep. of Chile) by differential pulse anodic stripping voltametry (DPASV). The radial distribution of the four metals investigated is similar in oaks from both regions. No significant differences in the average concentrations of the essential metals Cu and Zn in the oak tree rings from the two regions have been observed. For the toxic metals Cd and Pb, however, a significant increase of factor of 2 for Cd and of factor of 12 for Pb, respectively, has been found when the Königstein region and the Valdivia region are compared. This evidences a sensible pollution of the Königstein region by Pb, and to a lesser extent by Cd, since 1940 or earlier.

1. Introduction

Heavy metals originating from natural and anthropogenic sources have greatly contributed to environmental pollution over the past few decades (Nriagu 1984). The contribution of anthropogenic sources to the total element emission depends substantially on the element itself (Tab. 1). A substantial portion of the total amount of some toxic metals released from anthropogenic sources into the environment is emitted into the atmosphere. Therefore, a significant pathway for heavy metals pollution of terrestrial and aquatic ecosystems goes through the atmosphere whence after transport both wet and dry deposition occurs (Georgii et al. 1986). Some of toxic trace metals, e.g. Cd, can be directly dissolved in acid rain water and are therefore deposited in a form very suitable for subsequent uptake by vegetation (Nürnberg et al. 1982).

There exist various paths through which heavy metals can be uptaken by a tree, i.e. through leaves from the atmosphere and through roots from the soil. It has been found that the annual growth rings, especially those of the oak, can be used as a reliable means for the study of chronology of the heavy metal pollution in the forest (Wickern 1986, Baes 1985).

In the past few decades increasing damages have been observed in German forests which

---

*Attached from the Department of Chemistry, University of Norte, Antofagasta, Republic of Chile.
Taken in part from the Ph.D. Thesis, University in Bonn.

H.-W. Georgii (ed.), Mechanisms and Effects of Pollutant-Transfer into Forests, 193–202.
© 1989 by Kluwer Academic Publishers.

TABLE 1. Loads - given in 10^5 kg a^{-1} - from natural and anthropogenic pollution sources (Stumm and Keller 1984)

Element	Continental flying ash load	Volcanic flying ash load	Industrial particles emission	Load by fossile fuels	Total anthropog. emission	Interference factor
Co	40	30	24	20	44	0.63
Cr	500	84	650	290	940	1.61
V	500	150	1000	1100	2100	3.23
Ni	200	83	600	380	980	3.46
Cu	100	93	2200	430	2630	13.6
Cd	2.5	0.4	40	15	55	19.0
Zn	250	108	7000	1400	8400	23.5
As	25	3	620	160	780	27.9
Se	3	1	50	90	140	33.9
Hg	0.3	0.1	50	60	110	275
Pb	50	8.7	16000	4300	20300	346

Remark: Interference factor $= \dfrac{\text{total anthropogenic emission}}{\text{total natural emission}}$

are attributed to various contaminants including heavy metals as the tree can accumulate heavy metals for extended time periods. According to information provided by the government on forest damages in 1984, approximately 3.7 mil. ha., or about 50 % of the total forest area in Germany have been visibly damaged. Oak trees were those trees showing rapidly increasing damage (15 % in 1983, 43 % in 1984). A comparative evaluation of forest damage from 1984 to 1986 has shown that tree damages have substantially augmented in the reported period (BMELF 1984, HFV 1988).

Taking into consideration the problems mentioned above, we have undertaken a comparative study of the heavy metal concentrations in annual growth rings of the oak originating from two different regions. The region in Taunus (F.R.G.) is known to have been exposed for many years to contamination by heavy metals and, simultaneously, to acid rain (Tab. 2). On the other hand, the trees originating from the region of Valdivia (Rep. of Chile) have grown under natural conditions without any contamination by heavy metals from anthropogenic sources.

TABLE 2. Average monthly wet deposition of ecotoxic heavy metals Cd, Cu, Pb and Zn and of free acid in Königstein (Taunus, F.R.G.). Sampling period May 1983 - Jan. 1984 (taken from Kraemer et al. 1985)

Metals, resp. H$^+$	Average deposition (μg/m^2 month)
Cd	20
Pb	480
Cu	99
Zn	550
H$^+$	3100

2. Experimental

2.1 SAMPLING

Samples were taken from sound living oak trees. The oak tree exhibits ring-porous wood and other advantageous properties for this type of research. Moreover, the annual growth rings can be well separated and differentiated (Wickern 1986). Sampling was performed at the end of May 1985 in the Königstein region and in January 1987 in the Valdivia region.

      Soil sampling was performed at the same time as wood sampling. In order to not contaminate soil samples with heavy metals originating from the instruments used for sampling, the chosen soil levels were reached by removing soil horizontally and using plastic shovels up to the depth of 5 cm. The homogenized soil sample was dried at $100^{\circ}C$, milled and then aliquoted for the respective determination. The heavy metal concentrations and acidity in the soil samples taken in both regions are summarized in Tab. 3. An example of sampling of soil and wood is shown in Fig. 1.

Figure 1. Taking wood and soil samples in the Valdivia region

2.2. DETERMINATION OF HEAVY METALS

A sensitive and accurate differential pulse anodic stripping voltammetric procedure has been used for the simultaneous determination of Cd, Cu, Pb and Zn in wood samples. A contamination free mineralization of wood material was performed by wet digestion in an acid mixture of HNO_3 + $HClO_4$. The accuracy of the method has been tested with standard reference materials and by comparison with the electrothermal atomic absorption spectrometry (ETAAS) as an independent method. Using distilled HNO_3 in the wet digestion procedure, an extremely low concentration range of 0.1 µg/L of Pb and Cd could be attained with an r.s.d. of < 20 %. The blank of about 0.01 ppb (or about 0.15 ng absolute) of Pb is attainable. The high sensitivity of our method made possible the determination of heavy metals in individual oak growth rings. The experimental details of the method are given previously (Queirolo and Valenta 1987). An example of a simultaneous determination of Cd, Cu, Pb and Zn in an analyte of a digested wood sample is given in Fig. 2.

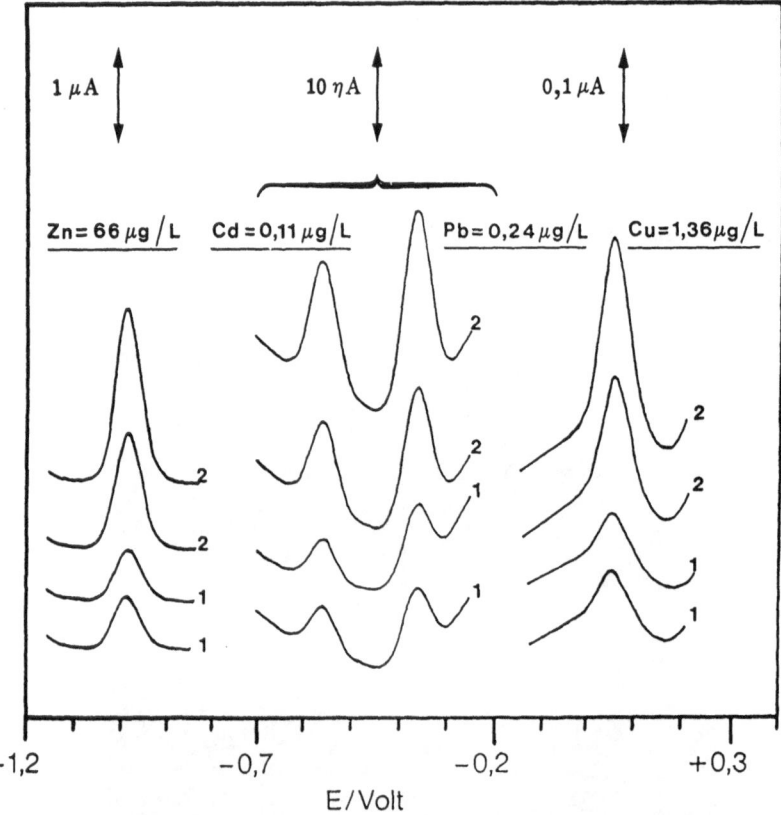

Figure 2: Simultaneous voltammetric determination of the heavy metals Cd, Cu, Pb and Zn in an analyte resulting from wet digestion of an individual growth ring of oak (Quercus Petraea). Oak Nr. 8 from Königstein, Taunus, 1987. Sample weight 0.01171 g; DPASV, Pulse height 50 mV, scan rate 10 mV s^{-1}, clock time 0.2 s, deposition time for Zn 30 s at -1.2 V, for Cd, Cu and Pb 300 s at -0.8 V. Curves 1 correspond to the sample, curves 2 to the first standard addition. Total determination time 45 - 50 min. Heavy metal concentrations in the sample (d.w.): Zn 51 µg g^{-1}, Cd 0.082 µg g^{-1}, Pb 0.18 µg g^{-1}, Cu 1.01 µg g^{-1}.

3. RESULTS AND DISCUSSION

In order to estimate typical concentrations of the heavy metals Cd, Cu, Pb and Zn in wood, samples from 7 oaks in the Valdivia region and 5 oaks in the Königstein region were taken and heavy metal concentrations determined in individual growth rings. An example of the radial distribution of the metals Cd and Pb in the oak Nr. 5 from the Valdivia region, corresponding to years 1971 - 1986 is given in Fig. 3 and Fig. 4, respectively. Radial distributions of Cu and Zn give a

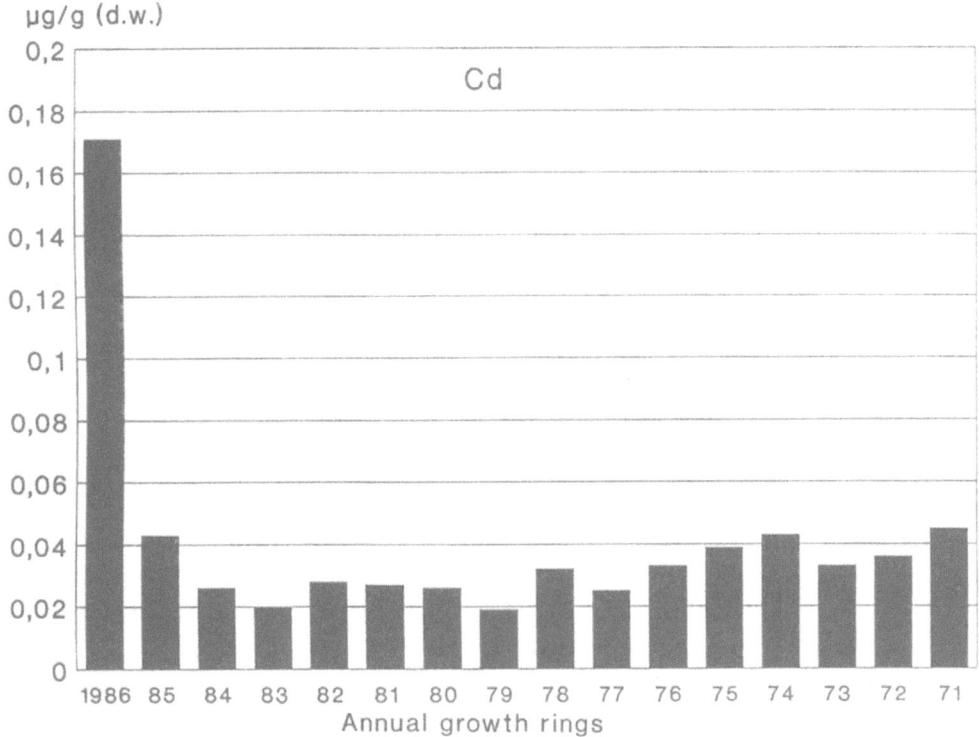

µg/g (d.w.)

Cd

Annual growth rings

Figure 3: Radial distribution of the Cd concentration in an oak wood sample (oak tree Nr. 5, Valdivia)

similar pattern. The steep rise of the metal concentration for the first annual ring, exhibited for all metals investigated, is not characteristic for the metal concentration in wood. This may be due to natural causes, as the principal transport of water and of nutrients proceeds in the most external part of the trunk. Alternatively, a part of the secondary phloem could be taken together with the adjacent ring so that the concentration determined corresponds to a mixture of both parts of the trunk. The course of the radial distribution of the metal concentrations in oak wood is discussed in detail elsewhere (Queirolo et al. 1989).

A major source of metals in oak wood is the soil. In Table 3 the values of the total concentration of the essential metals Cu and Zn in the soil of both regions investigated are given. The values do not represent typical values for the respective region. They serve only as an additional information about the metal content in the soil immediately adjacent to the oak tree from which wood samples are taken (see Fig. 1). The metal content is in the normal range observed for various types of the soil, i.e. 2 - 250 ppm with an average value of 30 ppm for Cu and 10 - 300 ppm with an average value of 50 ppm for Zn (Adriano 1986). In the Königstein region the major part of the metals Cu, Pb and Zn remains in the upper layer of the soil (depth 0 - 5 cm) formed by soil humic matter. These metals are firmly bound to functional groups of humic and fulvic acids with a high oxygen content, e.g. COOH, fenolic OH. On the contrary the soil in the Valdivia region is of volcanic origin which is very porous, has good internal drainage and possibly a low CEC (cation exchange capacity). Therefore these metals are uniformly distributed in the whole depth profile of the soil perhaps also due to the high average annual precipitation of about 2500 mm.

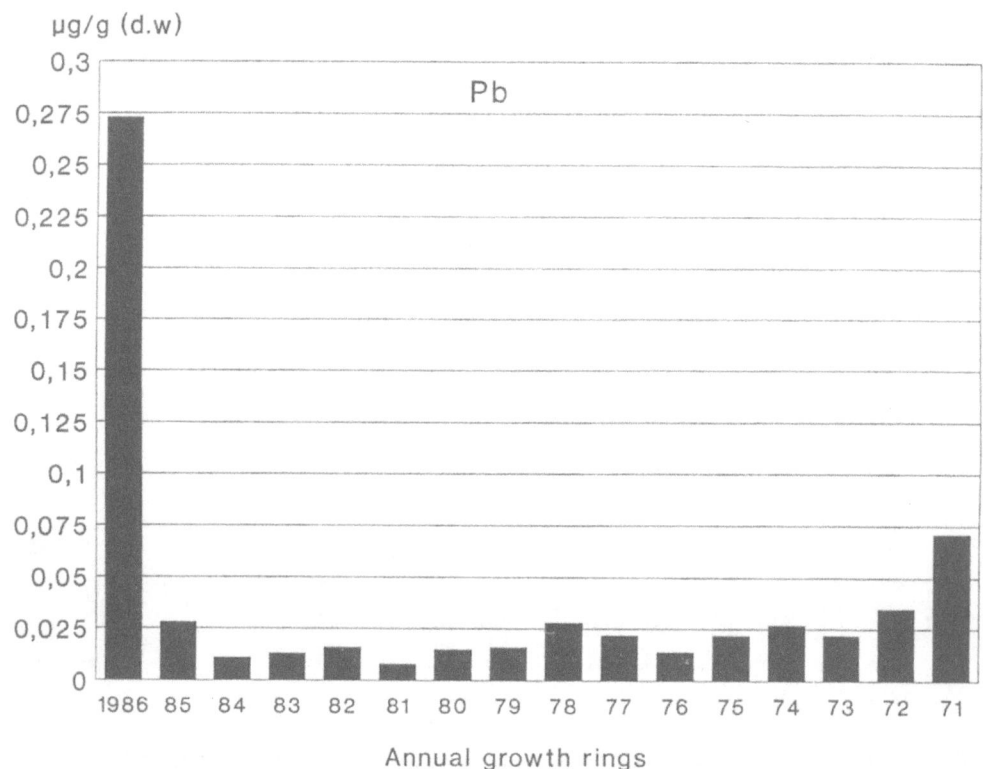

Figure 4. Radial distribution of the Pb concentration in an oak wood sample (oak tree Nr. 5, Valdivia)

TABLE 3. Heavy metal concentration (in $\mu g\, g^{-1}$, d.w.) and acidity in the soil in Valdivia and Königstein regions

Depth (cm)	Valdivia (Rep. Chile)					Königstein (F.R.G.)				
	Zn	Cu	Cd	Pb	pH (1M KCL)	Zn	Cu	Cd	Pb	pH (1M KCL)
0- 5	65	41	<0.1	8	4.4	51	16	<5	133	2.8
10-15	62	48	<0.1	8	4.5	36	8	<5	50	3.0
35-75	61	59	<0.1	10	4.8	37	7	<5	20	3.3

Remark: Heavy metal concentrations were determined after the wet digestion in aqua regia and correspond therefore to the total concentration.

On the other hand for Cd an average of 0.4 ppm is reported as a background level (Berrow and Reaves 1984). For Pb the mean value of 14 ppm emerges with the range 2.5 - 85 ppm. The average Pb concentration of 30 ppm in the upper organic layer is sensibly higher than the average concentration of 13 ppm in the mineral layer. It has been reported (Davies 1983) that concentrations greater than 110 ppm total Pb should not occur naturally in soils. Also the acidity of the soil in Königstein which is by about 1.5 pH unit lower than that in Valdivia cannot explain the differences in Cd and Pb concentration. Thus it can be concluded that the much higher Pb concentration and a fairly high Cd-concentration in the soil of Königstein with respect to those of Valdivia is caused by anthropogenic influences, e.g. atmospheric precipitates.

In Fig. 5 and Fig. 6 average values of metal concentrations in growth rings originating from 7 oaks from the Valdivia region with 5 oaks from the Königstein region are compared. It can be seen from Fig. 5 that in spite of the lower concentration of the essential metals Cu and Zn in the soil in Königstein their concentration in wood is a little higher. However, this difference is not significant and lays within a natural fluctuation. On the other hand significant differences in average concentrations of Cd and Pb can be observed in Fig. 6. The average Cd-concentration is about twice as high and that of Pb about 12 as high in the Königstein region compared to the Valdivia region. These differences can be explained by higher Cd and Pb concentrations in the soil of Königstein and by the lower pH in this region. It has been reported (Adrian 1986) that pH value of the soil enhances the solubility, mobility and phytoavailability of the heavy metals in the soil and thus increases their uptake by the plants. In fact a negative correlation has been found for some heavy metals between their uptake by the plant and the pH value of the soil.

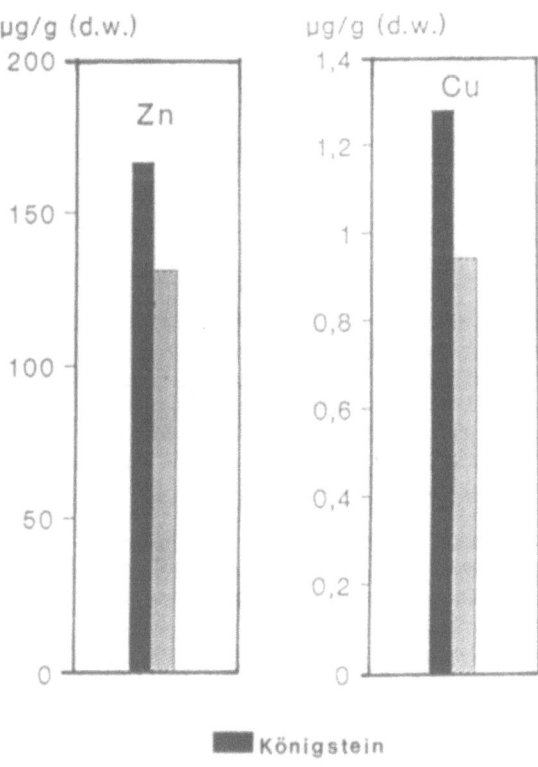

Figure 5. Comparison of average Zn and Cu concentrations in oak wood samples from Königstein und Valdivia regions, respectively.

200

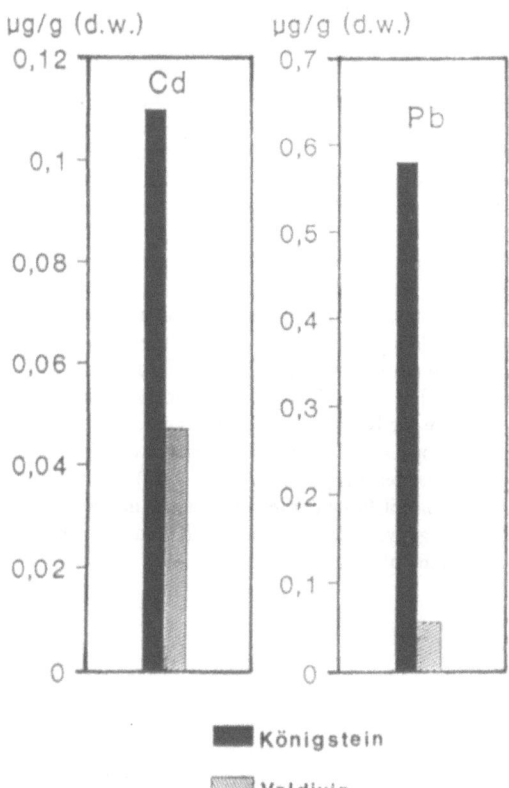

Figure 6. Comparison of average Cd and Pb concentration in oak wood samples from the Königstein and the Valdivia region, respectively.

4. CONCLUSION

The low pH value and the elevated concentrations of Cd and especially of Pb in the soil of Königstein, particularly in its superficial layer (depth of 0 - 5 cm), result to a major extent from the deposition of acid and heavy metals by atmospheric precipitates, as shown previously (Brechtel et al. 1986). This atmospheric pollution in the State Hessen is augmented in the forest regions because of the filter effect (interception) of the gases and particles in aerosols (Ulrich 1986, Rehfuess 1981, Block and Martels 1985). The trees investigated by us originate from the region of about 500 m above sea level where the effects of interception are more pronounced than in plane zones.

The low pH value can remobilize nutrients and heavy metals which are accumulated in the superficial zone of the soil and thus make them accessible for their uptake by the tree. The elevated concentration of Cd and Pb in the trees from the Königstein region compared to those of the Valdivia region may represent the pollution of this site by heavy metals since many decades. Nevertheless, the elevated metal levels in the wood cannot be denoted as dangerous at present as the trees investigated are classified as visibly sound (damage degree 0-1).

On the other hand the low concentrations of Cd and Pb encountered in the trees of the Valdivia region are by far lower than those hitherto indicated in the literature and may be considered as base levels of those metals in the oak tree grown under natural conditions without any pollution by heavy metals.

Acknowledgements: The authors are indebted to the Faculty of Forest Sciences of the Universidad Austral de Chile for permission and assistance in taking oak tree samples and soil samples in the Valdivia region. The assistance of the following institutions is also greatly appreciated: Hessische Landwirtschaftliche Versuchsanstalt, Landwirtschaftliches Untersuchungsamt Kassel-Harleshausen, Hessisches Forstamt Königstein und Central Laboratory for Chemical Analysis of the Nuclear Research Center Jülich. Financial support from the DAAD, Bonn, for the scientific excursion to Valdivia in 1986/87 is greatly acknowledged.

REFERENCES

Adriano, D.C. (1986) Trace elements in the terrestrial environment, Springer-Verlag, New York.

Baes, C.F. III (1985) "Elemental analysis of tree rings", Cambial Activities No. 11, 2-8.

Berrow, M.L., Reaves, G.A. (1984) in: Proc. Int. Conf. Environ. Contamination, CEP Consultants, Edinburgh, pp. 333-340

Block, J., Bartels, V. (1985) "Ergebnisse der Schadstoffdepositionsmessungen in Waldökosystemen in den Meßjahren 1982/83". Forschung und Beratung, Reihe C, 39, Landwirtschaftsverlag, Münster-Heltrup.

Brechtel, H.M., Balazs, A., Lehnardt, F. (1986) "Precipitation input of inorganic chemicals in the open field and in forest stands - results of investigations in the State of Hesse". in: H.W. Georgii (ed.), Atmospheric Pollutants in Forest Areas, Reidel Publishing Company, pp. 47-67.

Bücking, W. (1985) "Einfluß von Bestand und Standort auf einige Bioelemetgehalte im Sickerwasser". Reihe Tagungsberichte, 5, Nationalpark Bayerischer Wald, Grafenau, pp. 491-504.

Bundesministerium für Ernährung, Landwirtschaft und Forsten (BMELF). Waldschadenherhebung (1984). Schriftenr. Bundesminist. Ernährung, Landwirtsch. u. Forsten, Bonn.

Devies, B.E. (1983) Geoderma 29, 67-75 (cited from Adriano 1986).

Georgii, H.W., Grosch, S., Schmitt, G. (1986) "Feststellung der Schadstoffbelastung von Waldgebieten in der Bundesrepublik Deutschland durch trockene und naße Deposition". Abschlußbericht Teil A, im Auftrag des Umweltbundesamtes, Berlin.

Hessische Forstliche Versuchsanstalt (HFV) Hann. Münden (1988). Jahresbericht 1987. "Waldbelastung durch Immissionen". pp. 31-37, Hann. Münden.

Kraemer, H.W., Valenta, P., Nürnberg, H.W. (1985) in: Wald und Wasser. Prozesse im Wasser- und Stoffkreislauf von Waldgebieten. Nationalparkverwaltung Bayerischer Wald, Symposium "Wald und Wasser", Grafenau, pp. 175-189.

Lehnardt, F., Brechtel, H.M., Boness, M. (1984) "Ein Beitrag zur Quantifizierung der Versauerung ausgewählter Bäche im Bereich des Nordhessischen Buntsandsteingebietes". Materialien 1/84, Umweltbundesamt, Berlin, E. Schmidt Verlag, pp. 108-111.

Nriagu, J.O. (ed.) (1984) "Changing metals cycles and human health". Dahlem Konferenzen. Springer Verlag, Berlin.

Nürnberg, H.W., Valenta, P., Nguyen, V.D. (1982) "Wet deposition of toxic meals from the atmosphere". in: H.W. Georgiii, J. Pankrath (eds.), Deposition of atmospheric pollutants, Reidel Publ. Comp., Dordrecht-Boston, pp. 143-157.

Queirolo, F., Valenta, P., (1987) "Trace determination of Cd, Cu, Pb and Zn in annual growth rings by differential pulse anodic stripping voltammetry", Fresenius Z. Anal. Chem. 328, 93-98

Queirolo, F., Valenta, P., Stegen, S., Breckle, S.-W. (1989) "Heavy metal concentration in oak wook growth rings from the Taunus (F.R.G.) and the Valdivia (Rep. Chile) region", Trees, in press.

Rehfuess, K.E. (1981) "Über die Wirkung der sauren Niederschläge in Waldökosystemen". Forstw. Zentralblatt 100, 363-381.

Stumm, W., Keller, L. (1984)" Chemische Prozesse in der Umwelt. Die Bedeutung der Spezierung für die chemische Dynamik der Metalle in Gewässern, Böden und Atmosphäre", in: E. Merian (Ed.) Metalle in der Umwelt, Verlag Chemie, Teil I.3a, pp. 21-33.

Ulrich, B. (1984) "Deposition von Säure und Schwermetallen aus Luftverunreinigungen und ihre Auswirkungen in Waldökosystemen". in: E. Merian (ed.), Metalle in der Umwelt, Verlag Chemie, Weinheim, I.7e, pp. 163-170.

Wickern, M. (1986) "Jahrringchronologie des Bleis in Eichen am Autobahnrand". in: S.-W. Breckle, J. Hagemeyer (eds.), Bielefelder Ökol. Beitr., Vol. 2, pp. 25-40, Bielefeld.

INVESTIGATIONS ON FOG AND DEW

TRACE SUBSTANCE INPUT TO CONIFEROUS FORESTS VIA CLOUD INTERCEPTION

G. KROLL, P.WINKLER
Deutscher Wetterdienst
Meteorologisches Observatorium Hamburg
Postfach 650150
D-2000 Hamburg 65

ABSTRACT. The trace substance deposition to coniferous forests via cloud interception at higher elevations is assessed. This is achieved in three steps: (1) The deposition fluxes of cloud water are calculated according to the deposition model of Lovett using observed meteorological data. (2)At various mountain sites cloud water is collected and analyzed for selected trace substances. (3) By a certain procedure the deposition fluxes are combined with the measured concentrations in order to estimate the deposition of trace substances. Although large uncertainties have still to be overcome, the results show that at high altitudes the input of trace substances to coniferous ecosystems due to interception of cloud droplets can be as high or even higher than the deposition by rain. Especially the nitrate deposition via cloud interception is higher at most sites than the sulfate deposition whereas for wet deposition the sulfate input is higher than the nitrate input.

1. Introduction

Since the seventies a spreading forest decline has been observed in northern Europe as well as in northern America. Increasing air pollution has been identified as a major contributor for the forest damages (Schöpfer and Hradetzky 1984). Several pathways for the input of pollutants have been investigated in the past to find out the importance of their possible impacts on ecosystems besides the natural climatic stresses:
- dry deposition of aerosol particles and trace gases, including synergic effects. Especially ozone is believed to cause damages when occurring in high concentrations (Prinz et al. 1982);
- wet deposition ("acid rain") causes an acidification of the soil, hereby mobilizing toxic metals which then may be taken up by the plants and disturb the metabolism (Ulrich 1979);
- a third pathway is the deposition of trace substances by interception of cloud or fog droplets in forests (moist deposition). The input via this pathway is difficult to quantify, because it depends in a very complex manner on numerous meteorological, chemical and biological parameters.

Trace substances deposited via interception can effect trees in two ways: either directly during the relatively long contact times, because interception is a relatively slow process or indirectly via the soil, since the deposited fog water drips down after sufficient accumulation at the needles and leaves.

The relevance of fog interception for the hydrological cycle has been recognized early and first attempts to quantify this deposition flux of water have been made by Grunow (1957) and

205

H.-W. Georgii (ed.), Mechanisms and Effects of Pollutant-Transfer into Forests, 205–211.
© 1989 by Kluwer Academic Publishers.

its dependance on elevation by Baumgartner (1958). Various studies have meanwhile demonstrated, that the trace substance concentrations in fog or cloud water may be significantly higher than in rain, at least in polluted regions (e.g. Schmitt 1986). It can therefore be concluded that moist deposition may be an important pathway for input of trace substances to forest ecosystems at sites with high frequency of fog occurrence.

The moist deposition is mainly subjected to the following parameters:
– windspeed,
– fog frequency,
– liquid water content of fog/clouds (LWC),
– droplet size distribution,
– aerodynamic characteristics of the canopy, and
– pollutant concentrations of the fog water.

Ulrich et al. (1979) have developed the flux balance method where it is tried to separate wet from 'dry' deposition by collecting and analyzing precipitation beneath and apart of trees and calculating appropriate differences for selected ions. The 'dry' deposition, however, includes the moist deposition described above and therefore this method does not allow to quantify the deposition via fog interception.

The purpose of this paper is to present results on the assessment of trace substance deposition via cloud interception by coniferous forests. Our approach to this problem is:
– to measure the trace substance concentrations in fog or cloud water at different places for a certain period of time,
– to calculate the fog water deposition to coniferous forests at various sites using observed meteorological data,
– to combine both results in an appropriate manner in order to obtain the moist ion deposition by fog interception.

In the following sections we present some results to each of these three points, however, without going to much into details.

2. Deposition model

To calculate the cloud water deposition in a given coniferous forest we are using the cloud droplet deposition model by Lovett (1984). This is a one dimensional resistance model which describes the turbulent diffusion of cloud droplets into a 10 m high forest from above and the sedimentation as well as the impaction of droplets onto the vegetation components. Lovett calculates the resistances for each of three size classes of droplets whereas we use 20 size classes (1 - 20 µm radius). Because we do not make any use of Lovett's evaporation/condensation submodel, which may be allowed for our purpose, our input parameters include the windspeed, the LWC and the droplet size distribution in the air above the forest as well as the vertical distribution of the vegetation components, i.e. needles, branches, twigs and stems. In addition we insert the meteorological parameters as hourly mean values, since - as we have shown elsewhere (Kroll and Winkler 1989) - the use of observed hourly values gives a more realistic estimate of the deposition rate than the use of climatological average values. The impaction and sedimentation flux onto the vegetation components is calculated for each of the 20 size classes of droplets and the sum of the fluxes of all droplet sizes results in the deposition flux per hour.

In our simulation we use the hourly synoptic observations of 1985 as meteorological input parameters; the LWC values and drop size distributions were parameterized from the observed visibility. We obtained a cloud water deposition of 210 mm/a into a 10 m high coniferous forest at the Kahler Asten (KA) and of 1560 mm/a at the Grosser Arber (GA). These results amount to 15 % at KA and to 100 % at GA of the bulk precipitation of 1985. Although the number of fog hours is almost the same at both sites (ca. 3000 h/a), at the GA the deposition flux is so much higher because here very dense fogs and high windspeeds are occurring more often than at the KA. At both stations there is a pronounced seasonal cycle with a minimum of the deposition rate during the summer and a maximum during the winter.

The above calculated deposition fluxes can only be considered to give the order of magnitude. Major uncertainties originate from discrepancies in literature data on the capture efficiency and from visibility observations, which are available only as visibility range classes. In order to avoid unrealistic high LWC values at low visibilities we choose a parameterization that the LWC could not exceed 0,22 g/m^3. Considering the various error sources the overall uncertainty of the flux calculation for a distinct fog hour may be as large as 100 %. However, the uncertainty of the total yearly flux should be appreciably lower because some of the errors should compensate when summing the hourly fluxes over one year.

3. Cloud water composition

During the 18 month from October 1986 through May 1988 we have collected cloud water at five different stations of the German Weather Service (DWD) and the Umweltbundesamt (UBA) (Hohenpeißenberg [HP], Kahler Asten [KA], Wasserkuppe [WK], Grosser Arber [GA] and Schauinsland [SL]). This has been done with fog collectors which work according to the impaction principle and collect droplets with radii above 2,5 μm (Winkler 1986). Per fog event we took one sample which was analyzed for ammonium, sulfate, nitrate, chloride, lead, and manganese as well as for the electric conductivity and the pH-value. We have got a total of 401 samples, per station the numbers vary between 20 and 195 samples.

Station		HP	KA	WK	SL	GA	BJ
Conductivity	(μS/cm)	85	206	70	73	164	32
pH - value		4,82	3,70	4,21	4,07	3,83	4,40
Sulfate	(mg/l)	5,3	10,5	5,2	3,5	12,5	5,0
Nitrate	(mg/l)	6,3	14,2	5,1	2,9	13,8	4,4
Chloride	(mg/l)	1,1	2,1	1,7	0,9	2,6	0,5
Ammonium	(mg/l)	3,1	5,8	2,5	2,4	5,6	1,4
Lead	(μg/l)	77	86	57	116	115	
Manganese	(μg/l)	5	13	7	7	12	
Numbers		38	126	20	22	195	
LWC	(g/m^3)	0,025	0,057	0,055	0,042	0,127	

Table 1: Mean values of the trace substance concentrations in fog water at five mountainous sites and in rain water at one site. The concentrations in the fog water are standardized to a uniform liquid water content of 0,1 g/m^3; additionally the number of samples and the mean measured liquid water contents are given.

For reasons of brevity the arithmetic mean values of the trace substance concentrations are given here (cf. Tab. 1). Before averaging we standardized the concentrations to a uniform LWC of 0,1 g/m^3 in order to avoid introduction of additional scatter due to different LWC's. As can be seen in Tab. 1 the highest values could be ascertained for nitrate, ammonium, manganese, conductivity and pH-value at the KA and for sulfate, chloride and lead at the GA. Appreciable regional differences in the chemical composition of fog water are occurring while the composition of the rain water is more homogenious. The mean concentration values do not only show relatively high variations between the different stations but there are also strong variations between the samples of one and the same stations. Peak values exceed the reported average values by factors up to 20. It was, however, not possible to identify a seasonal cycle for any of the trace substances.

For reasons of comparison the last column in Tab. 1 gives the mean trace substance concentrations in rain water at the UBA-station Brotjacklriegl (BJ), which is only 30 km away from the GA. These rain values, which may be regarded as representative throughout the FRG (UBA 1986), are significantly lower than the fog water values at the GA, for which the enrichment factors lie between 2,5 (sulfate) and 5,2 (conductivity, chloride).

4. Estimation of trace substance deposition

The estimation of trace substance deposition into a coniferous forest can be achieved by combining the observed concentrations of pollutants in fog water with the calculated fog water deposition rates. This can be done in different ways and accordingly the results will vary (cf. Tab. 2 and Fig. 1).

Station	Mode	Deposition		Trace substance deposition (kg/ha·a)			
		(mm/a)		Ammonium	Sulfate	Nitrate	Chloride
		wet	moist				
	R	1500		21,0	75,0	66,0	7,5
GA	1		1460	64,2	143,1	159,1	29,2
	2		1460	37,1	88,7	101,4	20,7
	3		1460	38,3	91,6	104,8	21,3
	4		1460	24,8	58,2	61,0	15,5
KA	1		210	21,3	38,6	52,3	8,2
	2		210	21,8	41,9	60,0	9,7
	3		210	23,9	45,7	65,5	10,6
	4		210	14,5	28,6	37,8	6,5

Table 2: Pollutant deposition rates to a coniferous forest via fog interception at two mountain sites. The calculation modes are described in the text. For comparison also the deposition rates via rain (mode R) at the GA are given (for this the concentration values from BJ are used, cf. Tab. 1). The sulfate and nitrate deposition rates at the GA are shown in Fig. 1.

A first rough estimate is obtained when the calculated fog water deposition per year is simply multiplied by the mean trace substance concentrations at the site (which are standardized on the mean observed LWC values). At the GA for example the deposition of sulfate will be 143 kg/ha·a and that of nitrate 159 kg/ha·a (cf. mode 1 in Tab. 2). The corresponding deposition values by precipitation are much lower, for sulfate 75 kg/ha.a and for nitrate 66 kg/ha.a (cf. columns 1 and R in Fig. 1). Note, that the nitrate deposition via fog exceeds that of sulfate for all calculation modes while the situation is opposite for rain.

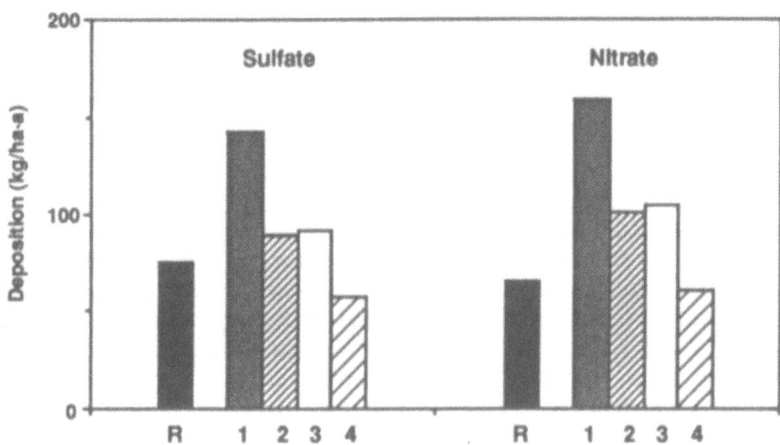

Figure 1: Comparison of different assessments of the sulfate and nitrate deposition to a coniferous forest at the GA; R: wet deposition via rain (the concentration values from BJ [cf. Tab. 1] are used); 1 - 4: different calculation modes for ion deposition via fog interception.

The results become more precise, when the hourly calculated deposition fluxes are combined with hourly observed concentration values. Since continuously working fog collectors are, however, not available we took the following approach: for each fog day one data set for the chemical composition was chosen from all concentration measurements. For each fog hour of this day the concentrations are standardized on the LWC that has been parameterized from the visibility for the same hour, and then they are multiplied by the hourly calculated water deposition flux. On the next day with fog a new concentration set was taken and again varied during the day inversely to the LWC. This was done with all fog days until the available concentration measurements have been used, and than we started with the first concentrations set another time. Thus we introduced concentration fluctuations in the course of fog events in such a manner that all peak values enter the calculations with their observed frequency. At the GA we calculate by this procedure a sulfate deposition of about 90 kg/ha.a and a nitrate deposition to about 100 kg/ha.a (cf. mode 2 in Tab. 2, columns 2 in Fig. 1). The deposited amount of pollutants is significantly reduced as compared to mode 1 because the extreme concentration values are specifically weighted by the calculation mode. The influence of the few peak concentration values (nitrate up to 250 mg/l) is compensated by the low frequency of their occurrance. At the KA the results for the pollutant deposition are slightly increased by this calculation procedure, because here such peak concentration values have not been observed (compare values for modes 1 and 2 for KA in Tab. 2).

The influence of high concentration values on the deposition flux can be even further demonstrated by assuming the following scenario. If the highest values (in this case 15 % of the measurements) are excluded from the calculations, the deposited amount of pollutants decreases by about 35 % at the GA (cf. mode 4 in Tab. 2, columns 4 in Fig. 1) and by about 30 % at the KA. This means that relatively few peak concentrations increase the deposition via fog over-proportionally.

The influence of a droplet size dependent concentration on the pollutant deposition can be neglected. It is assumed that small droplets are higher concentrated as larger ones (Trautner 1989; Schmitt 1986). At the same time they are less effectively deposited so that the pollutant deposition is increased by only 5 % at the GA and 10 % at the KA (cf. mode 3 in Tab. 2, columns 3 in Fig.1).

5. Conclusion

By combining model calculations of the cloud water deposition with measurements of the concentrations of trace substances in cloud water it can be demonstrated that the pollutant deposition into a high elevation coniferous forest via cloud droplet interception can be as high or even higher as that via precipitation. This is especially caused by the fact that trace substances are higher concentrated in fog and cloud water than in rain water. The concentrations of trace substances in cloud water vary stronger than in rain water, so that appreciable local variations in the input of trace substances into forests by interception of cloud droplets is to be expected which is not the case for rain.

Due to local variations of the wind profile appreciable local variations of the deposition via fog or cloud interception are to be expected even over relatively small distances (forest edges, clearings and so on). Another important difference between deposition via rain and fog interception is that the rain is falling more uniformly while the intercepted fog water drips down just beneath the trees so that the trace substance input to the soil occurs exactly in that area where trees are being rooted.

Our assessments show that forests in mountainous areas with high fog frequencies receive substantially higher ion deposition rates than was expected by mass balance calculations up to now. It is therefore not surprising that in these sites, which are often considered as clean area regions, the forests are especially damaged.

Acknowledgement: This work was supported by the Ministry of Research and Technology under grant 07431018.

References

Baumgartner, A. (1958) 'Nebel und Nebelniederschlag als Standortfaktoren am Gr. Falkenstein', Forstw. Cbl. 77, 257-272.

Grunow, J. (1957) 'Vergleichende Messungen des Nebelniederschlags', UGGI, Ass. Int. Hydr. Sci. Assemblee Toronto.

Kroll, G. and Winkler, P. (1989) 'Influence of meteorological parameters on interception of cloud droplets in a coniferous forest', accepted for publication in Beitr. Phys. Atmos.

Lovett, G.M. (1984) 'Rates and mechanisms of cloud water deposition to a subalpine Balsam fir forest', Atmos. Environ. 18, 361 - 371.

Prinz, B., Krause, G. H. M., and Stratmann, H. (1982) Waldschäden in der Bundesrepublik Deutschland, Bericht Nr. 28, Landesanstalt für Immissionsschutz des Landes Nordrhein-Westfalen, Essen.

Schöpfer, W. and Hradetzky, J. (1984) 'Der Indizienbeweis: Luftverschmutzung als maßgeblliche Verursachung der Walderkrankung', Forstw. Cbl. 103, 231-248.

Schmitt, G. (1986) 'The temporal distribution of trace elements concentration in fog water during individual fog events', in H. W. Georgii (ed.), Atmospheric pollutants in forest areas, Reidel, Dordrecht, pp. 129-141.

Trautner, F. (1989) 'Collection and Properties of fog water', in O. Lange.; E.D. Schulze, R. Oren (eds.), Acid Rain and Forest Decline in the Fichtelgebirge, Ecological Studies, Springer, Berlin (in press).

UBA (1986) Monatsberichte aus dem Meßnetz, Bd. 1, Berlin.

Ulrich, B., Mayer P. and Khanna P.K. (1979) Deposition von Luftverunreinigungen, Schriftenreihe Forstl. Fak. Univ. Göttingen, Bd. 58, Sauerlanders Verlag, Frankfurt.

Winkler, P. (1986) 'Observation on fog water composition in Hamburg', in H. W. Georgii (ed.), Atmospheric pollutants in forest areas, Reidel, Dordrecht, pp. 143-151.

Comparison of wet deposition inputs via rain and fog interception at
a high elevated forest site

G. Schmitt
DEKRA-Institut für Sicherheit, Umweltschutz und Energie
Schulze-Delitzsch-Straße 49
7000 Stuttgart 80

ABSTRACT

The absolute amounts of water and atmospheric trace subtances deposited
to forests due to the interception of fog and cloud droplets are a
function of meteorological parameters and geometrical structure of the
vegetation surfaces. This report examines a simple method of determining interception deposition during isolated instances of high-density
fog over long periods of time. The results indicate that fog interception is an important factor in the total input of trace elements into
forest ecosystems. In forested subdued mountain areas with a high instance of fog, interception deposition can cause inputs of pollutants,
containig specific elements, of up to 50 mg/m^2 per event.

INTRODUCTION
The interception of fog water by the vegetation plays a major role in
damage to forests in many parts of Central Europe. Considerable forest
damage was observed in regions with a high incidence of fog. Early interest in fog research concentrated only on the absolute amounts of the
water leached out by the vegetation.

At high-altitude forest sites the interception of fog water can reach
the same order of magnitude as that of normal precipitation. Linke
(1916) and Grunow (1955) show that the intercepted fog water reaches 35
- 150 % of the annual precipitation. Today, in view of the high concentration of pollutants, the main interst has turned towards the chemical
composition of fog water. Therefore, research into the causes of recent
damage to forests primarily requires knowledge of the element masses
deposited during fog events, in addition to their concentrations in
fog, in order to deduce the possible harmful effects to vegetation as a
direct result of the high element concentrations on the vegetation surface. Fog interception is therefore directly dependent on a series of
parameters which can mutually interact with it. The element interception depends on the liquid water content (LWC), the droplet size spectrum, the fog frequency and the actual wind speed. However the surface
structure and the physical condition of the vegetation also influence
the absolute amounts of pollutants impacted by the vegetation.

H.-W. Georgii (ed.), Mechanisms and Effects of Pollutant-Transfer into Forests, 213–220.
© *1989 by Kluwer Academic Publishers.*

For a long time, direct measurements of the element masses deposited on the vegetation by interception were not available. Gradient measurements or eddy correlations, which cannot be determined in the stand without defining micrometeorological variables such as LWC, the droplet size spectrum, wind, temperature and humidity distribution on several measuring levels, require considerable technical input. Furthermore, the different chemical composition of the fog droplets, as a function of their size, must be taken into account. In view of this, this report it intended to present a simplified method of determining the interception of fog droplets.

Sampling method

This method simply requires the fog concentration an the absolute quantity of fog water impacted by the vegetation. The fog ist collected outside the stand, i. e. in a clearing in the forest or above the crown area, and is analysed for its element content. The quantity of fog water deposited on the vegetation is collected as throughfall water in the stand using suitable collectors (Fig.: 1). A detailed description of the method used for fog water sampling is given in Schmitt (1987). The interception deposition is given by equation 1:

$$FID_E = C_E \cdot (T_f \cdot C_C) \tag{1}$$

FID_E = Fog interception deposition of element E (mg/m^2)
C_E = Concentration of element E in fog (mg/L)
T_f = absolute amount of throughfall water (L/m^2)
C_C = Capacity of the canopy (L/m^2)

Similar to dry and wet deposition, the interception deposition has the dimension mass/area and time. In the following description, the time always refers to the duration of an individual fog period. The amount of water gathered by the collectors must be corrected to include that remaining in the crown area. This comprises the amount of water required to wet the entire vegetation surface at the beginning of a fog period or the fraction which remains at the end of a fog period and which then evaporates again. This parameter, termed specific capacity, is a function of the density of the stand and can be calculated from the difference in the amount of precipitation on the clearing and below the canopy during individual fog periods.

The experiments were carried out at a forest stand on top of the "Kleiner Feldberg/Taunus" mountain (800 a. s. l.). The station is located 25 km northwest of Frankfurt and is characterized by a high fog frequency. Over a period of 30 years, fog was recorded on an average of 235 days per year (visible range less than 1000 m). High wind speeds, increasing the deposition efficiency by impaction, are also characteristic of mountain stations such as this. A mean value of 3 mm was recorded for the specific capacity of the spruce stand at the station. During the fall and winter months in 1986 throughfall and fog water were collected simultaneously.

SPRUCE STAND

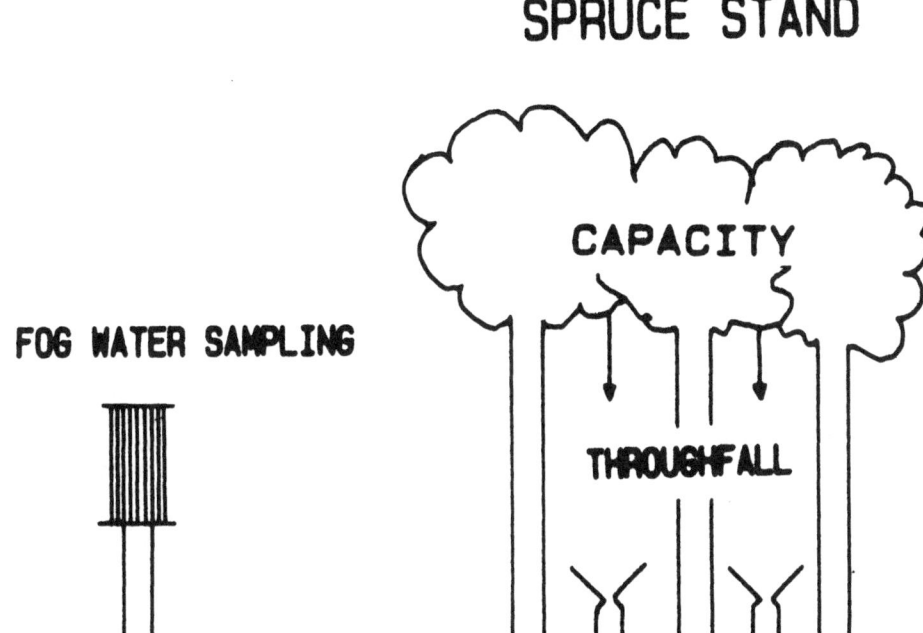

FOG WATER SAMPLING

CAPACITY

THROUGHFALL

Fig.: 1 Sampling method for determination of fog interception

The investigations only concentrated on instances of high-density fog with high liquid water contents and over extended periods (> 10 h). Sampling was begun when the visible range was less than 300 m. Fog periods with higher visible ranges showed low LWC. In this case no measurable throughfall rates were observed. The fog water was analysed of pH level, electrical conductivity, anions, cations and heavy elements. All data relating to $SO_4^=$ and NO_3^- were calculated as S and N respectively.

RESULTS

In order to assess the rates of interception, the mean concentrations (including maximum and minimum values) in the fog and rainwater must first be given. As shown in Table 1, the mean concentrations of the fog are clearly greater than those of the rain. The maximum and minimum values illustrate the wide concentration range of the fog. For example the pH values ranged from 2.3 to 7.9 in the individual samples. In some cases the minimum concentrations in the fog water are higher than the mean concentrations in the rainwater. Measurements of the fluctuations in the concentrations with time during individual fog periods show that the highest concentrations are recorded at low liquid water contents, which are characterisitc of the beginning and end of a fog period (Georgii et al. 1986, Schmitt 1986).

Tab.: 1   Arithmetic mean-, maximum- and minimum concentrations in
          fogwater samples compared to mean values in rainwater.
          Period: autumn 1983 - autumn 1986;
          station: "Kleiner Feldberg/Taunus"

	mean	fog max. (mg/l)	min.	rain mean (mg/l)
$SO_4^=$ - S	13.30	90	1.40	150
NO_3^- - N	10.90	99	0.40	0.95
Cl^-	8.60	80	0.60	1.17
NH_4^+	21.60	94	1.00	1.40
Pb	0.214	2.0	0.006	0.014
Fe	0.513	3.4	0.053	0.029
Cd	0.0041	0.020	0.0004	0.0005
Na	2.46	41.8	0.03	0.34
K	0.84	5.1	0.02	0.29
Ca	2.24	15.4	0.10	0.28
Mg	0.61	3.2	0.10	0.08
pH	3.8	7.9	2.3	4.3

The total interception quantities of anions, ammonia, calcium, H^+-ions
and liquid water with respect to time, calculated using the formula
described above (Equation 1), are shown in fig.: 2. A total of three
fractions of throughfall water was collected during the fog period. To
facilitate calculation, the element concentrations in the fog water are
determined with respect to the sampling intervals of the throughfall.
The figure shows that 13.5 mm of precipitation was registrated during a
fog period of 13 hours, including the specific capacitiy of 3 mm. As a
result of the concentrations already detected in the fog water, the
quantity of sulphate impacted by the vegetation is 62 mg/m². Similarly,
high desposition rates were also recorded for chloride (55 mg/m²) and
ammonia (48 mg/m²).
The values for nitrate and calcium (approx. 30 mg/m²) were somewhat
smaller. At pH values between 3.61 and 3.98, the total H^+-ion deposition is 2.5 mg/m². The distribution of interception rates for the fog period of 31.10.86 is similar (Fig.: 3). During this period 10 different
samples of the throughfall were collected.

The total amount of water was 11.7 mm which was almost identical to the
value obtained during the previous period. Due to the lower total concentration of trace elements in the fog water, in comparison to the period of 25.11.86, the resultant mass elements leached out of the vegetation is less. The highest values recorded were chloride (46 mg/m²)
and ammonia (43 mg/m²).

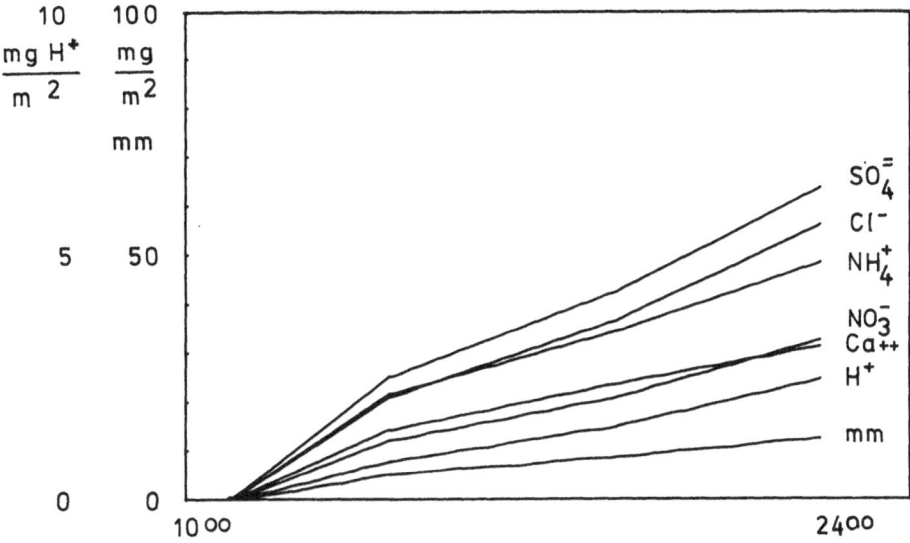

Fig.: 2 Cumulative interception of anions, Ca^{++}, NH_4^+, H^+
and water during a single fog event (25.11.1986).

A value of 37 mg/m^2 was determined for sulphate. The H^+-ion deposition
is approximately half that of the previous example. This is due to the
higher pH values which, in this case, were in a small range of pH 4.0.
The quantity of H^+-ions recorded was 1.3 mg/m^2. In this period, the
quantity of trace substances fed to the vegetation, as a result of the
interception of fog droplets, was also considerably high.

The result obtained from these individual periods alone, which are representative of all of the extended, high-densitiy fog periods, clearly
show the significance of fog interception with regard to atmospheric
input of trace substances. The mean value from fog interception is compared to the mean wet deposition per precipitation occurrence in order
to assess the test results recorded.

It has to be pointed out that the mean precipitation rate (11.2 mm) is
equal to the absolute amount of water deposited by fog interception
(12.6 mm). Due to the higher concentrations in fog, however, the ele-
ment-specific interception rates are significantly higher (up to a factor of 7). From the results it can be clearly seen that extremely large
quantities of trace substances and liquid water are leached out of vegetation when fog occurs and, as opposed to wet deposition from rain,
are absorbed by vegetation.

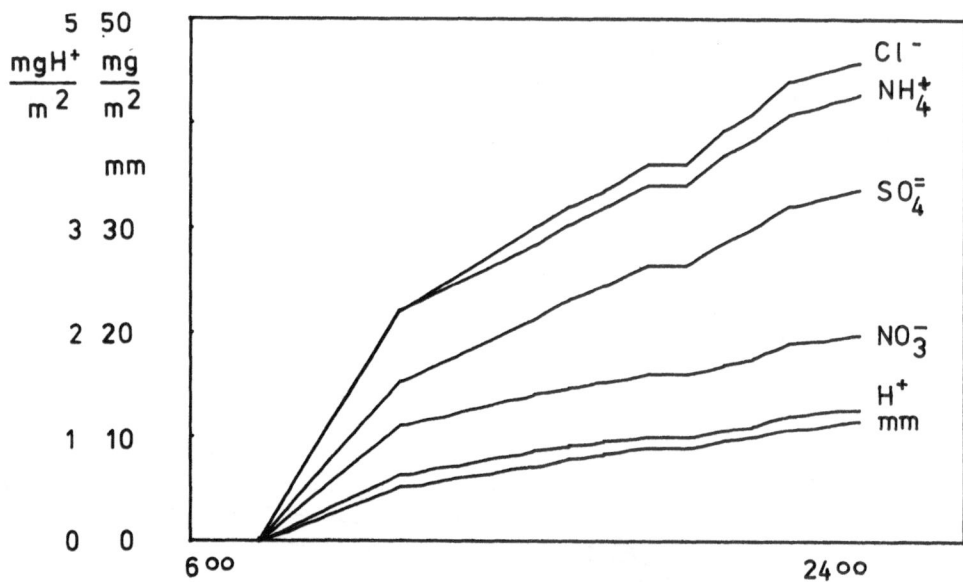

Fig.: 3 Cumulative interception of anions, NH_4^+, H^+
and water during a single fog event (30.10.1986).

Tab.: 2 Mean wet deposition by rain (1983 - 1986) and interception
deposition during single fog events (1986) at the station
"Kleiner Feldberg / Taunus"

	rain	fog
	(mg/m^2 . event)	
$SO_4^=$ - S	12.3	48
NO_3^- - N	5.9	25
Cl^-	7.3	51
NH_4^+	10.3	46
Ca^{++}	3.6	27
H^+-ion	0.6	1.8
	(mm/event)	
water	11.2	12.8

CONCLUSIONS

The interception deposition of fog must also be taken into consideration in order to provide a complete description of deposition processes in high-altitude forest areas. Using a simple pathway to calculate the absolute amounts of fog interception, the results show that interception deposition is one main method of atmospheric input of trace substances. During single fog events with high density and long duration, the intercepted element masses were greater than 50 mg/m^2 and per event, whereas the intercepted fog water amounted to 12 mm/event. Model calculations from Lovett (1982) and Kroll and Winkler (1987) show that the intecepted water amounts may be as high as 1 mm/h depending on the LWC of fog. The measured values are confirmed to satisfaction via this simple method. Measurements and model calculations demonstrate the important role played by fog interception deposition with respect to observed damages caused to forest ecosystems. In conclusion, it should also be mentioned that the results obtained from the small number of events are by no means representative of interception rates recorded over a long period of time. The recorded measurements should only be used as approximate values for initial estimations when quantifying the actual deposition inputs via interception on natural surfaces.

A detailed examination of the atmospheric input with regard to fog interception has yet to be carried out.

ACKNOWLEDGEMENTS

The investigations on fog chemistry and fog interception were carried out at the Institute for Meteorology and Geophysics, University of Frankfurt/Main. The work was supported by the Umweltbundesamt, Berlin under contract number 104 02715

LITERATURE

Georgii, H.W.,
S. Grosch,
G. Schmitt: Feststellung der Schadstoffbelastung
von Waldgebieten in der Bundesrepublik
Deutschland durch trockene und nasse
Deposition;
Abschlußbericht Teil A Forschungsprojekt
104 02715 im Auftrag des Umweltbundesamtes
Universitätsinstitut für Meteorologie und
Geophysik Frankfurt/M., 1986, p. 247

Grunow, J.   Der Nebelniederschlag im Bergwald, Nieder-
schlagszurückhaltung und Nebelzuschlag
Forstw. Zbl. 74, 1955, pp. 21 - 36

Kroll, G., P. Winkler   Estimation of wet deposition via fog
K. Grefen and J. Löbel (eds.), Environmental
Meteorology Kluwer Acad. Publ., 1988,
pp. 227 - 236

Linke, F.   Niederschlagsmessung unter Bäumen
Met. Zschr. 33, 1916 pp. 140 - 141 and
38, 1921, p. 277

Lovett, G.M.   Rates and mechanismus of cloud water
deposition to a subalpine balsam fir forest
Atm. Envir. Vol 18, No 2, 1984, pp. 361 - 371

Schmitt, G.   Methoden und Ergebnisse der
Nebelanalyse Dissertation,
Institut für Meteorologie und Geophysik,
Frankfurt, 1987, p. 199

Schmitt, G.   The temporal distribution of trace
element concentrations in fog water
during individual events, in Atmospheric
pollutants in forest areas, ed. by
H.W. Georgii 1986, Reidel Pub. Company,
Dordrecht pp. 129 - 141

DESIGN AND OPERATION OF A TWO-STAGE FOGWATER COLLECTOR

DIETER SCHELL and HANS-WALTER GEORGII
Institute of Meteorology and Geophysics
University of Frankfurt
Feldbergstr. 47
6000 Frankfurt / Main
Federal Republic of Germany

ABSTRACT. A sampling device has been developed, which allows to separate the fog droplet spectrum into two size fractions. The first stage collects droplets down to a diameter of 4 μm using a string collector technique. The second stage with a calculated 50 % cut-off diameter of 2 μm was constructed under application of numerically developed and experimentally proven design criteria. Depending on the actual liquid water content, the collector has been operated with a time resolution of 60-120 minutes during several fog events under different conditions. The samples have been analyzed for pH and total conductivity as well as for the major ions by Ion Chromatography. Concentration values in smaller droplets were found to be up to 14 times higher than the corresponding concentrations in larger droplets.

1. Introduction

The determination of the chemical composition of fogwater is a crucial point in the investigation of stress factors affecting forest ecosystems. Recent estimations on the absolute deposited amount of atmospheric pollutants by fog in forest areas at mountain sites show comparable dimensions to the deposited amounts by rainwater (Kroll 1988).

Up to now many techniques for the collection of fogwater have been developed and because of this variety, any data on the chemical composition of fogwater are only negotiable if information about the sampling device used is provided. Especially the cut-off diameter should be known because of a strong size dependence of the droplet solute concentration. Therefore it is possible that some methods underestimate important size fractions and do not make reliable predictions on the real chemical fogwater composition.

Thus, for an exact determination of its composition and for a better understanding of the fog microphysics it seems to be helpful to separate the fog droplet spectrum into different size fractions.

Most active fogwater collectors are based on the principle of inertial impaction at obstacles in an airstream. The impaction surfaces either can be screens as reported in Jacob et al. (1985), Fuzzi et al. (1988) or solid impaction surfaces (Winkler 1986, Däumer 1988).

Solid impaction surfaces commonly are used in cascade impactors to obtain different size fractions of ambient particles. Such cascade impactors are made of series of impaction plates. The diameters of the nozzles or slits above each impactor plate

221

H.-W. Georgii (ed.), Mechanisms and Effects of Pollutant-Transfer into Forests, 221–229.
© *1989 by Kluwer Academic Publishers.*

222

become increasingly smaller as the air moves through the impactor to allow an increasing velocity of the air through these orifices and a reduction of particle sizes impacted on the plates. The cut-off diameters can be accurately predicted by applying the design criteria of Marple and Willeke (1976) which are numerically developed and experimentally proven.

Applying this technique for the collection of fog droplets one has to pay attention to the following problems:

-It is not possible to collect and accumulate droplets on collection plates since they lose their shape by contacting any kind of obstacle. Therefore, and for the prevention of evaporation losses, one has to ascertain that the collected water runs immediately out of the airstream into a collecting vessel.

-Owing to the high mass of the larger droplets compared to smaller particles it is not possible to turn the airstream around as often as it is necessary in cascade impactors. This would result in immense wall losses.

-By collecting liquid samples one has to make sure that each stage delivers a feasible amount of water for chemical analysis. The available liquid water content which is normally within the range of 0.01 - 0.6 g m^{-3} (Georgii et al. 1987) prevents to separate the fog droplet spectrum into large numbers of size fractions.

The construction of a multistage fog collector can be managed by application of the technique of aerosol sampling by respecting the mentioned problems with actual fogwater collection.

2. Method

The device represents a combination of jet impactor and active string collector. This design has been chosen to avoid the mentioned problem of wall losses. By using strings for the first stage, larger droplets with diameters above 4 μm will be collected without any wall losses. The second stage is a rectangular jet impactor with a calculated cut-off diameter of 2 μm.

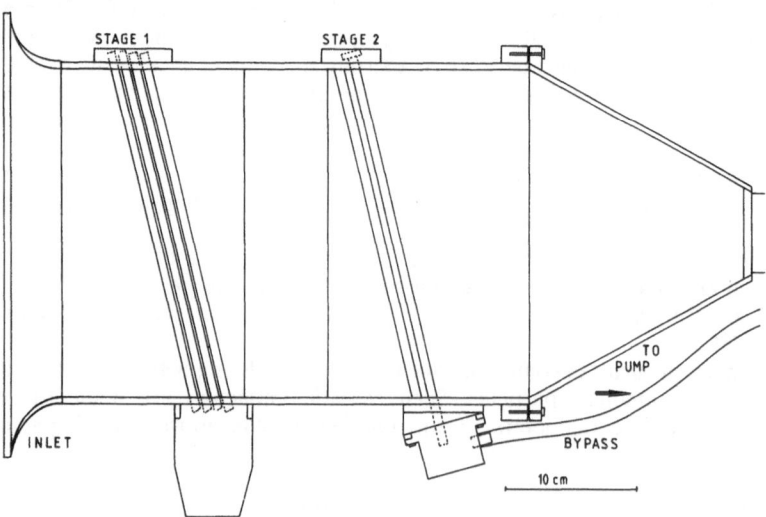

Figure 1. Two-Stage Fogwater Collector (side view)

Figure 1 shows the side view of the collector. The two stages are tilted forward at an angle of about 15 degrees against the direction of the airstream to support the downflow of collected fogwater into the sampling vessel. At the second stage this downflow is supported additionally by generating a low pressure using a bypass from the vesselholder to the pump. To render a representative fogwater collection, the inlet of the device is formed like a trumpet. During field measurements both collector and pump are mounted on a rotatable stand. A wind vane takes care, that the inlet of the fogwater collector is always directed towards the wind. This may not be satisfactory under conditions of low wind velocities e.g. of radiation fog in a valley. However, measurements of capping clouds on mountain sites can be strongly influenced by inlet losses of the bigger droplets if the device is not oriented into the wind at all times.

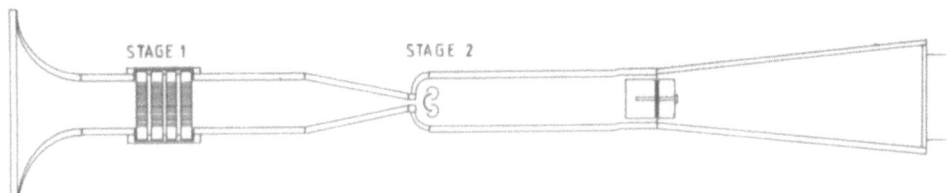

Figure 2. Two-Stage Fogwater Collector (top view)

The top view of the collector (fig. 2) shows the narrowing of the cross-section in front of the second stage. The flow accelerates to windspeeds which are necessary to collect droplets down to a diameter of 2 μm. The geometry at this part of the device, especially the nozzle width, its minimum length and the distance between impaction surface and nozzle outlet is calculated under application of the design criteria of Marple and Willeke (1976).

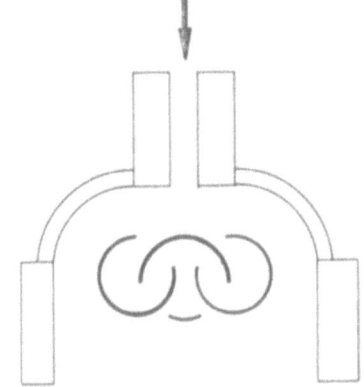

Figure 3. Detailed view of the second stage

The detailed view (fig. 3) of the second stage illustrates the technique to force impacted droplets out of the main airstream into the rear part of this tube system, where the collected water accumulates and runs down into the sample bottle.

TABLE 1. Technical specifications of the sampling device

Inlet height	23.8	cm
Inlet width	2.8	cm
Nozzle width 1. stage	2.32	cm
String diameter	0.028	cm
Number of strings per row	17	
Number of rows	8	
Distance between each string	0.13	cm
Air velocity 1. stage	6	m s^{-1}
Cut-off diameter 1. stage	4	μm
Nozzle width 2. stage	0.3	cm
Distance between nozzle and impaction surface	0.5	cm
Air velocity 2. stage	46.7	m s^{-1}
Cut-off diameter 2. stage	2	μm
Flow rate	120	m^3 h^{-1}

Pump: Staplex TF1A-2

Materials:	Strings:	Teflon
	Vesselholder:	Teflon
	Impaction surface:	Plexiglass
	Frame:	Plexiglass

3. Field Experiments

During 1987 and 1988 several field experiments have been carried out at three different sites, representing different types of fog :

On November 11, 17 and 18, 1987, three events of radiation fog where investigated at the *FISBAT* (Istituto per lo Studio dei Fenomeni Fisici e Chimici della Bassa e Alta Atmosphera) Field Station of San Pietro Capofiume in the eastern part of the Po Valley, near Bologna (BO 11, BO 17, BO 18). This site can be characterized by relative low wind speeds associated with high atmospheric loading of trace gas and particulate matter in the boundary layer, caused by the fact that the Po Valley is both the largest agricultural as well as industrialized area in Italy (Fuzzi 1988).

On May 6, 1988 a marine fog situation has been investigated during a campaign on the *Forschungsplattform Nordsee* (FPN 6), which is located about 50 sm northwest of Helgoland in the North Sea. Far away from major anthropogenic sources of pollutants, pH-values of 2.5 to 3.8 were found. Wind velocities in the range of 2.5 to 7.5 m s^{-1} were observed during this fog event.

Finally, another successful sampling period on November 10, 1988 at the mountain station *Taunus Observatorium* on top of *Kleiner Feldberg*, 20 km northwest of Frankfurt is described (TO 10). A slight rain started in the morning hours of November 11 and affected the samples taken during the last phase of the fog event.

Generally the collector was operated with a temporal resolution of 60-120 minutes during the whole fog period.

3.1. CHEMICAL ANALYSIS

Immediately after sampling some basic analysis were carried out:
-the sample volume was determined by differential weighing;
-the total conductivity was determined using a commercial conductometer along with a suitable detector cell for little sampling volume;
-for pH- measurements a glass electrode was used.
Then the samples were kept refrigerated until they were analyzed for the major ions (SO_4^-, NO_3^-, Cl^-, Na^+, NH_4^+, K^+) by Ion Chromatography.

4. Results

4.1. MEAN VALUES

To obtain a general view on the concentration range of trace components in individual fog events, mean element concentrations were calculated from consecutive fog water samples. By comparing these values (fig 4.) higher concentrations are generally found in the fraction of smaller droplets. Compared with the fraction of larger droplets maximum enhancement factors > 10 are realized (Tab. 2.). The highest solute concentration of 12 mmol/l was found on November 18 (BO 18) for nitrate in the fraction of the small droplets. The corresponding value of 3 mmol/l for large droplets is also the highest concentration of samples collected on the first stage.

The widest range of the concentration ratio stage 2/stage 1 was found for the H^+-Ion (Tab. 2.): extreme values from 14.9 (BO 11) down to an inverse value of 0.9 (TO 10) were observed.

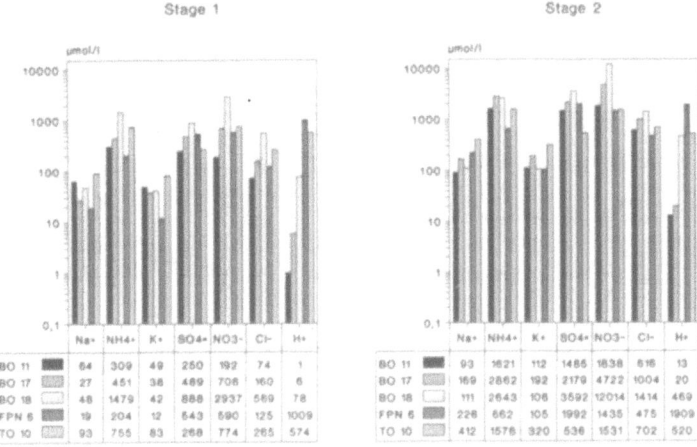

	Na+	NH4+	K+	SO4-	NO3-	Cl-	H+
BO 11	64	309	49	250	192	74	1
BO 17	27	451	38	489	708	160	6
BO 18	48	1479	42	888	2937	589	78
FPN 6	19	204	12	643	590	125	1009
TO 10	93	755	83	268	774	265	574

	Na+	NH4+	K+	SO4-	NO3-	Cl-	H+
BO 11	93	1621	112	1485	1838	515	13
BO 17	169	2862	192	2179	4722	1004	20
BO 18	111	2643	106	3592	12014	1414	469
FPN 6	228	662	105	1992	1435	475	1909
TO 10	412	1576	320	536	1531	702	520

Figure 4. Mean concentrations of each compound taken over different fog periods.

TABLE 2. Solute concentration ratio stage 2/stage 1 of different fog events

	Na⁺	NH₄⁺	K⁺	SO₄⁼	NO₃⁻	Cl⁻	H⁺
BO 11	1.4	5.2	2.3	5.9	9.5	8.2	14.9
BO 17	6.3	6.4	5.0	4.5	6.7	6.3	3.3
BO 18	2.3	1.9	2.5	4.1	4.1	2.5	5.9
FPN 6	11.6	3.3	8.6	3.7	2.4	3.8	1.8
TO 10	4.4	2.1	3.9	2.0	2.0	2.6	0.9

4.2. TEMPORAL VARIATION OF CONCENTRATIONS

Mean values derived from data observed during the complete fog period can only be a rough approximation since there is a wide time dependent variety of concentrations. Fig.5 shows the temporal variation of the ion concentrations as well as the liquid water content during the event of November 10 (TO 10). The highest concentrations, especially of ammonium and nitrate are found at the beginning, corresponding to the low liquid water content. As LWC rises in both size fractions during this event, the concentrations of the compounds analyzed simultaneously decrease. Obviously the LWC seems to influence decisively the ion-concentrations. Nevertheless there is an increase in the concentrations of potassium, sodium and chloride especially in the smaller droplets along with an increase of LWC towards the end of the measurements which points to LWC independent variations of concentrations (see also chapter 4.3.).

Because of a slight rainfall the measurements had to be stopped before the fog disappeared. Therefore the results obtained towards the end of the sampling period show the influence of an increasing amount of raindrops. The LWC decreases for larger but still increases for smaller droplets. This might indicate an effective removal of the larger fog droplets by collision with raindrops. The smaller fog droplets are apparently not influenced by this mechanism.

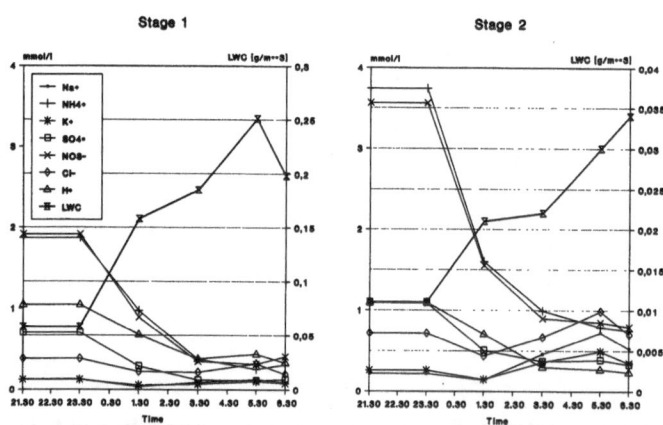

Figure 5. Temporal variation of concentrations and liquid water content of November 10, 1988 (TO 10).

4.3. NORMALIZED CONCENTRATIONS

As a first approximation, the LWC of each of the two dropsize fractions was estimated as the ratio of sample volume per total volume of air passed through the collector.

Solute concentrations normalized by LWC will lead to some information on ion concentrations per volume air. By plotting the temporal distribution (fig. 6), obviously these normalized concentrations are higher for samples of the first stage than for small droplets. This behaviour is opposite to the element concentrations of the fogwater solutions and is the result of higher LWC- values in the fraction of the large droplets which cannot be compensated by the higher solute concentrations in the smaller droplets (eq. 1).

$$K = C \cdot LWC \tag{1}$$

with :   K   : concentration per volume air $[\mu g\ m^{-3}]$
C   : solute concentration $[\mu g\ g^{-1}]$
LWC   : liquid water content $[g\ m^{-3}]$

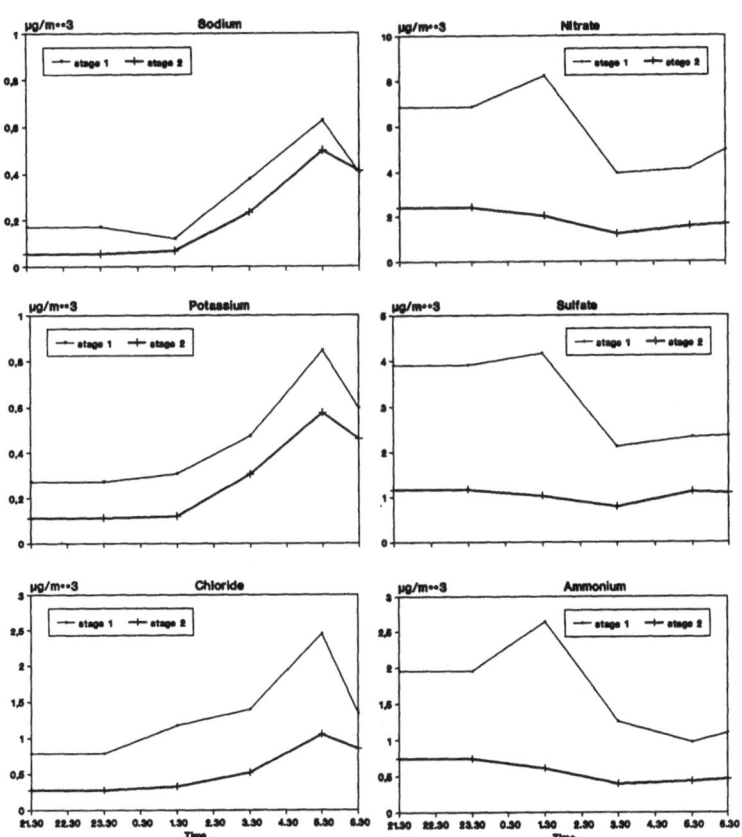

Figure 6. Normalized concentrations of November 10, 1988 (TO 10).

Following fig. 6, two groups of elements can be distinguished showing a generally different course:

-The temporal variation of sodium, potassium and chloride shows a parallel lapse for the concentrations of both stages. Low concentrations in the beginning are followed by an increase up to the maximum at 5.30 a.m. and a distinct decrease again with the last sample.

-The behaviour of nitrate, sulfate and ammonium is completely different. While concentrations remain more or less constant over the whole period for the second stage, concentrations analyzed on the first stage start at relative high levels, reaching a maximum at 1.30 a.m. . Then the concentrations sharply decreased to minimum values followed by a slight increase again within the last three hours of this sampling period.

This different general behaviour might be caused by different mechanisms concerning trace element uptake by fog droplets. Sodium and potassium only are present as particles (e.g. NaCl and KCl) while the species NH_4^+, NO_3^- and $SO_4^=$ also can be formed by uptake of gaseous compounds like SO_2, NO_x, HNO_3 or NH_3. According to the similar temporal distribution of Cl^- compared to Na^+ and K^+ a possible contribution from gaseous HCl seems to be negligible at least for this individual fog event.

5. Conclusions

The prototype of a two stage fogwater collector presented above was operated during several field experiments yielding first data on the chemical composition of two fog droplet size fractions. The main results of the investigations can be summarized as follows:

-The solute concentrations in the droplet size range >4 μm are generally lower than in the range $4> d >2$ μm.

-The normalized concentrations show an opposite behaviour.

-The variation of liquid water content during a fog event is not the only parameter affecting solute concentration variations in fog droplets.

There is no doubt, that this collector, despite of its good applicability, has to be improved in some details by evaluation of the experiences gained in the past. Consequently it would be favourable to increase the collection rate to render a better time resolution and to extend the chemical analysis to additional compounds. But also some design changes of the second stage and a shift of cut-off diameters to larger droplets might be useful to obtain higher sample volumes with this stage. At the moment, the total volume of collected water in the second stage is about 10-15 % of the volume collected in the first stage. This sampling rate seems to be too high compared to expected rates estimated from characteristic fog droplet size distributions. This must be caused by a collection efficiency of the first stage being less than 100 %.

Nevertheless, the results show that size fractionated fog sampling reveals some interesting additional information about size dependent concentration ranges and temporal distributions. It should contribute to a better understanding of fogwater chemistry as well as scavenging processes concerning fog droplets.

To verify the calculated properties of this collector, it will be unavoidable to perform a calibration procedure. An appropriate experimental setup is in the state of planning.

6. Acknowledgement

This investigation was sponsored by the *Umweltbundesamt* under contract no. 10402635

7. References

Däumer, B., R. Niessner, D. Klockow (1988) 'Design and calibration of a low volume fog sampler', J. Aerosol Sci., Vol. 19, No. 2, 175-181

Enderle, K.H., W. Jaeschke (1988) 'Sammlung und chemische Analyse von Nebelwasser unter Berücksichtigung der Mikrophysik des Nebels', Berichte des Zentrums für Umweltforschung der Johann Wolfgang Goethe Universität Frankfurt, No. 5

Fuzzi, S. G. Orsi, G. Nardini, M.C. Facchini, S. McLaren, E. McLaren, M. Mariotti (1988) 'Heterogeneous Processes in the Po Valley Radiation Fog', J. Geophys. Res., Vol 93, No. D9, 11141-11151

Georgii, H.-W., S. Grosch ,G. Schmitt (1987) 'Schadstoffbelastung in Waldgebieten durch Deposition und Interzeption', Forschungsbericht, Forschungsprojekt 104 02 635 im Auftrag des Umweltbundesamtes, Eigenverlag des Universitätsinstituts für Meteorologie und Geophysik, Frankfurt am Main

Jacob, D.J.,J.M. Waldman, M. Haghi, M.R. Hoffmann, R.C. Flagan (1985) 'Instrument to collect fogwater for chemical analysis', Rev. Sci. Instruments, 56, 1291-1293

Kroll, G., P. Winkler (1988) 'Estimation of wet deposition via fog', in K. Grefen and J. Löbel (eds.), Environmental Meteorology, Kluver Academic Publishers, 227-236

Marple, V.A., K. Willeke (1976) 'Impactor Design', Atmospheric Environment Vol. 10, 891-896

Munger, J.W., D.J. Jacob, J.M. Waldman, M.R. Hoffmann (1983) 'Fogwater Chemistry in an Urban Atmosphere', J. Geophys. Res., 88, No. C9, 5109- 5121

Schmitt, G. (1986) 'The temporal distribution of trace element concentrations in fogwater during individual fog events', in H.-W. Georgii (ed.), Atmospheric Pollutants in Forest Areas, D. Reidel Publishing Company, 129-141

Winkler, P. (1986) 'Observations on fogwater composition in Hamburg', in H.-W. Georgii (ed.), Atmospheric Pollutants in Forest Areas, D. Reidel Publishing Company, 143-151

THE INFLUENCE OF MANGANESE ON THE OXIDATION OF SULFITE IN DEW WATER

B.G.Arends, S.Eenkhoorn
Netherlands Energy Research Foundation (ECN),
P.O.Box 1
1755 ZG Petten, the Netherlands

ABSTRACT. High concentrations of manganese were found in dew water on grass which are believed to be catalytically active in the oxidation of sulfite. In order to describe this oxidation the reaction was studied in a fog chamber. In the concentration and pH range, normally found in dew on grass, the formation of sulfate is linearly dependent on the concentration of manganese and on the square root of the concentration of sulfite. The results indicate that the manganese catalysed sulfite oxidation is an important reaction in dew on grass.

1. INTRODUCTION

The concentrations of sulfate and manganese in dew water deposited on an artificial surface, a polyetheen plate, show large differences in comparison with the concentrations in dew water deposited on plants. Even if relative concentrations are compared, higher manganese and sulfate concentrations are found in dew sampled from plants. Manganese concentrations determined in dew from grass in the dunes of North Holland show values varying between 12 and 70 µmol/l. At the same location, manganese concentrations in dew taken from the polyetheen plate are about 2 µmol/l. Manganese can be leached from plants by dew or rain, especially when the plants are growing on acidic soil. The pH of dew samples obtained from plants measured at the location in North Holland is about 5.

Mean sulfate concentrations in dew from plants measured during the summer months are 1.2 mmol/l, while the concentrations measured in dew taken from the plate during the same period are 0.24 mmol/l.

This is an indication for the importance of the catalytic oxidation of sulfite by manganese in dew. Though many studies on the catalytic oxidation of sulfite by manganese have been performed (Martin, 1984 and 1987, Ibusuki, 1984 and 1987), the kinetics in this concentration regime is not yet known. The different kinetic descriptions for this reaction found in literature result in completely different rates of sulfate formation.

In order to present a better picture of the oxidation of sulfite catalyzed by manganese, the reaction is studied in a fog chamber. The advantage of using a fog chamber instead of a thin layer of water is,

231

H.-W. Georgii (ed.), Mechanisms and Effects of Pollutant-Transfer into Forests, 231–238.

that the uptake of gases in the small droplets is less limited by mass transport than in a thicker layer of water. For most reactions the rate of diffusion from the gas phase into the droplet is higher than the rate of reaction of the aqueous compound. This is an essential condition for the study of the reaction kinetics of aqueous phase reactions which receive the reactants from the gas phase.
The aim of the study is to measure the kinetics of sulfate formation in the presence of manganese in the concentration- and pH-range found in natural dew.

2. METHOD

The study is performed in a fog chamber described in detail elsewhere (Arends, 1988). The chamber with a volume of 1 m^3 consists of pyrex glass. Humidified clean air is continuously circulated through the chamber at a flow of 40 l/min. Fog is produced by a pneumatic nebulizer, producing fog droplets with a mass median diameter of 10 μm. The fog solution consists of a weak buffer of acetate (1 x 10^{-3} to 1 x 10^{-2} mol/l) in order to maintain a constant pH of 5. The liquid water content (LWC) of the chamber is 3-5 g/m^3, which is high enough to obtain a sample suitable for analyses within a reasonable period of time (30 minutes). The sample is collected by a polyetheen cyclone with an air flow of 20 l/min. This cyclone collects more than 99 % of all liquid water in the air flowing through the cyclone.
SO_2 is introduced into the chamber via a mass-flow controller. The concentration of SO_2 in the gas phase must be constant before fog is produced, but it is varied per experiment between 130 and 1300 $μg/m^3$. When fog production starts, the concentration decreases rapidly until a steady state is reached. Before the experiments start, a solution of $MnCl_2$ is added to the fog water. The concentration of manganese in the droplets varies from 9 to 90 μmol/l. Experiments without manganese are done as well. In all experiments, the temperature of the chamber is 23°C. The experiments are performed in the dark.
The water samples are analysed for sulfite, sulfate and manganese. Sulfite is analysed spectrophotometrically after stabilisation by formaldehyde. Sulfate is analysed by an ion chromatographic method and manganese by atomic absorption spectrometry.
The reaction rate is calculated using the concentration of sulfate measured in the sample divided by the mean residence time of the droplets (3 minutes). For the calculation of the rate constant, the concentration of sulfite is needed. The concentration of sulfite in a droplet is assumed to be constant during the whole residence time, as the concentration of SO_2 measured at different points in the chamber is the same, and the gas/water equilibrium can be established fastly. Various kinetic descriptions are tested for experiments with varying concentrations of sulfite, of manganese and for different concentrations of the buffer. Descriptions, in which the rate constant increased or decreased with varying concentration of one of the compounds are rejected.

3. RESULTS

The formation of sulfate is strongly increased by the presence of manganese.
Some oxidation of sulfite is observed in experiments without manganese as well, but this is a small amount compared to the catalytic oxidation of sulfite. The oxidation in the absence of manganese is assumed to be a wall effect of the cyclone. A correction for this oxidation can be made according to:

$$d[SO_4^{2-}]/dt = k_o \, [S(IV)] \tag{1}$$

Here, S(IV) is the sum of sulfite and bisulfite. The value of k_o remaines constant for different concentrations of sulfite, but it increases with decreasing concentrations of the buffer, for example k_o = 0.0034 s^{-1} with a buffer concentration of 2×10^{-3} mol/l and 0.0018 s^{-1} with a buffer concentration of 4×10^{-3} mol/l.
The remaining formation of sulfate is considered to be an oxidation reaction catalysed by manganese. Figure 1 shows that there is a linear relation between the concentration of manganese and the formation of sulfate. The data in figure 1 are obtained by a series of experiments in which the concentration of sulfite is 62 μmol/l.
The catalytic reaction is not first order with respect to S(IV). In figure 2 the formation of sulfate at different sulfite concentrations is shown for three different concentrations of Mn^{2+}. There is no linear relation between the concentration of sulfate and the concentration of sulfite at constant Mn^{2+} concentration. The curves are fitted lines for the kinetic equation

$$d[SO_4^{2-}]/dt = k \, [S(IV)]^{0.5} \tag{2}$$

The order of reaction can be determined more accurately by a log/log plot of the sulfate formation versus the concentration of sulfite. This results in the best fit for the measured values:

$$d[SO_4^{2-}]/dt = k_1 \, [Mn^{2+}] \, [S(IV)]^{0.5} \tag{3}$$

with k_1 = 3.7 ± 1.2 $(mol/l)^{-0.5} \, s^{-0.5}$ at pH = 5.
The results of all experiments performed at buffer concentrations 2×10^{-3} and 4×10^{-3} mol/l are shown in table 1. The sulfate concentrations given in the table are already corrected for the oxidation in the absence of manganese using equation (1). The table shows that the calculation used here is the best fit for the measured values.

4. COMPARISON WITH OTHER STUDIES

The constants k_2, k_3 and k_4 given in table 1 are values calculated according to kinetic descriptions used in other studies.
Martin (1984) assumes that the equation:

Table 1: Measured results of experiments at different buffer
concentrations

nr	pH	Mn	SO_3^{2-}	SO_4^{2-}	k_1	k_2	k_3	k_4
		μmol/l	μmol/l	μmol/ls	$M^{-0.5}s^{-1}$	s^{-1}	$M^{-1}s^{-1}$	$M^{-1}s^{-1}$
buffer 1 x 10^{-3} mol/l								
1	4.95	9.09	22.25	.141	3.28	.0191	696	1703
2	4.55	9.09	53.12	.094	1.42	.0322	195	1142
3	4.3	45.45	16.50	.763	4.13	.0185	1017	369
buffer 2 x 10^{-3} mol/l								
4	4.8	9.09	74.00	.189	2.42	.0363	281	2289
5	5.2	18.18	33.50	.309	2.94	.0059	508	936
6	4.85	18.18	62.75	.503	3.49	.0215	441	1520
7	4.75	18.18	66.62	.630	4.24	.0339	520	1905
8	4.72	18.18	129.75	.754	3.64	.0435	320	2282
9	4.5	18.18	117.62	.735	3.73	.0703	344	2224
10	5.07	45.45	9.00	.388	2.85	.0016	949	188
11	4.5	41.82	26.62	.950	4.40	.0172	853	543
12	4.86	43.64	72.25	1.227	3.31	.0089	389	644
13	5	90.91	14.75	1.784	5.11	.0022	1331	216
14	4.4	90.91	16.50	2.365	6.40	.0114	1576	286
15	4.3	90.91	30.50	2.495	4.97	.0151	900	302
buffer 4 x 10^{-3} mol/l								
16	4.85	8.55	191.62	.294	2.48	.0568	179	4021
17	4.98	8.73	217.75	.166	1.29	.0228	87	2175
18	5	17.09	67.12	.486	3.47	.0166	424	1663
19	4.83	16.55	130.62	.453	2.40	.0245	210	1656
20	4.87	43.64	69.62	1.447	3.97	.0103	476	760
21	5.1	92.73	17.38	1.448	3.75	.0013	899	168
22	4.5	90.91	11.12	1.621	5.35	.0062	1603	196
23	4.8	98.18	15.12	1.492	3.91	.0025	1005	155
24	4.8	92.73	24.38	2.495	5.45	.0046	1104	290
buffer 1 x 10^{-2} mol/l								
25	5.05	17.45	374.62	.316	.94	.0093	48	1039
26	4.96	87.27	109.38	2.242	2.46	.0032	235	294

Descriptions for the kinetic constants, $(R = d[SO_4^{2-}]/dt)$
$R = k_1 \, [Mn^{2+}] \, [S(IV)]^{0.5}$
$R = k_2 \, [Mn^{2+}]^2 \, / \, [H^+]$
$R = k_3 \, [Mn^{2+}] \, [S(IV)]$
$R = k_4 \, [Mn^{2+}]^2$

S(IV) = 62 μmol/l

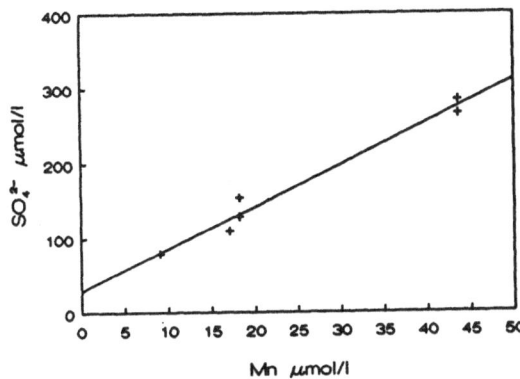

Figure 1.
The effect of the manganese
concentration on the
catalytic formation of
sulfate at pH = 5.

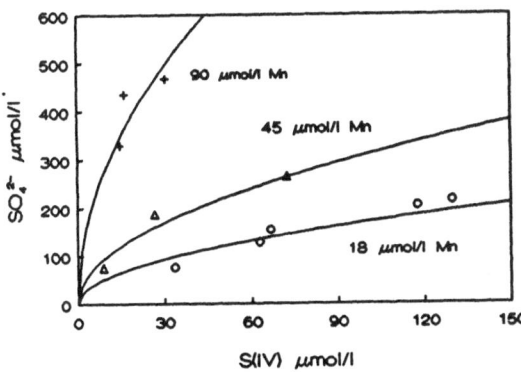

Figure 2.
The effect of the con-
centration of S(IV) on the
catalytic formation of
sulfate at three different
manganese concentrations.

$$d[SO_4{}^{2-}]/dt = k_2 \ [Mn^{2+}]^2/[H^+] \tag{4}$$

is the best description for the relation between sulfate formation and
manganese concentration up to pH 5. But this study shows that the rate
of formation of sulfate increases with increasing sulfite concentration.
Our results can be compared to the kinetic description found by
Ibusuki et al. (1987). They measure a linear correlation between the
formation of sulfate and the concentrations of sulfite and manganese
according to:

$$d[SO_4{}^{2-}]/dt = k_3 \ [Mn^{2+}] \ [S(IV)] \tag{5}$$

When the reaction rate constant k_3 is calculated from our experiments
the value increases at lower sulfite concentrations. This means that a
first-order dependence on sulfite cannot be used to describe the
results of our experiments.
If the absolute amount of sulfate formed during one hour is calculated
according to the kinetic expression of the present study, to that of

Martin (1984) and to that of Ibusuki (1987), the sulfate formation according to Martin is about a factor of thousand higher than the amount measured in the present experiments. The sulfate formation calculated according to Ibusuki is about a factor of 2 - 3 higher than the value measured here at high concentrations of sulfite and a factor of 2 lower at low concentrations of sulfite (table 2). The difference in sulfate formation according to Ibusuki compared to this study may be caused by impurities in the water. Even two rate constants measured by the same person at two different laboratories (Ibusuki, 1984 and 1987) are different by a factor of 2.5 due to a different treatment of the water. Another possibility is that the formation of sulfate at low concentrations of sulfite is linear with respect to sulfite, while at higher concentrations the sulfate formation depends on the root of the sulfite concentration. Compared to Ibusuki's experiments the range of the concentration of manganese in this study is higher by a factor of 10.

Table 2:
Sulfate formation according to three different models at pH = 5, a) at low concentrations of sulfite, b) at high concentrations

	S(IV) μmol/l	Mn^{2+} μmol/l	SO_4^{2-} μmol/lh this study	μmol/lh Ibusuki (1987)	μmol/lh Martin (1984)
a)	1	10	130	61	170000
	1	40	530	245	2700000
b)	50	10	940	3060	170000
	50	40	3770	12200	2700000

In a more recent publication, Martin (1987) suggests that the catalytic formation of sulfate is dependent on ionic strength. For sulfite concentrations in the range below 1 μmol/l, Martin measured a second-order kinetics according to (5) with a reaction rate constant of 630 ℓmol⁻¹s⁻¹ for an ionic strength of 2 x 10⁻³ mol/l, and k = 374 ℓmol⁻¹s⁻¹ for an ionic strength of 1 x 10⁻² mol/l. This is of the same order as the second-order rate constants k_3 calculated from our experiments (table 1). But from these experiments a dependence on ionic strength cannot be concluded.
For sulfite concentrations higher than 100 μmol/l Martin (1987) assumes that the kinetic relation depends on the concentration of manganese only, according to:

$$d[SO_4^{2-}]/dt = k_4 \ [Mn^{2+}]^2 \qquad (6)$$

However, k_4 should be depending on ionic strength as well. From Martins experiments, values for k_4 are 450 ℓmol⁻¹s⁻¹ (ionic strength 2 x 10⁻³ mol/l) and 290 ℓmol⁻¹s⁻¹ (ionic strength 1 x 10⁻² mol/l). If

k_4 is calculated from our experiments, the value is about a factor of 2-4 higher. There are not sufficient data at higher sulfite concentrations available to determine a possible dependence on ionic strength at these sulfite concentrations.

5. DISCUSSION

The mechanism of the catalytic oxidation of sulfite is not yet completely clear. Hoffmann (1984) made a review about different reaction mechanisms which have been proposed. In all cases sulfite is oxidized by oxygen. During the chain reaction, sulfite and sulfate radicals are formed. The initiation occurs possibly via a transition complex of sulfite with manganese. It is not yet known whether Mn(II) is transformed to Mn(III) in this transition complex. The kinetic expression found here does not lead to a clear reaction mechanism.
The importance of the manganese catalyzed oxidation of sulfite in dew and rain is shown in table 3.

Table 3:
Importance of the catalytic oxidation of sulfite and the oxidation by ozone for different concentration ranges in dew and rain at pH 5.
a) low concentrations of sulfite and ozone

	S(IV)	O_3	Mn^{2+}	SO_4^{2-} cat	SO_4^{2-} by O_3
	µmol/l	µg/m³	µmol/l	µmol/h	µmol/h
dew	1	40	20	300	8.3
rain	1	40	.04	.5	8.3

b) high concentrations of sulfite and ozone

dew	50	160	20	1900	1700
rain	50	160	.04	3.7	1700

In table 3 the production of sulfate is calculated using the kinetic relation determined in this study. The catalytic oxidation is compared to the oxidation by ozone, taken from Erickson (1977). The calculation is performed for cases in which the concentrations of SO_2 and O_3 are low as well as for cases in which they are high. The rate of reaction is calculated for a pH of 5. Table 3 shows that the catalytic oxidation is very important for the formation of sulfate in dew, while it is less important in rain water. The relative importance of the catalytic oxidation is highest at low concentrations of ozone and sulfite. However, the absolute amount of sulfate formed by catalytic oxidation is higher at high concentrations of S(IV). This means that the thin water layer which forms on plants during the night can be a very reactive mixture.
It should be noted that the pH of rain decreases with continuing oxidation of sulfite, which leads to a decrease in the rate of reaction. Dew on plants is buffered and the pH will remain constant for a longer time. At a higher pH, more sulfite can be dissolved in

238

the water layer and as a consequence, the reaction in dew will
generally be faster than in rain.
As the catalytic oxidation involves sulfite and sulfate radicals, a
correlation between the reaction rate of the catalytic sulfite
oxidation and plant damage could be possible. In forest areas in the
Federal Republic of Germany, a relation between damage of spruce trees
and the concentration of manganese in the needles has been found
(Gaertner, 1985). As the concentration of manganese in the needles
does not seem to be toxic for the trees, the oxidation of sulfite by
manganese leached from the needles might be an explanation for the
observed damage.
This reaction is an example of a clear difference in the chemistry and
reactivity of a rain droplet and a layer of dew on a plant. It should
be considered that reactions occurring in a dew layer can be different
from those known from the chemistry in rain or clouds. In addition,
the reactive water layer may be very important for the damage by
water-soluble atmosperic compounds on plants.

6. REFERENCES

Arends, B.G., Mallant, R.K.A.M., van Wensveen, E., Gouman, J.M. (1988)
'A fog chamber for the study of chemical reactions.' Netherlands
Energy Research Foundation, ECN-210.
Erickson, R.E., L.M.Yates, R.L.Clark, D.McEven (1977) 'The reaction of
sulfur dioxide with ozone in water and its possible atmospheric
significance'. Atmos. Environ. 11, 813-817.
Gärtner, E. (1985) 'Mangangehalte in Altfichten, Boden und
Kronendurchlass an jeweils gleichen Standorten.' VDI-Berichte 560,
VDI Kommission Reinhaltung der Luft: Waldschaden, Einflussfaktoren
und ihre Bewertung, 559-573.
Hoffmann, M.R., D.C.Jacob. (1984) 'Kinetics and mechanisms of the
catalytic oxidation of dissolved sulfur dioxide in aqueous solution:
an application to nighttime fog water chemistry.' In J.G.Calvert
(ed.), SO_2, NO and NO_2 oxidation mechanisms: Atmospheric
considerations. Butterworth Publishers, Boston, London.
Ibusuki, T., H.M.Barnes. (1984) 'Manganese(II) catalysed sulfur
dioxide oxidation in aqueous solution at environmental
concentrations.' Atmos. Environ. 18, 145-151.
Ibusuki, T., K.Takeuchi. (1987) 'Sulfur dioxide oxidation by oxygen
catalyzed by mixtures of manganese II and iron III in aqueous
solutions at environmental reaction conditions.' Atmos. Environ. 21,
1555-1560.
Martin, L.R. (1984) 'Kinetic studies of sulfite oxidation in aqueous
solution.' In J.G.Calvert (ed.), SO_2, NO and NO_2 oxidation
mechanisms: Atmospheric considerations. Butterworth Publishers,
Boston, London.
Martin, L.R., M.W.Hill. (1987) 'The effect of ionic strength on the
manganese catalyzed oxidation of sulfur (IV).' Atmos. Environ. 21,
2267-2270.

MEASUREMENT OF DEW AND FOGWATER DEPOSITION IN FOREST STANDS

F. Dröscher, J. Nickel and E. Mikisch [*]

Institut für Verfahrenstechnik und Dampfkesselwesen
Abt. Reinhaltung der Luft, Universität Stuttgart
Pfaffenwaldring 23, D-7000 Stuttgart

ABSTRACT. In this study canopy weighing and leaf-washing were combined for the analysis of dew water deposition on natural spruce and surrogate test trees. This method yields the time fraction of canopy wetting, the quantity of wetting water on the tree canopy by ordinary and occult precipitation continuously. Leaf washing with definite dilution allows the calculation of material deposition during dew as well as fog episodes.

1. WETTING WATER LAYERS ON THE TREE CANOPY SURFACE

Dew and fog water deposititon do not only contribute to the water balance of forest ecosystems, they also play an important part in the deposition of trace materials on forest stands. These two forms of occult precipitation interfere with the material budget in different ways. Significant quantities of water deposition can be found during episodes of fog and cloud water interception on mountain tops or western slopes of mountain ranges. Sometimes they are associated with canopy through fall, i.e. fog water drips off the canopy. For a grown-up spruce stand this equals a precipitation height in excess of the canopy water retention capacity of 2-3 mm. At exposed summit locations fog interception can amount to 10mm in 12 hours [1]. At these mountain locations frequently advection fog occurs with fog that underwent longer distance transport (e.g. up to 200 fog days per year at the kl. Feldberg near Frankfurt/M. or the gr. Arber in the Bavarian Forest). There, fog water interception is likely to match ordinary precipitation by rain and snow with respect to the yearly precipitation heights.

With respect to the chemical input of trace elements, fog water deposition can well exceed the wet deposition by rain and snow at many locations with pronounced fog exposition due to generally increased concentrations of most fog water constituents compared to rain water [1,2,3].

The importance of light fog or fog with reduced wind speeds that do not result in canopy through-fall is the generation of wetting water covers on the canopy surfaces rather than the direct chemical input. Fog water constituents remain in the canopy, no material is washed off the needles or leaves. Fog water interception extends the times of canopy wetting. It then acts in much the same way as dew water deposition.

---

[*] The authors gratefully acknowledge a research grant of the German Federal Minister of Science and Technolgy

H.-W. Georgii (ed.), Mechanisms and Effects of Pollutant-Transfer into Forests, 239–247.
© 1989 by Kluwer Academic Publishers.

Dew water deposition occurs in a regular pattern during clear nights due to the condensation of water vapour from the ambient air on cold surfaces.

The wetting water cover improves deposition conditions and affects plant-atmosphere interactions:

- Dry deposition of gases and particles is improved on wet surfaces. Water-soluable acidic gases like SO_2 or NO_2 are readily absorbed in water films as well as the only alkaline gas NH_3. Moreover, improved sticking coefficients for particle deposition due to sedimentation and eddy diffusivity result in a better canopy interception efficiency and an improved retention of the deposited material in the canopy.

- Wetting water films provide a space for heterogeneous chemical reactions involving dissolved matter from particles and plant leachate as well as absorbed gases. Due to the diverse origin of its components, a wide variety of chemical compounds is present in the wetting water film including potential catalysts for the oxidation of SO_2 to sulphate. Especially manganese and to a less extent iron act as effective catalysts. Manganese in particular is expected to occur in high concentrations due to its part in plant physiology /4/. These time-dependent reactions rely on long durations of canopy wetting.

- Wetting water films enhance the material transfer across the plant-atmosphere interface. Depending on the concentration gradient between the intercellular solution and the wetting water layer, dissolved material is either taken up by the plants or is leached from the plant. Proton buffering of the canopy is such a process: Protons in the wetting water layers (e.g. from rain) are exchanged for cations in the plant solution like K^+, Ca^{2+}, Mg^{2+}, and Mn^{2+}. Since these diffusion processes are highly time dependent, the duration of canopy wetting is important for the leaching.

2. MEASUREMENT OF FOG AND DEW WATER DEPOSITION

Experimental approaches to assess fog water deposition on forests generally use fog water analyses from the atmosphere obtained by fog water sampling. Fog water deposition is then either calculated with the help of a fog droplet impaction modell /5,6/, by the measurement of canopy through-fall (adding the water retention capacity of the canopy) or by tree weighing /7/, where a tree is placed on a balance and the weight increase due to canopy wetting is measured.

All three methods assume a uniform distribution of fog water constituents over the entire drop size spectrum. The chemical composition of the canopy wetting water itself remains unknown. This can be determined either from canopy through fall analysis in the case of heavy fog water interception or from leaf washing experiments during light fog. In order to exclude leaching of plant materials from natural trees, chemically inert surrogate trees can be used. The quantity of fog water deposition and the times of canopy wetting are conveniently measured by weighing the test trees.

Little is known about dew water deposition on forests. Attempts to quantify dew water deposition were restricted mainly to the gravimetric determination of water volumes on surrogate surfaces like porous plates (Leick-plate) or a steel wire mesh (Hiltner's dew balance) as standard receptors. From these methods no information on the dew water composition is available.

On the other hand the chemical processes in dew water are studied using artificial dew formation on chilled surfaces from a defined atmosphere /8/. This approach excludes the influence of particle deposition and of plant leachates on the dew chemistry. Direct sampling of dew water from the forest canopy is not feasible due to the small size of dew droplets on needles or leaves. Therefore leaf washing has to be applied. Leaf washing of natural trees is interfered by leaching processes. The use of surrogate trees excludes this biological influence.

3. COMPARISON OF NATURAL SPUCE AND SURROGATE SURFACES AS DEW WATER RECEPTORS

Whereas for fog water deposition the geometrical arrangement of a surrogate receptor is of prime importance, for dew deposition two additional parameters have to be considered the IR-absorption spectrum of the surrogate surface and the heat conductivity of the material.

Therefore a comparison experiment was conducted for 11 dew nights on an open grass field on the University campus in Stuttgart-Vaihingen, involving pairs of natural spruce trees and of surrogate trees built to represent the natural trees as well as a plane dew plate with a surface temperature sensor as reference along with meteorological and air analytical parameters. The experimental set up is depicted in Figure 1.

In the afternoon each receptor was rinsed with destilled water to remove particulate impurities on the surfaces and allow to dry. In the evening the weight was taken as well as at sunrise in the morning. In the morning each receptor was rinsed with a known volume of water (definite dilution) to wash off the dew water. The dripping water was then sampled and analysed for ionic species. The average terminal concentrations in the dew water were calculated from these samples and the dilution ratio.

Also the average deposition to the surface water layer was determined using the canopy or plate surface area. The terminal dew water quantity of the various receptors is plotted in Figure 2 along with relevant dew parameters.

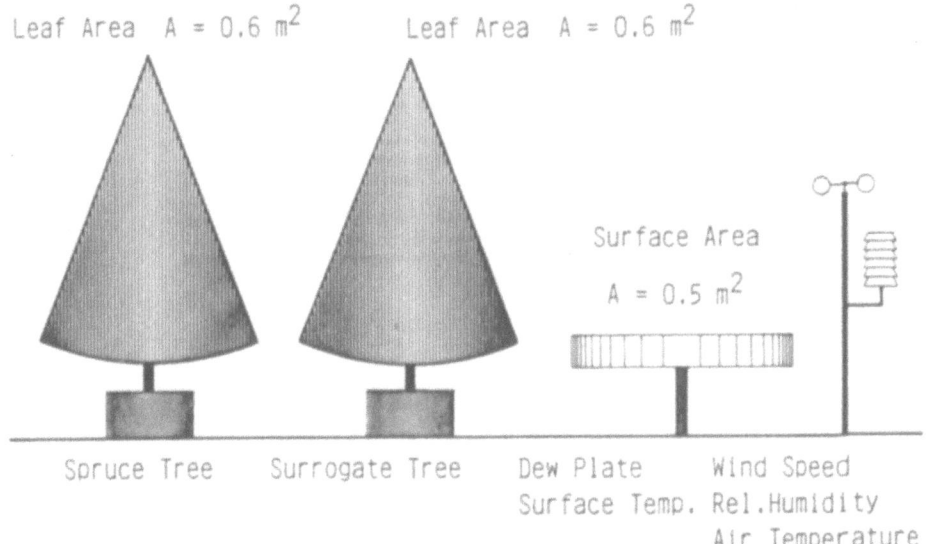

Figure 1: Comparison of natural spruce tree, surrogate tree, and plane dew plate as dew water receptors on an open grass field on the University campus in Stuttgart-Vaihingen.

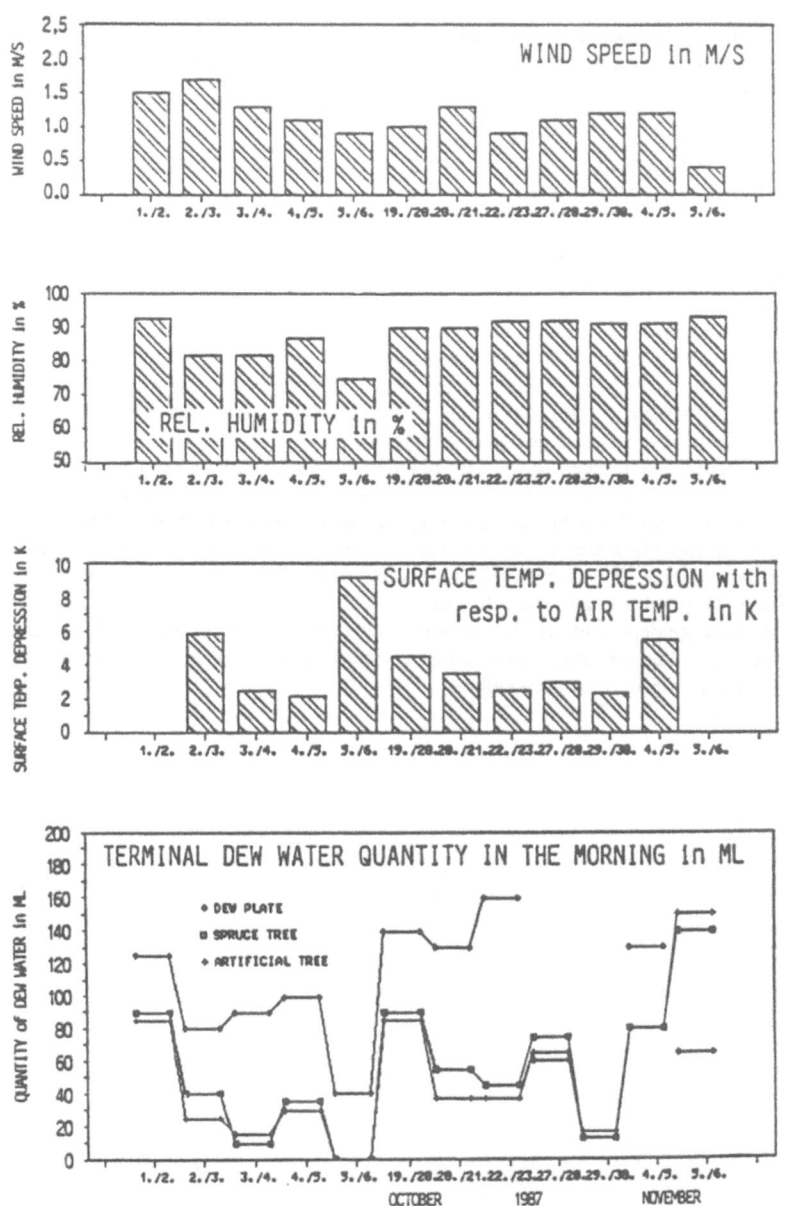

Figure 2: Terminal dew water quantity in the morning on dew plate, natural spruce tree and surrogate tree (bottom graph) and average values of the dew parameters wind speed, relative humidity, and the surface temperature depression on the dew plate with respect to air temperature for the duration of each dew episode. Stuttgart-Vaihingen, Oct./Nov. 1987.

The correlation of the terminal dew water deposition on natural and surrogate trees are given in Figure 3 (top). The high correlation coefficient indicate the close relation of the two quantities.

Figure 3 also plots the correlations of calculated average pH-values in the dew water as well as the sulphate deposition per needle area. These correlations are fair with increasing deviations at high dilution ratios of above five to eight.

Except for that limitation of the method of definite leaf washing surrogate trees can adequately replace natural trees not only in fog, but also in dew water depositon studies to exclude leaching and evapotranspiration interference.

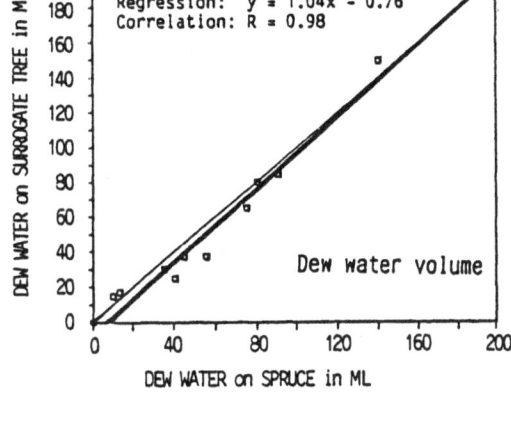

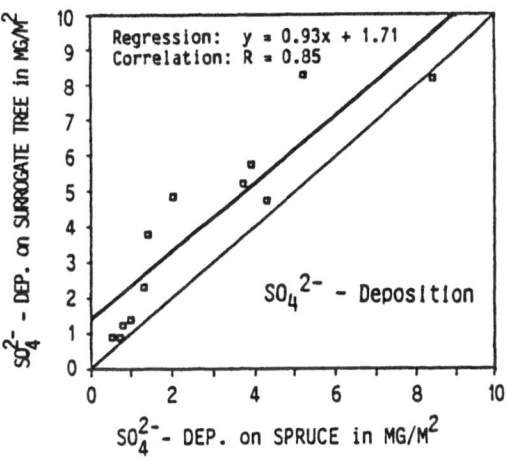

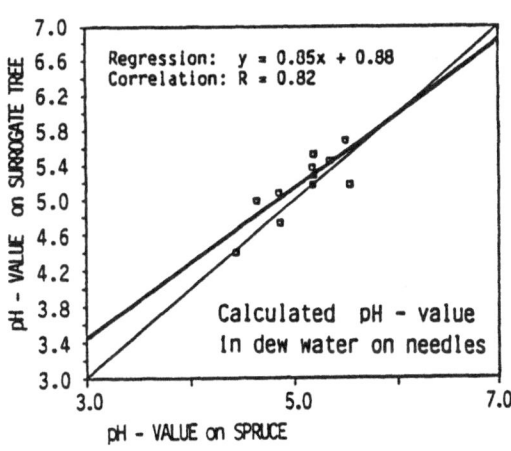

Figure 3: Correlation of dew water deposition, sulphate deposition, and terminal dew water pH-value calculated from definite dilution by leaf washing between commercial surrogate tree and natural spruce tree. 11 dew nights, Stuttgart-Vaihingen, Oct./Nov. 1987.

244

4. CONTINUOUS RECORDING OF CANOPY WETTING

In order to investigate canopy wetting in the forest canopy the method was partly automated and the dew and fog water receptors moved into the canopy of a mature spruce stand in the Black-Forest (see Figure 4). The experimental setup was mounted 20m above the ground on the IVD measuring tower at Freudenstadt-Schöllkopf. It comprises a surrogate tree canopy installed on an electrical balance and sampling facilities for drip-off water from the canopy allowing for fractional sampling during heavy fog water interception as well as for integrative sampling.

Two examples for the continuous recording of canopy wetting are presented in Figure 5. It plots half-hour mean values of canopy wetting along with relevant meteorological parameters for a seven day period in Sept. 1988 (left column) and for a dew night Nov. 5th - 6th 1988 (right column).

The first 4 days in the seven days' period in September 1988 were characterized by the absence of rain. Strong radiation indicates clear skies on Sept. 19th, 21st (after fog dissipation) and 22nd. This is also reflected by the temperature of the dew plate surface

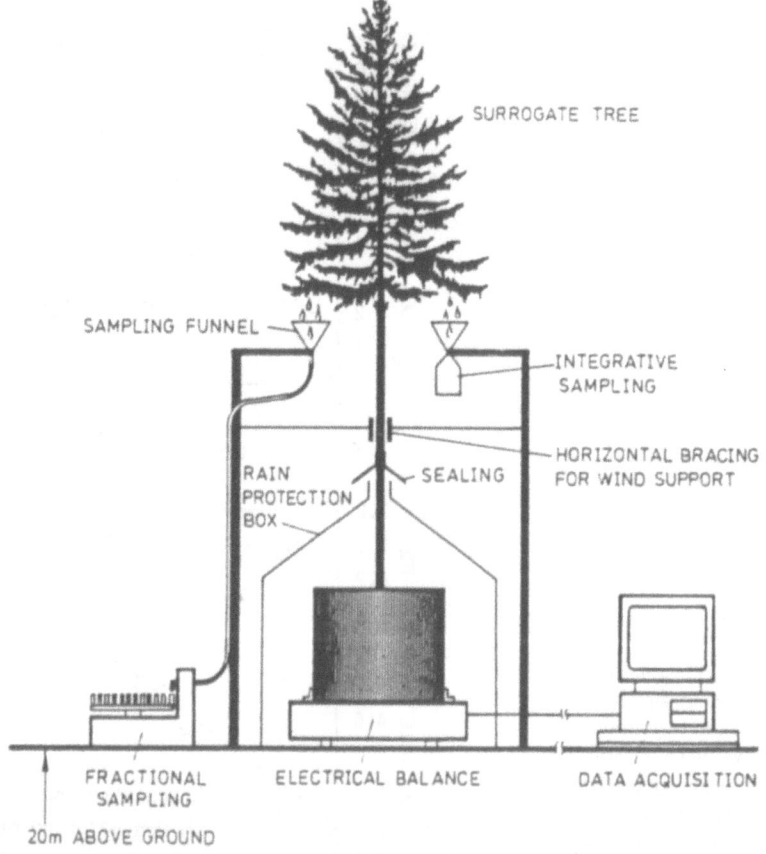

Figure 4: Experimental Set-Up for Continuous Fog and Dew Water Weighing and Sampling of Canopy Dripping Water.

which is much above the air temperature during the day. In the first two nights no radiational cooling can be detected due to elevated fog formation.

Only in the night of 21st to 22nd, radiational cooling is detected. Consequently dew fall is noticed on a wetness sensor giving the time fraction of wetness on a slightly inclined plastic sheet. Gentle dew fall amounts to 80g dew fall at the end of that night corrosponding to a wetting film of 0.03mm. In the following days a storm system moves across Freudenstadt associated with heavy rain fall on two days. This leads up to the dynamic canopy water retention capacity of 1100g, or 0.4mm water layer thickness respectively.

A single dew episode is analysed using the data for Nov. 5-6, 1988. During an anticylonic period marked by clear skies (see radiation) a gentle surface temperature depression on the dew plate compared to the ambient air evolved in the evening up to 5 a.m.. During that period dew formed at a constant rate of about 10g/h. After 5 a.m. up until sunrise 8-8.30a.m. elevated fog occurred and stopped the surface chilling by IR reflexion. This results in an unchanged dew water volume on the surrogate tree canopy. The dew fall amounts to 125g or 0.045mm. After sunrise the dew layer rapidly dissipates.

5. CONCLUSIONS

In order to assess dew and fog water deposition on coniferous trees the methods of canopy weighing and leafwashing were applied to natural trees as well as surrogate surfaces. For dew and fog water deposition surrogate trees can well represent natural trees both with respect to the quantity and to the chemical composition of the wetting water layer on the canopy. Calculations of the average deposition rate on the needle surface are generally more accurate than those for the species concentrations. Uncertainties become large when huge amounts of rinsing water is required for leaf-washing. Canopy weighing allows to record the development of dew and fog water deposition precisely. Therefore the method is a helpful tool in the analysis of dew and fog water deposition on forests.

6. REFERENCES

/1/   Schmitt, G.: Methoden und Ergebnisse der Nebelanalyse. Diss. Universität Frankfurt. Berichte des Inst. f. Meteorologie und Geophysik der Univ. Frankfurt Nr. 72, 1987.

/2/   Trautner, F.: Entwicklung und Anwendung von Meßsystemen zur Untersuchung der chemischen und pysikalischen Eigenschaften von Nebelwasser und dessen Deposition auf Fichten. Diss. Universität Bayreuth, 1988.

/3/   Chamides, W.L. (1987): Acid dew and the role of chemistry in the dry deposition of reactive gases to wetted surfaces. J. Geophys. Res., 92, p. 895-908.

/4/   Arends, B.: The effect of manganese on the oxidation of SO_2 in dew water. This volume.

/5/   Kroll, G. and P. Winkler: Estimation of Wet Deposition via fog. In: Grefen, K. and J. Löbel: Environmental Meteorology, Kluwer Acad. Publ. Dordrecht 1988.

/6/   Lovett, G. (1984): rates and machanisms of cloud water deposition to a subalpine balsam fir forest. Atm. Environ. 18 p. 361-371

/7/   Trautner, F.: Collection and properties of fogwater. In: Lange, O. and E.D. Schulze (Eds); Acid rain and forest decline in the Fichtelgebirge. Ecological Studies Ser. Springer,Heidelberg in press

/8/   Slanina, J.: Untersuchung des Zusammenhangs von Schadstoffkonzentrationen in Luft, Tau und Nebel. VDI-Berichte, Nr. 608. VDI-Verlag Düsseldorf 1987

246

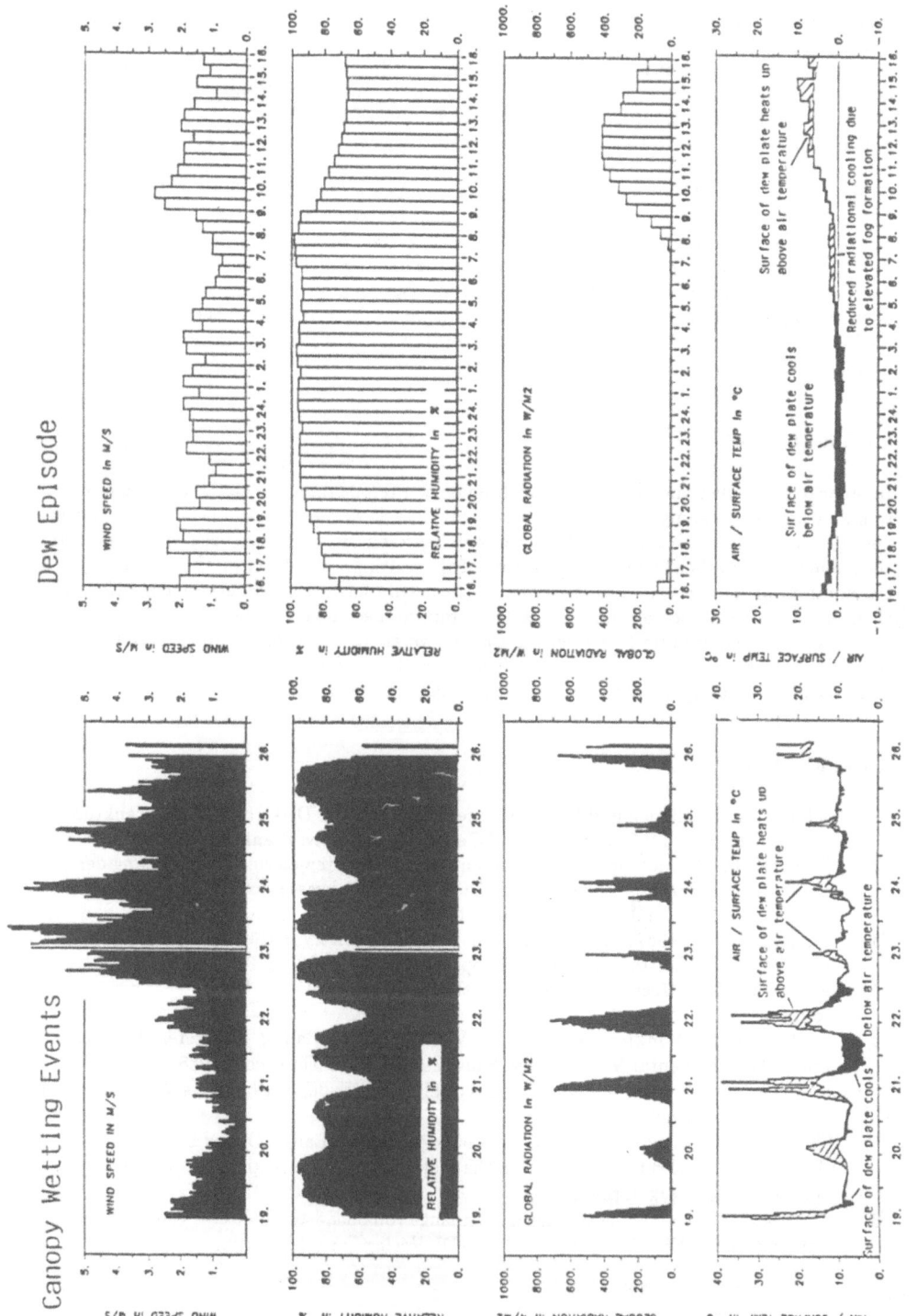

Canopy Wetting Events

Dew Episode

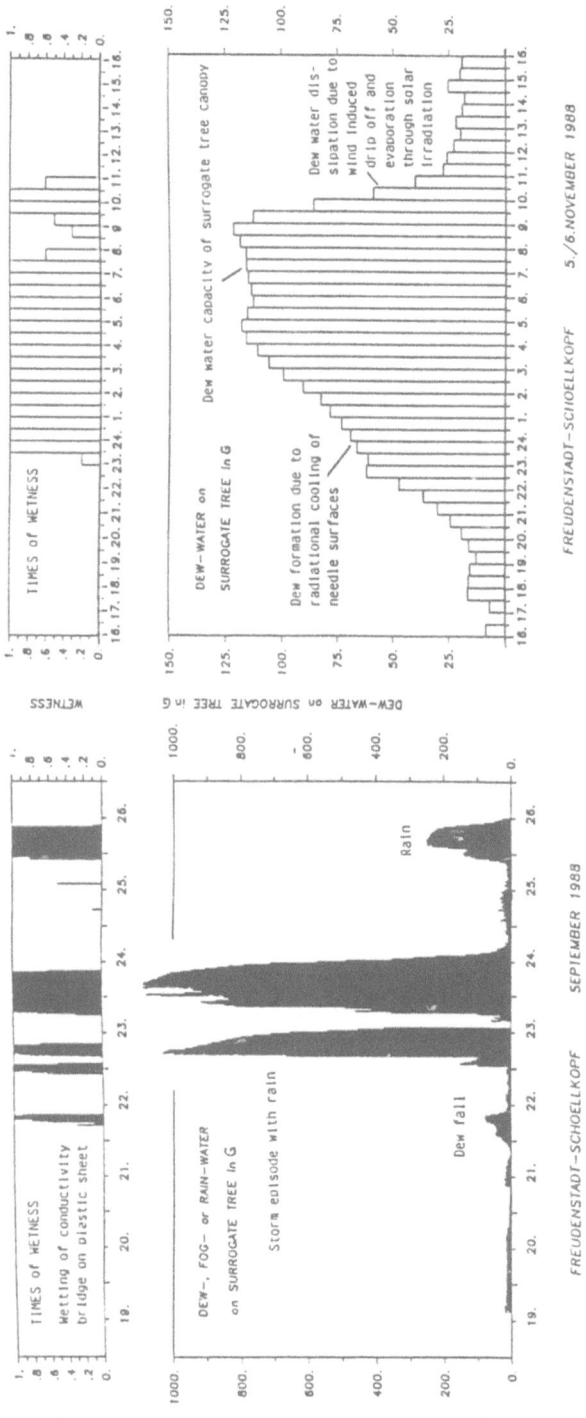

Figure 5: Half-hour mean values of dew-, fog or rainwater deposition on surrogate tree (A = 2.8m² leaf area) (bottom graphs) during a 7 day period in Sept. 1988 (left) and during the dew night of Nov 5th-6th 1988 (right).

For ease of interpretation dew parameters are arranged in the resp. columns representing wind speed, relative humidity, global radiation, temperature of ambient air and the surface of a dew plate, as well as time fraction of wetness. Freudenstadt-Schöllkopf.

EFFECTS
OF ATMOSPHERIC POLLUTANTS ON
VEGETATION

EFFECTS OF PHOTOOXIDANTS ON PLANTS

HEINZ RENNENBERG AND ANDREA POLLE

Fraunhofer Institut für Atmosphärische Umwelt-
forschung
Kreuzeckbahnstr. 19
D-8100 Garmisch-Partenkirchen, FRG

ABSTRACT. Photooxidants such as ozone or hydroperoxides are
potentially damaging to plants, because they can rise the
intracellular amount of toxic oxygen species. Since aerobic
organisms are adapted to live with the threat of oxidation,
protection mechanisms exist in plant cells that can remove
injurious agents in the symplastic as well as in the
apoplastic space. However, when the stress applied exceeds
a critical threshold of the adaptation capacity of the
metabolic responses, cellular injury, reduced vitality or
even death of the plants may result.

The term "photooxidants" comprises a group of air
pollutants that are produced in the atmosphere under the
influence of light with nitrogen oxides and reactive
hydrocarbons being the major precursors [1]. In
industrialized regions these precursors are predominantly
of anthropogenic origin, whereas biogenic emissions
contribute to a major part of these compounds in the remote
atmosphere. Most of our knowledge on the effects of
photooxidants on plants is based on experiments with ozone
[2]. However, this compound is only one relatively stable
reaction product among many other instable compounds
produced in complex series of reactions [3]. Therefore,
other photooxidants, like oxygen radicals and peroxides, so
far not investigated in any detail, may be much more
injurious to plants.

To evaluate the toxicity of photooxidants to plants, it has
to be considerd that living in an oxygen containing
atmosphere is unequivocally connected with an exposure to
photooxidants. Toxic oxygen species are consistently
produced in numerous metabolic pathways and, therefore,
belong to the every day's life of a plant. At high light
intensities, e.g., the availability of CO_2 in the

251

H.-W. Georgii (ed.), Mechanisms and Effects of Pollutant-Transfer into Forests, 251–258.
© *1989 by Kluwer Academic Publishers.*

chloroplasts becomes the rate limiting step in photosynthesis. Under these conditions more NADPH is produced by photochemical reactions than can be used for the synthesis of carbohydrates in the Calvin-Cycle. As a consequence, $NADP^+$ as electron acceptor becomes scarce. To prevent damage to the photosynthetic electron transport system from an insufficient electron flow, excited electrons are transferred either directly from photosystem I or indirectly via reduced ferredoxin to molecular oxygen (fig. 1).

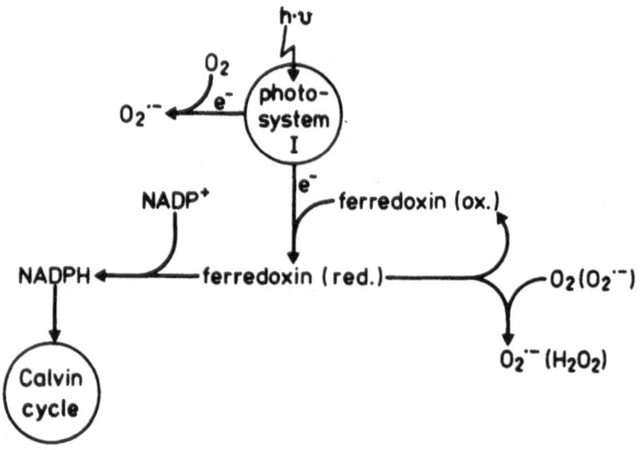

Figure 1.   Generation of toxic oxygen species during light-
            saturation of photosynthesis

The products of these reactions are superoxide radicals and hydrogen peroxide [4]. The same oxygen species, but also hydroxyl radicals, may accidentally be released during respiration, when electrons are successively transferred to molecular oxygen during ATP synthesis [5]:

$$O_2 \xrightarrow{e^-} O_2^{\cdot -} \xrightarrow[H^+]{e^-} H_2O_2 \xrightarrow{e^-} 2\ HO^{\cdot} \xrightarrow{e^-} H_2O$$

The generation of toxic oxygen species during essential metabolic processes like photosynthesis and respiration implies that the development of detoxification mechanisms for these compounds is a matter of survival in an oxygen containing atmosphere. The strategies evolved in plants for the removal of toxic oxygen species can be divided in (1) enzymatic processes and (2) chemical reactions with antioxidants. Hydrogen peroxide, organic peroxides, and superoxide radicals, e.g., can be detoxified enzymatically, whereas ozone or hydroxyl radicals can only be removed in plant cells by chemical reactions with antioxidants [6]. With these strategies not only injury by toxic oxygen species per se can be prevented, but also chain reactions initiated by these radicals can be interrupted. Still the distinction between enzyme catalized and chemical processes is artificial, since many photooxidants detoxified enzymatically can also be removed chemically by reactions with antioxidants, and antioxidants like ascorbate and glutathione are co-substrates in enzyme catalyzed detoxification reactions.

An important factor in the protection of plants from the action of atmospheric photooxidants is the subcellular localization of detoxification mechanisms. An atmospheric photooxidant may first get into contact with the cuticle and the stomata of the leaves. We presently do not know, at which extent photooxidants can penetrate the cuticle and at which extent they will react with the cuticle's structures. Reactions of photooxidants with the cuticle can only be called detoxification reactions, if they do not result in changes in the permeability of the cuticle for gaseous compounds and if the reacting structures of the cuticle can be renewed. Independent of whether a photooxidant has entered a leaf via the cuticle or via the stomata, it may be subjected to detoxification reactions in the apoplastic space of the leaf tissue (fig. 2).

To date these reactions have not attracted too much interest, although they may be extremely important for the protection of plant cells from photoxidative damage by air pollutants. Photooxidants already detoxified in the apoplastic space can not cause any damage inside the cells. E.g., it is known since a long time that peroxidases can not only be found inside plant cells, but are also associated with the cell walls [7], still total peroxidase activity and not the activity associated with the cell walls has been determined in many experiments on the action of atmospheric photooxidants on plants. To evaluate the conversion of atmospheric hydrogen peroxide and organic peroxides in the apoplastic space intensive studies on the activity and the substrate specificity of apoplastic peroxidases are urgently needed. The presence of ascorbate has recently been shown in cell walls of white mustard [8]. Because of its mobility [9-11] it would not be surprising, if the antioxidant glutathione can also be found in the apoplastic space.

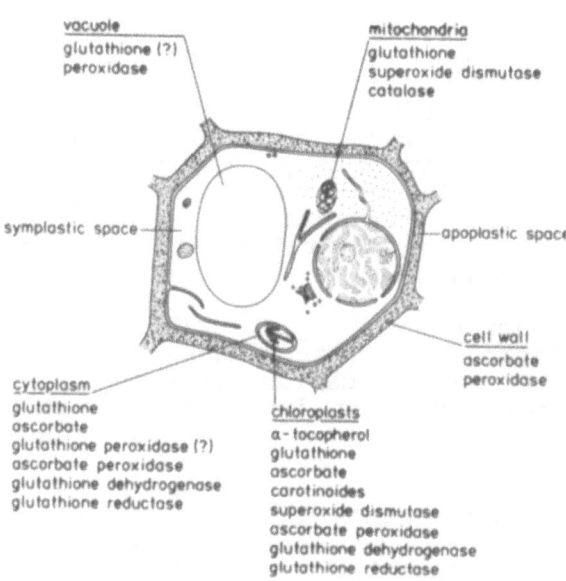

Figure 2   Subcellular localization of   detoxification mechanisms for toxic oxygen species in plant cells

Atmospheric photooxidants that have passed the apoplastic space will get into contact with the plasmalemma, an achillis heal for the damage of plant cells by these compounds mainly because lipid peroxidation may occur in this membrane. Inside the cells, different detoxification mechanisms for photooxidants are available in different compartments (fig. 2). In the mitochondria, superoxide radicals and hydrogen peroxide can be detoxified by a superoxide dismutase and a catalase, respectively [5].

Besides high concentrations of antioxidants, esp. ascorbate and glutathione, the chloroplasts contain a complex enzymatic detoxification system for superoxide radicals and hydrogen peroxide [4]. It is not clear whether this system or part of it also operate in the cytoplasm [12]. Since many years it is a matter of controversy whether cytoplasmatic glutathione peroxidase, an enzyme converting hydrogen peroxide and organic peroxides via oxidation of reduced glutathione, is present in the cytoplasm of plant cells. A conclusive proof for the existence of this enzyme in plants is still lacking. Therefore, it is not clear whether or not enzymatic detoxification mechanisms for organic peroxides do exist in the cytoplasm or inside plant organelles.

These examples show that in the apoplastic as well as in the symplastic space of plant cells a whole set of defense mechanisms is available for the detoxification of photooxidants. Significant injury to the cells will only be observed, when the capacity of these mechanisms exceeds the influx of an individual photooxidant. Therefore, the following conceptual model for the action of photooxidants (and other air pollutants) in plants can be proposed (fig. 3): The influx of photooxidants into plants will not affect the physiological state of the cells, as long as it does not exceed the capacity of the detoxification mechanisms present constitutively. Increasing influx of photooxidants will result in changes in the physiological steady state of the cells and may lead to an induction of individual defense mechanisms. This induction will enhance the detoxification capacity that may be determined as an increased content of an antioxidant or an increased activity of an enzyme. Superoxide dismutase activity, e.g., is constitutive in spruce needles; the activity of the enzyme decreases with increasing age of the needles in apparently healthy trees, but is maintained at its maximum level independent from needle age in damaged trees [13]. When the influx of photooxidants exceeds the maximum of inducible detoxification capacity and of repair and compensation mechanisms of the cells, injury events will reduce the cellular detoxification capacity.

This model is capable of explaining many discrepancies in the literature from reports on the action of photooxidants on plant metabolism. Under the influence of photooxidants like ozone the concentrations of antioxidants, e.g. ascorbate and glutathione, and the activity of detoxification enzymes are shown to increase in one study and to decrease in the other (cf. 14). The model implies that induction phenomena will enhance, and injury events may reduce antioxidant concentrations or enzyme activities.

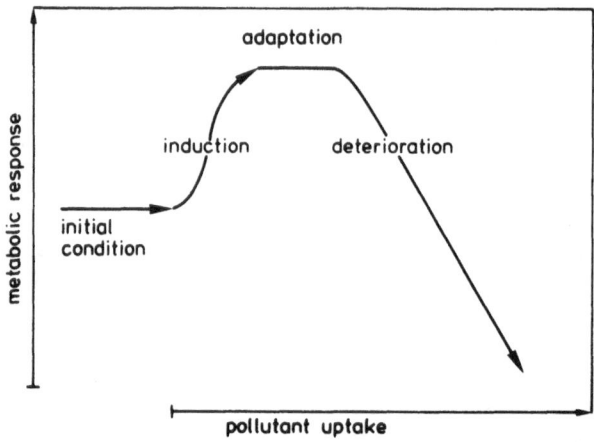

Figure 3 Physiological mechanisms for the action of
air pollutants on plants

To evaluate reports on the action of photooxidants on
plants it is important to know where in the series of
events occurring upon deposition of photooxidants in plants
the actual measurements have been performed. E.g., when the
superoxide dismutase activity of spruce needles was plotted
versus the damage of the trees expressed as the loss of
needles (fig. 4), superoxide dismutase activity initially
increased with increasing loss of needles, but declined in
needles of trees with high losses of needles. As a
consequence, low superoxide dismutase activities were
observed in the needles of healthy trees and the needles of
heavily damaged trees as well.

Connections between physiological parameters in plants and
damage by photooxidants can, therefore, only be elucidated,
if phenomenological measurements are replaced by process-
oriented studies. The observation that a physiological
parameter is enhanced or reduced in the presence of a
photooxidant is without any meaning in an effect-study as
long as the physiological state of the plant at the time of
analysis is unknown.

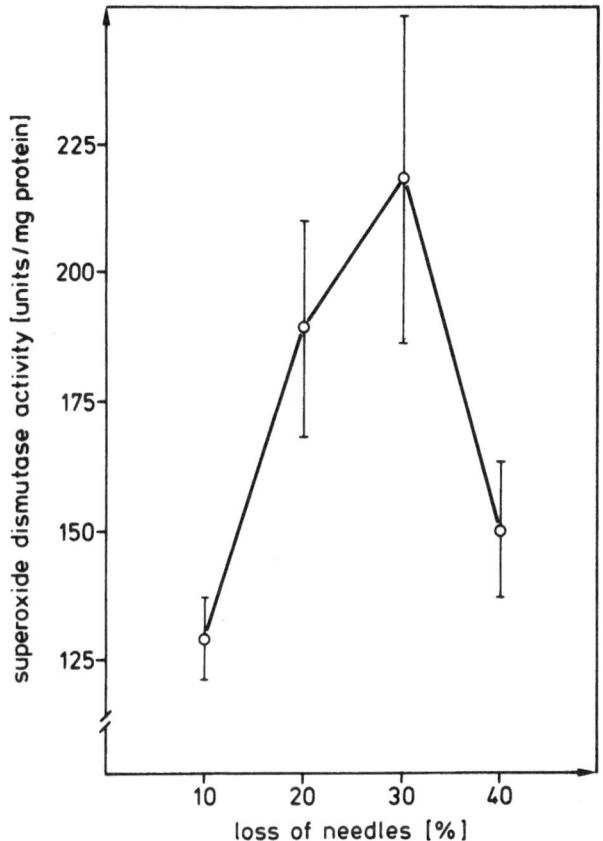

Figure 4   Relationship between superoxide dismutase
            activity and the degree of injury in spruce trees

References

[1]   Hough, A.M. and Derwent, R.G. (1987) 'Computer
       modelling studies of the distribution of
       photochemical ozone production between different
       hydrocarbons', Atmos. Environ. 21, 2015-2033.

[2]   Guderian, R., Tingey, D.T., and Rabe, R. (1985)
       'Effects of photochemical oxidants on plants', in:
       Air Pollution by Photochemical Oxidants, (ed.) R.
       Guderian, pp. 129- 296, Springer Verlag, Berlin.

[3]   Grennfelt P. and Schjoldager, J. (1984) 'Photochemical
       oxidants in the troposphere: a mounting menace',
       AMBIO 13, 61-67.

258

[4]   Halliwell, B. (1984) 'Chloroplast metabolism. The structure and function of chloroplasts in green leaf cells', Clarendon Press, Oxford.

[5]   Elstner, E.F. (1982) 'Oxygen activation and oxygen toxicity', Ann. Rev. Plant Physiol. 33, 73-96.

[6]   Rabinowitsch, H.D. and Fridowich, J. (1983) 'Superoxide radicals, superoxide dismutases and oxygen toxicity in plants', Photochem. Photobiol. 37, 679-690.

[7]   Gaspar, Th., Penel, C., Thorpe, T., and Greppin, H. (1982) 'Peroxidases 1970-1980, A survey of their biochemical and physiological roles in higher plants', Université de Genève, Genève, pp. 73-79

[8]   Castillo, F. and Greppin, H. (1988) 'Extracellular ascorbic acid and enzyme activities related to ascorbic acid metabolism in Sedum album L. leaves after ozone exposure', Environm. Exp. Bot. 28, 231-238

[9]   Rennenberg, H. (1984) 'The fate of excess sulfur in higher plants', Ann. Rev. Plant Physiol. 35, 121-153.

[10] Rennenberg, H. and Thoene, B. (1987) 'The mobility of sulphur in higher plants', Proceedings of the Internat. Symp. Elemental Sulphur in Agriculture Vol. 2, pp. 701- 707, Nice.

[11] Bergmann, L. and Rennenberg, H. (1978) 'Efflux und Produktion von Glutathion in Suspensionskulturen von Nicotiana tabacum', Z. Pflanzenphysiol. 88, 175-185.

[12] Gillham, D.J. and Dodge, A.D. (1986) 'Hydrogen-peroxide-scavenging systems within pea chloroplasts. A quantitative study', Planta 167, pp. 246-251.

[13] Polle, A. Krings, B., and Rennenberg, H. (1989) 'Superoxide dismutase activity in needles of Norwegian spruce trees (Picea abies L.)', Plant Physiol., in press

[14] Smith, I., Rennenberg, H., and Polle, A. (1989) 'Glutathione', in: Stress Responses in Plants, Eds. R. Alscher and J.R. Cumming. A. Liss. Inc, in press

EFFECT OF ACIDIC FOG ON EPICUTICULAR WAX LAYER AND WATER
STATUS OF PICEA ABIES KARST.

A.M.R. HOGREBE and K.MENGEL
Institute of Plant Nutrition
Südanlage 6
D - 6300 Giessen

ABSTRACT.   Experiments carried out with 5-year-old Norway
spruce trees showed that acidic fog (pH 3) had a detrimental
influence on the epistomatal and the epicuticular waxes. The
wax layer of the "pH 3" treated plants was damaged and had a
melted appearance whereas the needle surface of the control
trees that had received a fog of pH 5 showed a dense stand
of small wax threads. Treatment with acidic fog of pH 3
caused a decreasing water holding capacity of these trees
compared with those treated with control fog. The acidic
treatment resulted in higher transpiration rates of the
whole trees under water stress, in higher water loss of
excised twigs, and in lower values of the xylem water
potential during the day compared with the control fog of pH
5. The osmotic potential was not affected by the different
fog treatments. It is suggested that decline symptoms, now
widely found in Central Europa, are mainly caused by
combined effects of acidifying pollutants in the atmosphere
and unfavourable weather, e.g. periods of drought in summer
and winter.

1. INTRODUCTION

The pattern of fog precipitation and the so called "new
type" of forest damages are significantly correlated (Maier
1984). Air pollutants (e.g. SO_2 and NO_x) are dissolved in
fogwater in 10 to 100 times higher concentrations than in
rainwater because of their large relative surface area so
that strong acids are formed, e.g. H_2SO_4 and HNO_3 (Sigg et
al. 1987). Acidic fogs can reach extreme pH values of 1.8
(Sigg et al. 1987). It is known that acids in rain and fogwater cause lesions in needles of coniferous plants
(Huttunen et al. 1985). Magel and Ziegler (1986) and Mengel
et al.(1987) found a disintegration of the epicuticular wax

259

H.-W. Georgii (ed.), Mechanisms and Effects of Pollutant-Transfer into Forests, 259–265.
© 1989 by Kluwer Academic Publishers.

layer of Picea abies that had received a treatment with acidic fog.
It is believed that this damage should have an influence on the water relations of plants. Therefore stomatal and cuticular transpiration as well as the xylem water potential and the osmotic potential of young Norway spruce trees were measured.

2. MATERIALS AND METHODS

2.1. Plant material

Five- and six-year-old seedlings of Picea abies were grown as described by Mengel et al. (1989).

2.2. Treatments

Acidic fog of pH 3 and control fog of pH 5 was applied for a period of 8 weeks (21 May - 14 July, 1987) three times a week for 2 min at 06.00 h in the morning. The fog solutions were prepared from two parts of sulphuric acid and one part of nitric acid. Fog application was brought about by means of nozzles that produced a fine fog. The soil was covered by a plastic foil in order to protect it from the acid.

2.3. Water stress

After this period of fog application both types of treatments were subdivided into two further treatments differing in water supply. One received a full water supply, whereas in the water stress treatment the soil was gradually depleated of water. Hence, the total design comprised 4 treatments:
Acidic fog + full water supply
     "   "   + water stress
Control fog + full water supply
     "   "   + water stress

2.4. Measurements

2.4.1. Transpiration. Transpiration was measured gravimetrically as water loss of trees by means of weighing the pots and the trees each morning at 07.00 h and at 2 h intervals over a one day period. The evaporation loss was measured in separate containers with the same soil and amount of water but no trees. These values were substracted from the overall water loss.

2.4.2. Cuticular transpiration. Detached twigs of the

current year were exposed to the dry atmosphere of a desiccator containing silicagel. The water loss was determined gravimetrically at intervals over 48 h of drying.

2.4.3. Xylem water potential. Small twigs of the current year were cut off and immediately used for the measurement of the xylem water potential with the pressure bomb technique according to Scholander (1965).

2.4.4. Osmotic potential. Green needles of the current year were sampled, put in a small plastic syringe, and immediately frozen in liquid nitrogen. Sap was expressed with a mechanical press and used for the cryoscopical measurement of the osmotic potential by means of a micro-osmometer.

3. RESULTS

3.1. Transpiration (Fig. 1)

The transpiration rates of plants with ample water supply did not differ between the different fog treatments. Under conditions of water stress the transpiration rates were much lower and declined during the period of drought. Moreover, pretreatment with acidic fog (pH 3) resulted in higher transpiration rates compared to the control treatment (pH 5).

3.2. Cuticular transpiration (Fig.2)

Acidic fog of pH 3 caused a higher water loss by cuticular transpiration (less steep curve) compared with the control treatment. Moreover, the steep initial rates of water loss by stomatal transpiration of both treatments were different.

3.3. Xylem water potential (Fig.3)

Twigs that were pretreated with acidic fog of pH 3 had significantly lower water potentials (more negative values) than those that were pretreated with control fog (pH 5).

3.4. Osmotic potential

Osmotic potentials showed no differences between the fog treatments.

3.5. Ultrastructure of the needle surface (Fig.4)

The epicuticular wax layer of the trees treated with acidic

fog of pH 3 showed numerous lesions and had a melted
appearance. This contrasted considerably with the epicuticular wax layer of the control needles, characterized by
a dense stand of small wax threads.

4. CONCLUSION

The results obtained confirmed our assumption that the
damaged epicuticular wax layer of needles caused by application of acidic fog has a detrimental effect on the water
holding capacity of the trees. The clear tendency that under
water stress the trees that had received a pH 3 pretratment
showed higher transpiration rates compared to those pretreated with control fog of pH 5 was also found in the 1988
experiments. In this experiments the differences in transpiration between treatments were all highly significant and
therefore confirmed our earlier findings. For further
discussion in more details see Mengel et al. (1989).
        During the last few years several researchers have
found that the epicuticular wax layer of forest trees is
damaged in areas where the so called "new type" of forest
damage occurs (Parameswaran et al. 1985, Karhu and Huttunen
1986). It can therefore be assumed that the so called "new
type" of forest damage is to a large extend caused by
cuticle damage that leads to water stress, particularly in
dry years. This assumption is in good agreement with the
observations that the decline symptoms are especially
evident in, and after, years of drought (Rehfuess 1985).

5. FIGURES

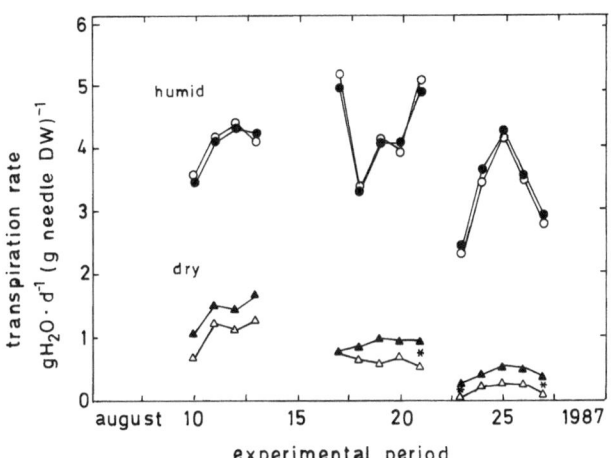

Figure 1. Transpiration rates of Picea abies under humid and dry conditions of the soil after the trees had been pre-treated with acidic fog of pH 3 (closed symbols) or with control fog of pH 5 (open symbols). Each point represents the mean of 12 replicates. *, significant differences between the fog treatments (P < 0,05).

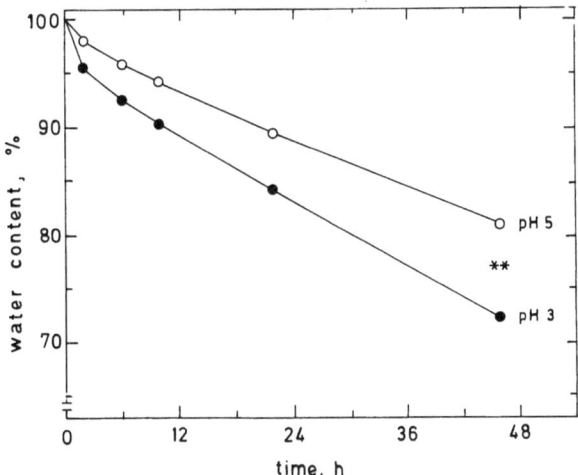

Figure 2. Water content of excised twigs of the current year (Picea abies) under the dry atmospheric conditions of a desiccator over a drying period of 48 h after the twigs were pretreated with acidic fog of pH 3 (closed symbols) and with control fog of pH 5 (open symbols). Each point represents the mean of 10 replicates. **, P < 0,01.

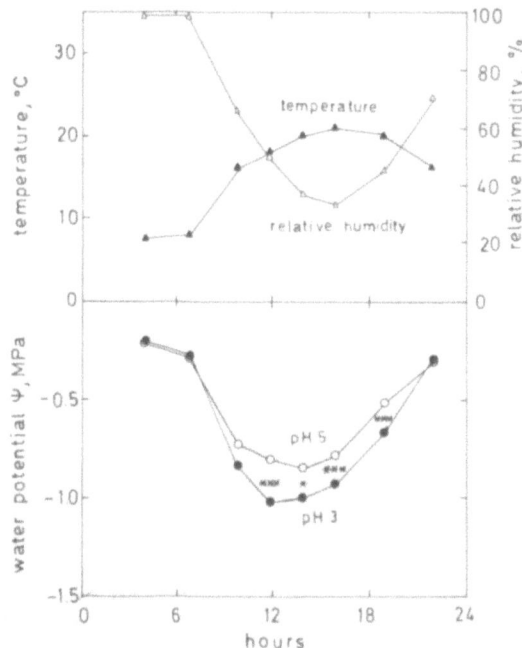

Figure 3. Top: Course of temperature and relative humidity of air during 7 August, 1987. Bottom: Changes in the xylem water potential of Picea abies during 7 August, 1987 under humid conditions of the soil after the trees had been pretreated with acidic fog of pH 3 (closed symbols) and control fog (open symbols). The points at 4, 7 and 22 h represent the mean of 10 replicates, and the other points represent the mean of 5 replicates each. *, P < 0,05, ***, P < 0,001.

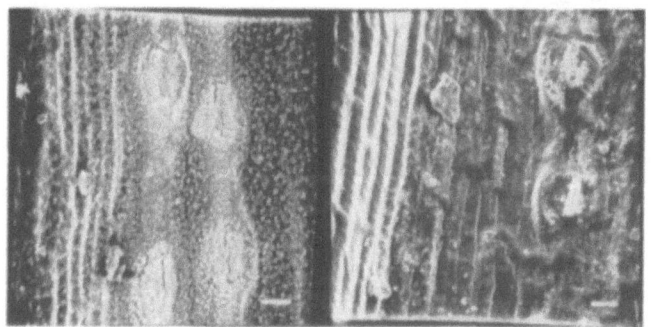

Figure 4. Scanning electron micrographs of the structure of epicuticular waxes in the stomatal area of Norway spruce needles of the current year after application of control (left) and acidic fog (right). Scale bars represent 5 um.

6. REFERENCES

Huttunen, S., Mäkelä, M. and Karhu, M. (1985) "Effects of acidic deposition on needle surfaces", in C.Troyanowsky (ed.), Air Pollution and Plants, Verlag Chemie, Weinheim, pp. 218-221.

Karhu, M. and Huttunen, S. (1986) "Erosion effects of air pollution on needle surfaces", Water Air Soil Pollut. 31, 417-423.

Magel, E. and Ziegler, H. (1986) "Einfluß von O₃ und saurem Nebel auf die Struktur der stomatären Wachspfropfen in den Nadeln von Picea abies (L.) Karst.", Forstwiss. Cbl. 105, 234-238.

Maier, G. (1984) "Aufnahme und Auswertung von Fichtenbeobachtungsflächen in Bayern", Schriftenreihe der Forstwiss. Fak. München und der Bayer. Versuchs- und Forschungsanstalten. Forstl. Forschungsbericht 62.

Mengel, K. Lutz, H.-J. and Breininger, M.-T. (1987) "Auswaschung von Nährstoffen aus jungen intakten Fichten (Picea abies)", Z. Pflanzenern. Bodenk. 150, 61-68

Mengel, K., Hogrebe, A.M.R. and Esch, A. (1989) "Effect of acidic fog on needle surface and water relations of Picea abies", Physiol. Plant. 75, 201-207.

Parameswaran, N., Fink, S. and Liese, W. (1985) "Feinstrukturelle Untersuchungen an Nadeln geschädigter Tannen und Fichten aus Waldschadensgebieten im Schwarzwald", Eur. J. For. Path. 15, 168-182.

Rehfuess, K. (1985) "On causes of decline of Norway spruce Picea abies (L.) Karst. in Central Europe", Soil Use and Management 1, 20-31.

Scholander, P.F., Hammel, H.T., Bradstreet, E.C. and Hemmingsen, E.A. (1965) "Sap pressure in vascular plants", Science 148, 339-346.

Sigg, L., Stumm, W., Zobrist, J. and Zürcher, F. (1987) "The chemistry of fog: factors regulating its composition", Chimia 41, 159-165.

EFFECTS OF H_2O_2 ON SPRUCE NEEDLES
- RESULTS OF OUTDOOR FOG CHAMBER EXPERIMENTS

G. MASUCH, H.G. KICINSKI, A. KETTRUP
Universität Paderborn
Fachbereich Chemie
D - 4790 Paderborn
Federal Republic of Germany

ABSTRACT. Three years old spruce trees (*Picea abies* Karst.) were exposed to acidic fog (pH 4) containing 864 ± 250 ppb H_2O_2 in outdoor fog chambers. We observed decreases in all tissue parameters of current-year needles. Phenols accumulated in the central vacuoles of the mesophyll cells. The amounts of piceatannolglycoside, kaempferol-glycoside, catechin as well as quinic and shikimic acids increased.

1. Introduction

Hydrogen peroxide is expected to be a major contributor to the acidification of cloud and rainwater because it is considered the primary oxidizer of SO_2 to SO_4^{2-} at low pH (<5.0) (Penkett et al. 1979, Calvert et al. 1983).

Dissolution of gaseous H_2O_2 from the air (Schwartz 1984) and solid surface generation of H_2O_2 (Chameides 1984, McElroy 1986) are dominant sources of aqueous phase H_2O_2.

The concentration of H_2O_2 in air is recognized to be relatively low, < 5 ppb (v/v) (Sakugawa et al. 1988). High concentrations of H_2O_2 can occur in the liquid phase, ranging from 850 ppb to 3 ppm at 150-200 m above sea level (Römer et al. 1985). The frequency of fog in elevated areas is about 120 days per year at 60 m and goes up to more than 200 days per year at locations over 800 m (Georgii et al. 1985).

As we have reported formerly (Masuch et al. 1986) H_2O_2 affects serious changes in the internal structure of beech leaves and spruce needles. We have repeated our experiments with acidic fog containing H_2O_2 in outdoor fog chambers under natural climatic conditions, using young spruce trees as test material.

267

H.-W. Georgii (ed.), Mechanisms and Effects of Pollutant-Transfer into Forests, 267–275.
© *1989 by Kluwer Academic Publishers.*

2. Materials and Methods

Experimental outdoor fog chambers were used to nebulize 3 years old spruce trees (*Picea abies* Karst.) with acidic fog (pH 4.0) containing 864 ± 250 ppb H2O2.

Fog containing H_2O_2 was sprayed four days a week from 4 a.m. to 7 a.m. over a period of eight weeks during June and July 1987. During the remaining time, the chambers were opened and the trees were exposed to natural climatic conditions.

Preparation and fixation of spruce needles are included in earlier works (Masuch et al. 1986). Sample preparation for the determination of phenolic acids in needles and the analytical methodology of needle phenols were presented elsewhere (Kicinski et al. 1988, Masuch et al. 1989).

3. Results

Significant decreases in the length of primary twigs (reference series: 70 ± 13 mm, H_2O_2-series: 46 ± 12 mm) and primary needles (reference series: 10 ± 1.7 mm, H_2O_2 series: 8 ± 2 mm) occured after exposure to H_2O_2. The tissue areas of the needle transverse sections are reduced in those needles exposed to H_2O_2 (Figure 1). The highest reduction rates were found in the vascular bundle (-29%), the hypodermis (-28%) and the mesophyll tissue (-17%); the area of the intercellular space also reduced by 22% (Figures 3, 4).The reduction of tissue area is due to a reduced amount of cells per needle. All the tissues of the vascular bundle decreased significantly after exposure to H_2O_2 (Figure 2) because of a reduced cambial activity (Figures 5 and 6). The number of stomata per surface area increased by 13%

Current-year needles polluted by H_2O_2 showed remarkable differences in the structural waxes, occluding the stomatal antechamber. In the reference group wax rods cover the guard cells and the outer vestibule; the epicuticular wax above the stomata forms a thee-dimensional network of anastomosing wax rodlets (Figure 7). After exposure to H_2O_2 cracks were built up in wax plugs and wax rods were fused together (Figure 8).

Mesophyll cells of spruce needles usually contain variable amounts of phenolic droplets in the central vacuoles. In the reference series, nebulized with acidic fog (pH 4.0) only, some large droplets can be found (Figure 9).

Current-year needles, treated with acidic fog containing H_2O_2, are closely filled up with small phenolic droplets knotted together like strings of pearls (Figure 10).

After exposure to acidic fog containing H_2O_2 the amounts of piceatannolglucoside, kaempferolglucoside, catechin, quinic and shikimic acids increased (Table 1), but the total amounts of quantified phenols did not change significantly as detected from methanolic extracts by an HPLC system with a photodiode-array detector.

Table 1. Amount of phenolic compounds (μg/g dry weight) in primary Norway spruce needles.

	Reference	864 ppb H_2O_2
Piceatannolglucoside	500	1.150
Kaempferol-glucoside	200	1.000
Catechin	5.000	9.000
Quinic acid	3.000	4.500
Shikimic acid	17.500	29.000

A summary of further results is given in Table 2.

Table 2. Summary of results for spruce needles; changes in tissues exposed to acidic fog containig 864 ± 250 ppb H_2O_2. Significant ($p<0.05$) decrease denoted by - ; high significant ($p<0.01$) decrease denoted by --.

length:		volumes:	
needle	-	total needle	--
primary twig	--	mesophyll cells	--
ray axis	--	intercellular space	--
cambium axis	-		
		cell number per transverse section:	
tissue areas:		mesophyll	--
total area	--	epidermis	--
epidermis	--	endodermis	-
hypodermis	--		
endodermis	-	cell rows:	
vascular bundle	--	cambium	--
intercellular space	-	xylem	--
mesophyll cell area	-	phloem	-
		cell numbers per longest cell row:	
tissue area of vascular bundle:		cambium	--
xylem	--	xylem	-
phloem	--	phloem	--
cambium	--		
strasburger cells	-	surface areas:	
sclerenchyma	--	internal	--
transfusion tissue	--	external	--

4. Discussion

Reductions in leaf tissues, nebulized with H_2O_2-containing acidic fog, we also observed in growing beech leaves (Masuch et al. 1985) and in leaves of agricultural grasses (Masuch, unpublished results). Therefore we can conclude that tissue reduction may be a main effect of H_2O_2-pollution.

As illustrated by SEM-pictures phenolic material increases in the vacuoles of the mesophyll cells, as a function of H_2O_2-pollution. As we did not find increased amounts of total phenols in methanolic extracts in the H_2O_2-series, we conclude that a special amount of phenols is not soluble in methanol. The function of accumulated piceatannolglucoside, kaempferol-glucoside, catechin and quinic and shikimic acids is not known until now. It is discussed that piceatannolglucoside may affect the osmotic system negatively (Harborne 1987). Production of phenols may increase the plant's resistance to transpiration (Fahn 1982).

The increased density of stomates is a symptom of water stress. H_2O_2 is considered to be one cause of degradation in epicuticular waxes (Elstner et al. 1984). Fused epicuticular wax and cracks in wax plugs may be causes for uncontrolled diffusion of water vapours (Huttunen et al. 1983). The evaporation rate of spruce needles exposed to H_2O_2-containing acidic mist is reduced (Dülme 1985). This may be a result of accumulated phenols in the central vacuoles. Phenols are able to bind water because of the periphere OH-groups. Histological and metabolic findings point to a disturbed water balance as a consequence of exposure to H_2O_2. Our findings make it obvious that acidic fog containig H_2O_2 is an important factor in damaging forest trees.

5. Acknoledgements

We are indebted to Mr J.T. Franz for SEM assistance and to J. Slanina and R.K.A.M. Mallant, ECN, Petten, The Netherlands, for expert help in creating fog and analyzing H_2O_2.

6. References

Calvert, J.G. and Stockwell, W.R. (1983) 'Acid generation in the troposphere by gas-phase chemistry', Environ. Sci. Technol. 17, 428A-443A.
Chameides, W.L. (1984) 'The photochemistry of a remote marine stratiform cloud', J. Geophy. Res. 89, 4739-4755.
Dülme, W. (1985) 'Wasserstoffperoxid als Luftverunreinigung - Analytik und Wirkungsuntersuchung', Diploma Thesis, Universität Paderborn.

Elstner, E.F. and Oβwald, W. (1984) 'Fichtensterben in "Reinluftgebieten": Strukturresistenzverlust', Naturw. Rundschau 37, 52-61.

Fahn, A. (1982) 'Plant Anatomy', Pergamon Press, Oxford, p. 42.

Georgii, H.W. and Schmitt, G. (1985) 'Methoden und Ergebnisse der Nebelanalyse', Staub-Reinhalt. Luft 45, 260-264.

Harborne, J.H. (1987) 'Biochemistry of Phenolic Compounds', Academic Press, London, p. 129-166.

Huttunen, S. and Laine, K. (1983) 'Effects of air-borne pollutants on the surface wax structure of *Pinus sylvestris* needles', Ann. Bot. Fenn. 20, 79-86.

Kicinski, H.G., Kettrup, A., Boos, K.-S. and Masuch, G. (1988) 'Single and combined effects of continuos and discontinuos O_3 and SO_2 immission on Norway spruce needles. II. Metaboloc changes', Intern. J. Environ. Anal. Chem. 32, 213-241.

Masuch, G., and Kettrup, A. (1985) 'Wirkungen von wasserstoffperoxidhaltigem sauren Nebel auf die Laubblätter junger Buchen (*Fagus sylvatica* L.)', VDI-Berichte 560, 761-776.

Masuch, G. Kettrup,A., Mallant, R.K.A.M. and Slanina, J. (1986) 'Effects of H_2O_2-containing acidic fog on young trees', Intern. J. Environ. Anal. Chem. 27, 183-213.

Masuch, G., Kicinski, H.G., Dülme, W. and Kettrup, A. (1989) 'Hydrogen peroxide dissolved in acidic fog as air pollutant - effects on spruce needles', Intern. J. Environ. Anal. Chem. 33, in press.

McElroy,W.J. (1986) 'Sources of hydrogen peroxide in cloudwater', Atmos. Environ. 20, 427-438.

Penkett, S.A., Jones, B.M.R., Brice, K.A., and Eggleton, A.E.J. (1979) 'The importance of atmospheric ozone and hydrogen peroxide in oxidizing sulphur dioxide in cloud and rainwater', Atmos. Environ. 13, 123-127.

Römer, F.G., Viljeer, J.W., van den Beld, L., Slangewal, H.J., Veldkamp, A.A. and Reijnders, H.F.R. (1985) 'The chemical copmpositions of cloud and rainwater. Results of preliminary measurements from an aircraft', Atmos. Environ. 19, 1847-1858.

Sakugawa, H. and Kaplan, J.R. (1988) 'Factors controlling the formation of gaseous H_2O_2 and O_3 and their control on rate of SO_2 oxidation to sulfate in Los Angeles air', J. Geophys. Res.

Schwartz, S.E. (1984) 'Gas- and aqueous-phase chemistry of H_2O_2 in liquid-water clouds', J. Geophys. Res. 89, 11589-11598.

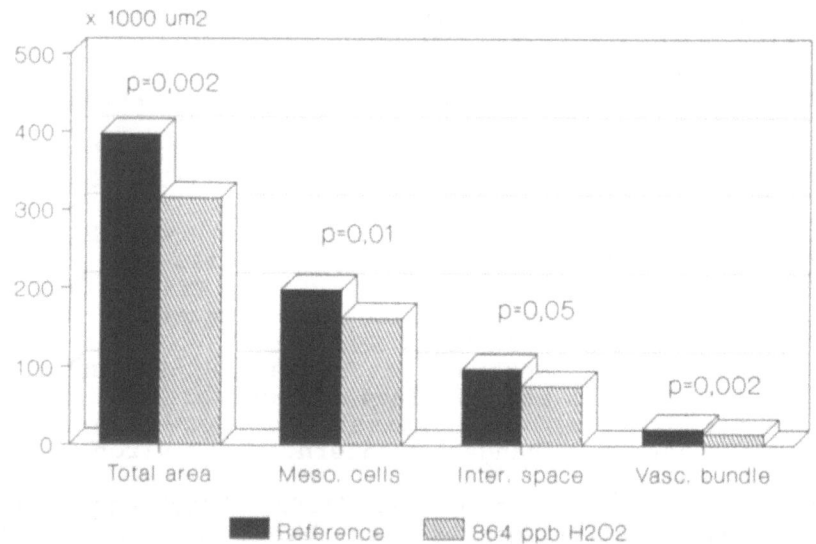

Figure 1. Areas of transverse section of primary spruce needles.

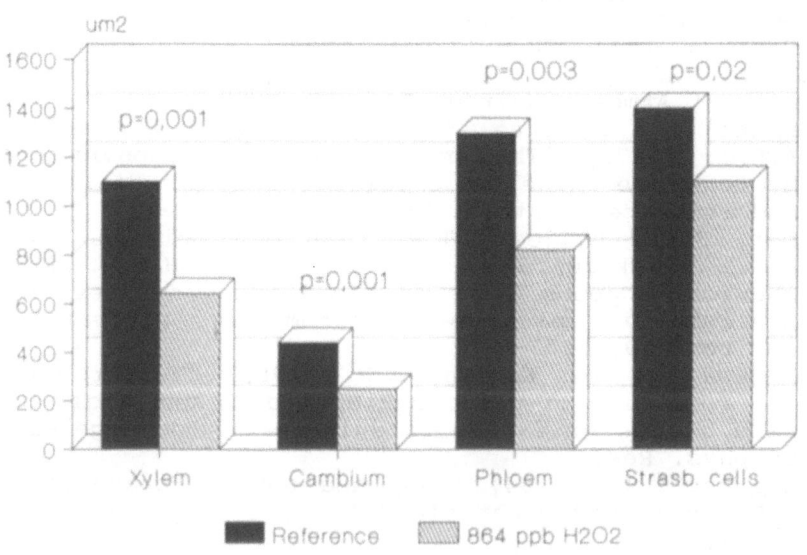

Figure 2. Tissue areas of transverse sections of the vascular bundle of primary Norway spruce needles.

Figures 3, 4. Transverse section of primary needles of Norway spruce taken from the reference series (Fig. 3) and the experimental series exposed to acidic fog containing H_2O_2 (Fig. 4). Bar = 50 μm.

Figures 5, 6. Transverse section of vascular bundles taken fom Norway spruce primary needles of the reference series (Fig. 5) and the H_2O_2-series (Fig. 6). Bar = 50 μm.

Figures 7, 8. Stomata of primary spruce needles with a three-dimensional network of crystalline wax from the reference series (Fig. 7) and with melted wax covering the stomata after exposure to acidic fog (pH 4) containing 864 ± 250 ppb H_2O_2 (Fig. 8). Bar = 5 μm.

Figures 9, 10. Mesophyll cells of Norway spruce primary needles. In the reference series packed chloroplasts can be seen in the background and some phenolic droplets in the central vacuoles (Fig. 9). Mesophyll cells of the experimental series exposed to acidic fog (pH 4) containing 864 ± 250 ppb H2O2 contain densly located small phenolic droplets filling the central vacuole (Fig. 10). Bar = 5 μm.

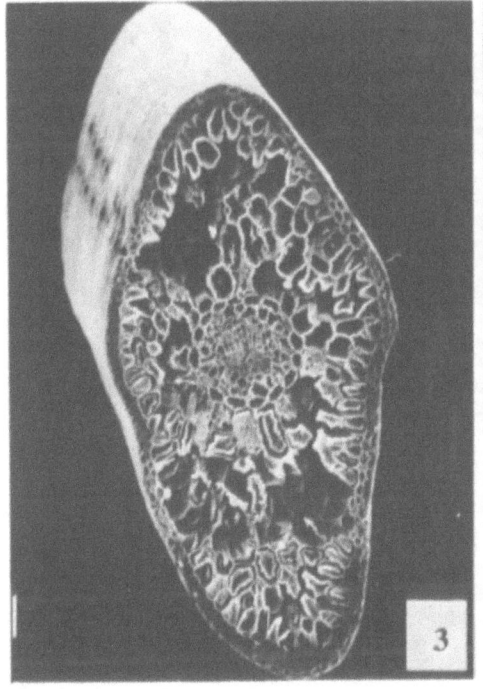 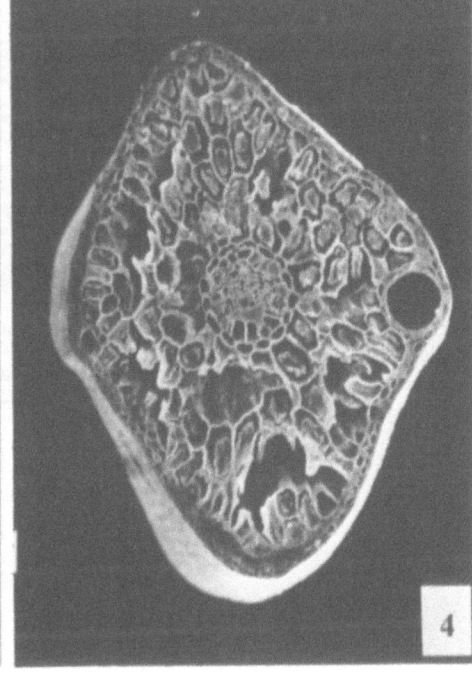

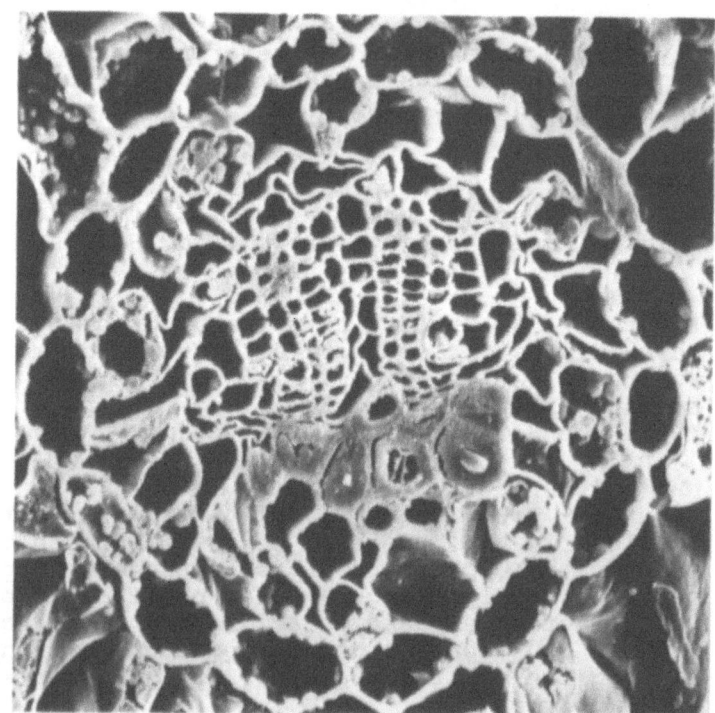

5

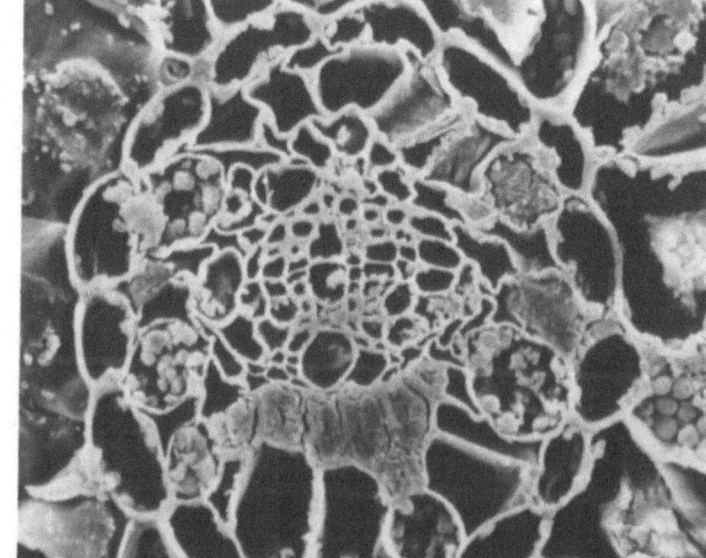

6

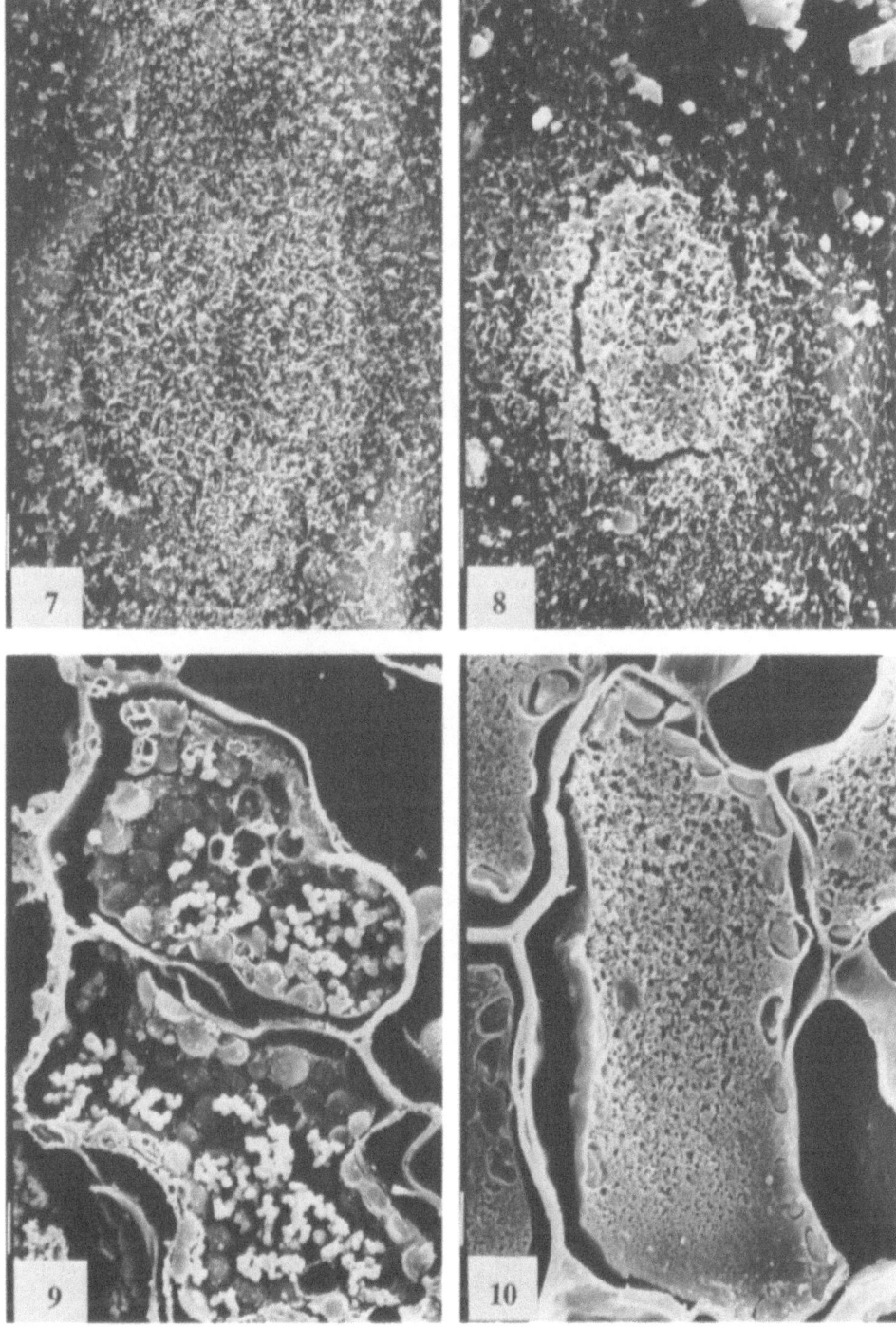

INTERACTIONS BETWEEN ROOT AND CROWN IN 90-YEAR OLD SPRUCE UNDER
DOCUMENTED CONDITIONS OF DEPOSITION

J. M. EICHHORN
Hessian Forest Research Station
Prof. Oelkersstr. 6
3510 Hann. Münden
Federal Republik of Germany

ABSTRACT. Relations between soil chemistry, fine roots and needle loss
are investigated on 90-years old Norway spruce trees at one of the main
forest decline measuring stations in Hesse (Kaufunger Wald, 550 m asl.).
Acid deposition causes at this site considerable soil acidity as well as
wash-out of nutrients. Under these conditions spruce are developing a
flat fine root system. The relative root tip frequency shows differences
between damaged and undamaged trees. Even in relation to crown surface
undamaged trees have far more root tips than damaged ones. This causes a
high sensitivity of damaged trees towards waterstress. Furthermore the
carbohydrate provision of fine roots is on the surveyed area not endangered by needle loss.

1. INTRODUCTION

Surveys on the effect of atmospheric pollution under open-field conditions can be supported considerably by a precise recording of the atmos-
pheric-chemical situation. Regular data on meteorology and airpollution
(HLfU-Bericht, 1989) have been available to the Main Measuring Station
Witzenhausen in the Kaufunger Forest since 1982. They are supported by
calculations of the flux ballance which are based mainly on both analysis of open-field and stand precipitation, and on surveys of seepage-
and creekwater. The work that has been undertaken as part of the Hessian
Research Programme ("Forest Pollution by Emmissions"; Gärtner et al.,
1989) offers a particularly good opportunity for researching the effects
of air pollutants on forest trees.
          Since the beginning of this century, the deposition of anthro-
pogenic air pollutants has caused far-reaching changes in the chemical
soil conditions on many Northern German and Hessian sites (Ulrich, 1989).
A cause for this is often found in the crass disparity between a high
acid input under spruce (3 kmol - 6 kmol/ha and yr), and the comparatively low buffering capability of typical forest sites.
          Representative surveys under spruce aged 60 - 80 years showed
pH values (KCl) of usually between 2.7 and 3.3 in the top soil (Gärtner

277

H.-W. Georgii (ed.), Mechanisms and Effects of Pollutant-Transfer into Forests, 277–284.
© *1989 by Kluwer Academic Publishers.*

minerals become soluble. This indicates a disintegration of the clay minerals.

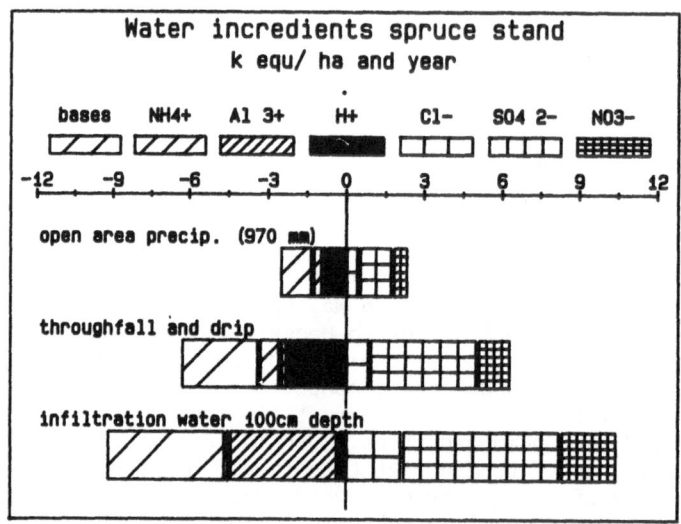

Figure 1. Water incredience in a 90-year old spruce stand of the Main Measuring Station Witzenhausen; each is the mean result of fortnightly samplings between 1984 and 1987 (according to BALAZS, 1989).

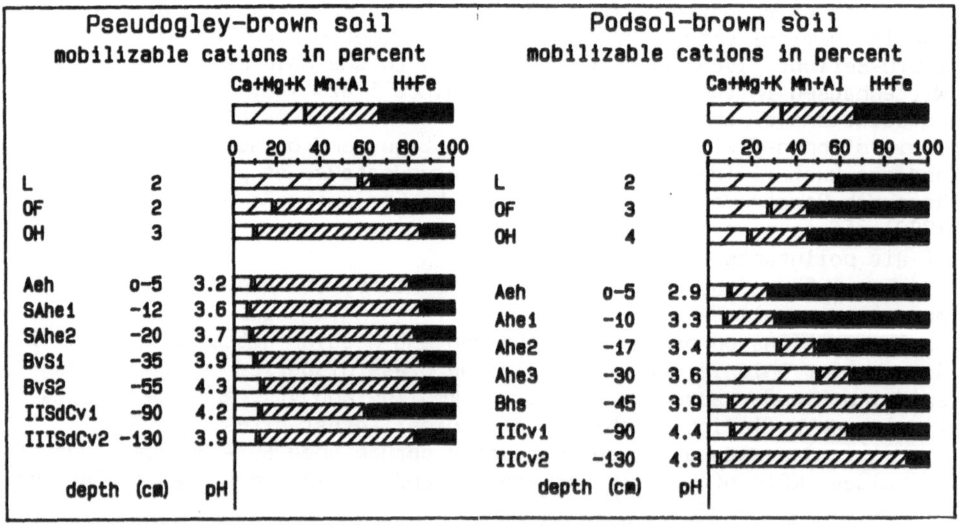

Figure 2. Shares of exchangeable cations in the effective exchange capacity. NH_4Cl-percolation

et al., 1989). This degree of acidification is evident for the extensive leaching of the nutrients calcium and magnesium. However, even at increased soil depth, there is no considerable increase in base saturation in many Hessian spruce stands at first. The stands sampled by Ulrich (1989) show that the acidification front on which a rapid increase in base saturation due to buffered neutral salts (e.g. $CaSO_4$) can be noticed lies on new red sandstone sites usually deeper than 3 meters. From this it follows that in the actual main root space the soluble Al-ions dominate in the soil solution. Thus, the interaction between soil condition and root development is of great significance when examining the physiology of trees subjected to pollution by emission.

Within the framework of acid soil conditions, the question of whether correlations can be found between root damage and needle loss on 90-year old spruce is now discussed below (Eichhorn, 1987).

2. MATERIAL AND METHOD

2.1 Survey area

The forest area observed is positioned in the compartment 83 A of the Hessian Forest District Witzenhausen at an altitude above sea level of 550 m. The general situation of the site results from the area belonging to the lower beech zone and being of moderate sub-atlantic climate moisture. Raw material of the soil formation is medium new red sandstone. On this developed in the southern part of the area a podsol brown soil type and in the northern part a pseudogley brown soil type. Regarding deposition and chemical characterisation, refer to subchapter 3.1.

2.2 Sampling

The characteristics of root and crown of 17 single trees were observed. The sampled trees were chosen according to the degree of damage to their crowns and to both sub sites.

The small-scale variablitity of the rooting of a forest soil makes it difficult to trace back the roots to each corresponding tree. Prior to the actual start of the survey, therefore, the development of the fine roots was assessed in different distances to the stand. In the surveyed site the most intensive fine root density is found at about 2 m distance from the butt end. At a distance of 6 m and 8 m, however, fine roots could be found only sporadically. the same pattern of distribution could be observed also with trees grouped closer together. Following this preliminary investigation, the roots can usually be traced back to the respective adjacent tree. The survey is thus based upon the sampling of each single tree and its respective roots. However, when choosing the sample trees, careful consideration was always given to the fact that they belonged to groups of either undamaged or damaged trees on a small-scale homogenous site.

The borings of the soil were done in a radial distance of 2 m around each spruce with a special soil gimlet (8 cm interior diameter).

The number of samples per date results from 17 trees, 10 bore holes per tree, and 12 depth gradiations per soil column. Samples up to a depth of 48 cm were taken. The results originate from three samplings in the autumn and one in May of consecutive years.

2.3. Parameters of Roots and Crown

The utilization of the samples took place following treatment in the "Autoclav" (vacuum treatment) and gentle rinsing under the "Binocular".
Live (vital) and dead (subvital) fine roots could firstly be distinguished as characteristic features. Under the "Binocular", theses differ in particular with regard to the oxidative state of the primary bark, its tearing strength and the existence of active root tips.
Active mycorrhiza are light meristems, that is light root tips with turgescent barkcells and a closed fungus mantle.
The starch determination was made using the enzymatic UV starch test by BOEHRINGER-Mannheim on fine roots which were frozen immediately in liquid nitrogen. Thereafter, the starch was determined photometrically using an EPPENDORF-photometer.
A quantitative measure of the size of tree crown is the surface of crown mantle (Kramer and Dong, 1985). For this feature, close correlations with the growth behaviour of spruce could be detected. A cone model of the trees, based on the basal area, is used for this purpose and its mantle area is calculated.
The definition of the degrees of crown damage corresponds to the guidelines for the annually recurring Forest Decline Survey agreed between the Federal States and the Ministry of Food, Agriculture and Forestry.

2.4. Measurements of Deposition

The recording of depositions is carried out using bulk precipitation samplers "Münden" which have a catchment area of 100 cm². Twenty samplers were used and analysed on the site in fortnightly rotation periods. Vacuum suction plugs were used to obtain samples of soil solution. Results from a soil depth of 100 cm are presented. The method developed by Ulrich was used for the flux balance (Ulrich, 1979).

3. RESULTS

3.1. Pollution of Stand by Emission

Analyses of the water nutrients found on the site show that, compared to open-field precipitation, site precipitation is clearly accumulated with nutrients (e.g. accumulation factor protons = 3.1; see fig. 1). When seeping through the top soil at a depth of up to 100 cm, the H+ ions of the site precipitation are buffered by aluminium. This indicates that under the given conditions the acid output of the ecosystem occurs mainly in the form of Al-ions. Hereby the aluminium compounds of clay

Soil analyses from the surveyed site indicate that the Al/Fe buffer range predominates. In particular the covering of the exchanger with magnesium and calcium is very low. Looking at Table 2, however, it emerges that there are also considerable differences between the two sub sites. At pH values in KCl of around 2.9, protons predominate in the composition of exchangeable cations in the top soil of podsol brown soil type. This points to a Fe buffer range.

In the top soil of pseudogley brown soil type, however, the domination of Al-ions and thus of the Al buffer range becomes already evident.

3.2. Fineroot Biomass

A distinct depth gradient of the fine root appeared in all spruce. The live (vital) fine roots are always particularly found in the upper 12 cm of the soil, that ist in the humus cover and in the humic top soil. In podsol brown soil type, 62 % of the total fine root biomass can be found in up to 32 cm depth in this layer, while on pseudogley brown soil type this is 55 %. Neither undamaged nor damaged spruce produce descending roots.

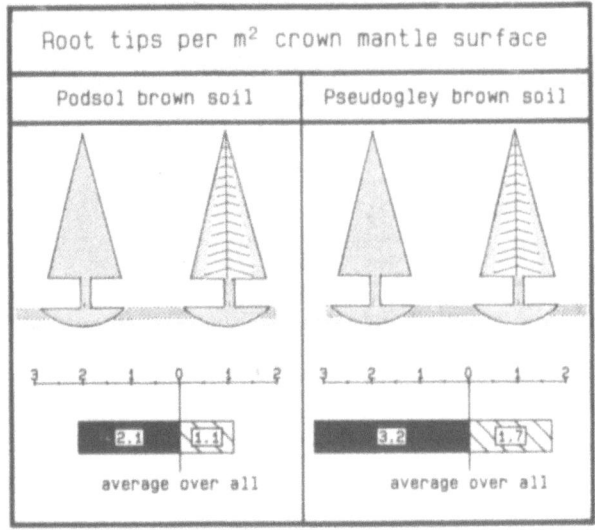

Figure 3. Root tip frequency in relation to crown mantle surface. Undamaged spruce trees have on both soil types far more root tips per crown mantle surface than damaged ones.

3.3. Mycorrhization

Undamaged spruce usually have a distinctly higher relative root tip frequency than damaged spruce (Fig. 3). This can be observed on both sub sites. In contrast, only slight differences can be detected between podsol and pseudogley brown soil type regarding the relative degree of mycorrhization.

For the interaction between root and crown particularly the root tip frequency in relation to each crown mantle area is important. Undamaged and damaged spruce differ particularly distinctly. Healthier trees possess considerably more root tips per unit crown mantle area than the check trees.

Using analyses of variance, it was examined whether connections exist between crown parameters as well as chemical element contents in needles, roots and soil and the relative root tip frequency. It emerged that the frequency of appearence of root tips on live (vital) fine roots happens in a positive context with the provision of the soil and roots with calcium and magnesium; spruce with a higher content of these bases usually have a higher relative root tip frequency. In contrast, there is a distinctly negative correlation between root tip frequency and Al content.

3.4. Starch Content

In the autumn of 1985, after completion of the vegetative season, samples were taken three times at weekly intervals in order to determine the starch content in live (vital) fine roots. Here distinctly higher starch contents were found in the fine roots of undamaged trees than in those of damaged trees. The higher the starch content of live (vital) fine roots, the lower the molar Ca/Al ratio (Fig. 4) of the soil.

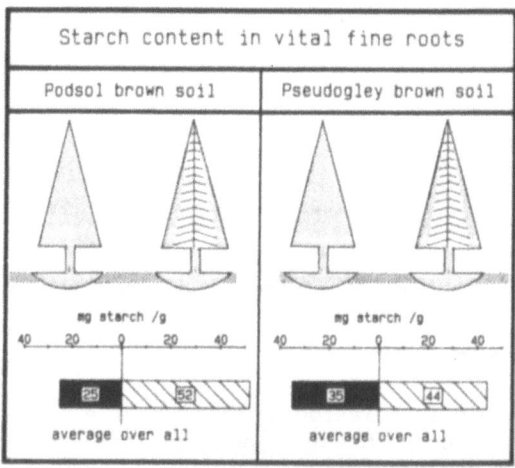

Figure 4. Starch content in live (vital) fine roots. Damaged spruce trees have on both sites higher starch contens than undamaged ones.

4. DISCUSSION

Basically, it can be assumed that the soil acidification has been in existence for a long time reached all trees in the physiology of their roots. This becomes particularly evident when observing how the roots run only close to the surface, whereas the physics of the soil would allow roots to penetrate considerably deeper. In Witzenhausen spruce have approximately three times more fine roots in the humic top soil than they do in the sub soil. The cause for such flatly developed root systems on physically deep sites is defined by Ulrich (1989) as a more favourable chemical level in the top soil. Available nutrients are more common in the top soil than they are in the mineral soil, due to litter decomposition. The aluminium which is harmful to plants becomes more significant in the humus-free mineral soil, whereas ist is found complexly-bound to organic matter in humus-rich layers.

If, as in the previous case, the roots of spruce are mainly limited to the top 12 cm of the soil, then these trees are subjected far more to risks from extreme temperatures and frost (Hüttermann and Godbold, 1985). But there are especially fluctuation in precipitation which effect the water budget of these trees more.

On the surveyed site, crown thinning dominates as a symptom of damage. With regards to the mechanism of needle loss, Gruber (1985) documents that the seperating zone which controls the loss is formed already six weeks after shoot-formation of the needles. In the case of high water stress, the structure of the seperating zone starts off the seperation mechanism. According to Gruber, in Witzenhausen the predominant percentages of needles is lost at the zone of seperation.

Turgescent root tips are active absorbing roots and as such are crucial for the uptake of water and nutrients. The number of root tips is, in relation to the tree's crown mantle surface, an important indication for the ratio between water-absorbing roots and transpiring needles. The distinct differences undamaged and damaged spruce with regards to this feature show that probably the waterbudget of the spruce is of basic importance in the formation of crown thinning.

Furthermore, these results do not prove that the provision of carbohydrates is restricted by the loss of older needles. Rather, the starch content of the live (vital) fine roots was, after completion of the vegetative season, and on both sub sites respectively, distinctly higher in those spruce with considerable needle loss than in the densely-foliaged check trees. The surveys in Witzenhausen document further an increased starch content in fine roots under unfavourable chemical soil conditions. The molar Ca/Al ratio is an indicator for this. The lower this value in the top soil, the more starch is accumulated in the fine roots. It can therefore be assumed that unfavourable chemical soil conditions are a major cause for an increased carbohydrate requirement of the roots.

From these experiments it can be concluded that on the surveyed area an unfavourable soil condition reduces the depth of root of spruce, lowers fine root biomass and fine root tip frequency. This leads to an increased sensitivity of spruce to waterstress. In contrast, the carbohydrate provision of fine roots is on the surveyed area not endangered by needle loss. Rather, the soil condition seems to trigger "sink" - characteristics for carbohydrates in the root.

5. ACKNOWLEDGEMETS

My particular thanks to the leader of the Hessian Forest Research Station Prof. Dr. Gärtner and the leader of the Institute for Forest Biology, University of Göttingen Prof. Dr. Hüttermann for their support of my work. Dr. Balaźs of the Hessian Forest Research Station kindly let me have data regarding the flux ballance of the surveyed stand.

6. REFERENCES

Balaźs, A. (1989) Immissionsbelastung des Waldes und seiner Böden - Gefahr für die Gewässer, Säurebilanz eines Fichtenbestandes im Hessischen Forstamt Witzenhausen, Brechtel, DVWK-Bericht zur 4. Wiss. Tagung "Hydrologie und Wasserwirtschaft", in press.

Eichhorn, J. (1987) Vergleichende Untersuchungen von Feinwurzelsystemen bei unterschiedlich geschädigten Altfichten (Picea abies KARST.), Diss., Forschungsber. der Hess. Forstl. Versuchsanstalt, 3, 179 p.

Gärtner, E. J., Balaźs, A. and Eichhorn, J. (1989) Waldschäden und Bodenschutz, Forst- u. Holzwirt, 1, 3-5.

Gruber, F. (1985) Morphologische Abweichung bei der Entwicklung der Triebe von Fichte, Ber. Forsch. Zentr. Waldökosysteme, Exk. Führer pp. 14-20.

HLfU-Bericht (1989) Waldbelastungen durch Immissionen, Forschungsbericht der Hess. Landesanstalt f. Umwelt 4. Zwischenber. in press.

Hüttermann, A, and Godbold, D. L. (1985) Was wir über den Wald wissen, Nießlein, Voss, Pflanzenphysiologische Mechanismen beim Waldsterben, Köln, pp. 190-196.

Kottke, I. and Agerer, R. (1983) Untersuchungen zur Bedeutung der Mykorrhiza in älteren Laub- und Nadelwaldbeständen des Südwestdeutschen Keuperberglandes, Mitt. d. Ver. f. Forstl. Standortskunde und Forstpflanzenzüchtung 30, pp. 30-39.

Kramer, H. and Dong, P. H. (1985) Kronenanalyse für Zuwachsuntersuchungen in immissionsgeschädigten Nadelholzbeständen, Forst u. Holzwirt, pp. 115-118.

Ulrich, B., Mayer, R. and Khanna, D. (1979) Deposition von Luftverunreinigungen und ihre Auswirkungen in Waldökosystemen im Solling, Schriftenr. Forstl. Fakultät Göttingen, 291 p.

Ulrich, B. (1989) Effects of acid Deposition on Forest Ecosystems in Europe, Adv. Environm. Sci. (Springer), 4: in press.

DIFFERENT CATION EXCHANGE IN THE CANOPY OF TWO NEIGHBOURING BEECH STANDS
WITH REGARD TO SOIL CHEMISTRY

N. Neikes, M. Kazda and R. Wittig
Institut für ökologische Pflanzphysiologie
und Geobotanik, Abt. Geobotanik
Universitätsstrasse 1
D–4000 Düsseldorf 1

ABSTRACT. Rates of proton buffering and cation leaching in the canopy
were determined in two neighbouring beech stands (Fagus sylvatica L.), a
Luzulo–Fagetum on sandstone and a Melico–Fagetum on limestone. Nearly
the same buffering was found in both stands. Particular cations,
however, showed different rates of foliar leaching. Higher amounts of K
and Mn were found in the throughfall of the Luzulo–Fagetum, whereas Ca
came up to higher rates in the Melico–Fagetum. These differences
concerning the turnover of cations in the canopies were related to the
nutrient content of the leaves and to the mineral element supply from
the soil on both sites.

1. INTRODUCTION

Acid deposition affects the element cycling in forest ecosystems in
several ways (Johnson et al. 1985). The processes of H^+-buffering and
cation exchange in the canopy may alter the nutrient status of the trees
(Ulrich 1983a, Cronan and Reiners 1983, Matzner and Ulrich 1984). These
proposed changes are discussed to be dependent on atmospheric load of H^+
as well as on site characteristics (Ulrich 1983b, Matzner 1988). In
artificial acidification experiments, however, low pH–values of the rain
did not decrease element contents of cationic nutrients (Kelly and
Strickland 1986, Mengel et al. 1987).
    Comparing nutrient contents in foliage from a beech stand on
limestone and on sandstone, respectively, relations between buffering
capacity and cation supply were found (Waraghai and Beese 1985). There
are also indications for increased buffering capacity in the canopy of
beech forests on limestone compared with other stands on acid soil in
north–west Germany (Ulrich 1983b, Gehrmann 1987). Since different geographical regions exhibit also different pollutants load, canopy buffering reflects not only the nutrient supply from the soil but also the
different composition of atmospheric deposition.
    In order to eliminate the influence of different deposition rates,
two neighbouring beech forests were selected for this investigation.

285

H.-W. Georgii (ed.), Mechanisms and Effects of Pollutant-Transfer into Forests, 285–294.

Since these two stands show similar characteristics except for the soil, differences concerning the cation exchange in the canopy could be explained by soil chemistry and mineral nutrition of the trees.

2. METHODS

The investigated sites are located at the west slope of the Egge mountains (FRG, 8°95'E, 51°42'N) exposed to winds from the Westphalian Bight and the Ruhr Area. Both stands were selected from a large mature beech forest of about 20 ha. Depending on soil conditions and refering to the different herb layers the forest communities can be described as a Luzulo-Fagetum (LF) and a Melico-Fagetum (MF). The main characteristics of these two stands are given in table 1 and 2.

TABLE 1. Main characteristics and soil conditions of the two compared beech stands (Wagner-Hucke 1986)

forest community	LUZULO-FAGETUM	MELICO-FAGETUM
elevation inclination exposition	355 m 8° NE	360 m 5° NE
age height stem diameter	108 years 25 m 33 cm	108 years 30 m 42 cm
parent material soil (FAO-classi- fication) humus	loess over sandstone cambisol 'Braunerde' mould	loess over limestone luvisol 'Parabraunerde' mull
rooting depth	about 50 cm	about 80 cm

In each stand the throughfall (TF) was collected weekly with eight "Münden 100"-bulk precipitation gauges (sampling area: 100 cm^2 each, Brechtel et al. 1986). The volume of the stemflow (ST) from four trees (about 200 m^2 canopy projection area) was determined and aliquot samples for chemical analysis were taken. The precipitation in the open (PD) was collected with four gauges, situated 200 m away from the stands. Soil solution was obtained in each stand by ten tension lysimeters (P 80 ceramic-cups) at continuing low pressure between -50 and -60 kPa. They were installed in depths between 60-80 cm (LF) and 80-90 cm (MF) in the lower C soil horizon.

The samples were analyzed for H^+ (glass electrode), NH_4, Na, K, SO_4, Cl, NO_3 (ion chromatography) Ca, Mg, Mn, Fe, Al (atomic absorption spectrophotometry). The HCO_3 in the soil solution from the beech stand on limestone (MF) was determined titrimetrically. In early September 1988, samples of the foliage were taken from the crown of each stand and

TABLE 2. NH$_4$Cl–exchangeable cation content (mg.kg^{-1} dry soil) and pH (H$_2$O) of the soils (Wagner-Hucke 1986). (* Due to the higher amounts of CaCO$_3$ in these horizons NH$_4$Cl–extraction may not reflect the actual exchangeable cation contents)

depth	LUZULO–FAGETUM						
	Ca	Mg	K	Mn	Fe	Al	pH
0 – 5 cm	170	17	66	16	53.0	295	4.2
5 – 10 cm	68	8	39	24	5.0	279	4.3
25 – 30 cm	30	3	21	13	0.6	251	4.5
45 – 50 cm	20	2	14	4	0.7	134	4.5
65 – 70 cm	50	6	31	4	0.9	345	4.5

depth	MELICO–FAGETUM						
	Ca	Mg	K	Mn	Fe	Al	pH
0 – 5 cm	1020	41	116	49	2.1	238	4.9
5 – 10 cm	455	10	68	13	1.2	324	4.9
25 – 30 cm	100	4	54	7	0.7	426	4.6
45 – 50 cm	230	8	87	9	0.1	721	4.6
65 – 70 cm*	11600	83	189	20	11.0	687	5.0
95 –100 cm*	4200	102	168	0	0.0	1	7.7

analyzed after HNO$_3$–digestion for cationic nutrients by atomic absorption spectrophotometry.

The annual rates of interception deposition (ID) and total deposition (TD = PD + ID) were calculated by using the particulate interception deposition of Na as reference. The rates of foliar leaching and cation exchange in the canopy as well as the gaseous deposition of SO$_4$ and H$^+$ were obtained by flux balances. A detailed description of the calculation method is given by Ulrich (1983a) and Meiwes et al. (1984).

3. RESULTS

The results are available for a period of one year from October 1986 till September 1987. Precipitation during that period exceeded the average for the region by 25 percent. The rates of total deposition (TD) in the two stands, calculated from the flux below the canopy (throughfall TF and stemflow ST) and the precipitation deposition (PD), are given in table 3. The values of the total input (TD) are validated already by subtraction of leaching as well as of the increased nutrient loss during leaf senescence. The total deposition is nearly the same in both stands.

The interception deposition (ID) is evident from the difference between total deposition and precipitation deposition the ID amounting 60 to 70 % of the latter. Due to gaseous SO$_2$–deposition this contribution of the ID additionally increases for about 20 % for SO$_4$ and H$^+$. The total acid deposition calculated by summing up H^+ and NH_4 was 1.8 kmol eq.ha^{-1}.a^{-1} in the open and 3.2 kmol eq.ha^{-1}.a^{-1} in each stand, respectively.

TABLE 3. Annual rates of precipitation deposition in the open (PD) and total deposition (TD) in the two stands

	volume $1.m^{-2}$	H^+ $kmol.ha^{-1}.a^{-1}$	NH_4-N	NO_3-N	SO_4-S	Cl
				$kg.ha^{-1}.a^{-1}$		
PD	1390	0.79	14.0	10.6	18.6	21.3
TD (LF)	1209	1.61	22.6	17.2	35.4	39.8
TD (MF)	1259	1.52	24.0	18.2	34.4	41.6
	Na	Ca	Mg	K	Mn	Fe	Al
		$kg.ha^{-1}.a^{-1}$				$g.ha^{-1}.a^{-1}$	
PD	10.7	6.0	1.5	3.0	73	612	595
TD (LF)	17.2	9.6	2.4	4.8	120	990	960
TD (MF)	18.3	10.2	2.6	5.1	130	1050	1020

The average monthly pH-values for the precipitation deposition, the stemflow and the throughfall are shown in figure 1. In the winter period (Nov.- Apr.), the interception deposition causes lower pH-values in the stands, as compared with the precipitation in the open. During the vegetation period (May - Oct.), the interception of pollutants is compensated by H^+-buffering in the canopy. Thus an increase of pH in the two subsamples of the precipitation below the canopy (TF, ST) can be observed. There are only little differences between the two stands, whereas higher pH-values occur in the throughfall of the Melico-Fagetum. In coincidence with the seasonal pattern of the pH-values, an enrichment of the throughfall with cationic nutrients (K, Mn, Ca, Mg) has been found. In particular, the increase of fluxes of K and Mn during the vegetation period is evident (fig. 2).

Table 4 shows the calculated quantity of leaching and buffering processes within the canopy of the two beech stands. In both the rates of proton buffering are with about 50 % of the total H^+-deposition almost identical. Regarding the annual rates of cationic nutrient leaching, the two compared stands exhibit some differences. Considerable higher amounts of Mn were found in the throughfall of the Luzulo-Fagetum. Here, also K occurred to a higher extent. In contrast, more Ca was leached in the canopy of the beech forest on limestone. The differences for K and Ca are significant at 99 % level, those for Mn at 99.9 % level (Wilcoxon signed rank test for weekly data).

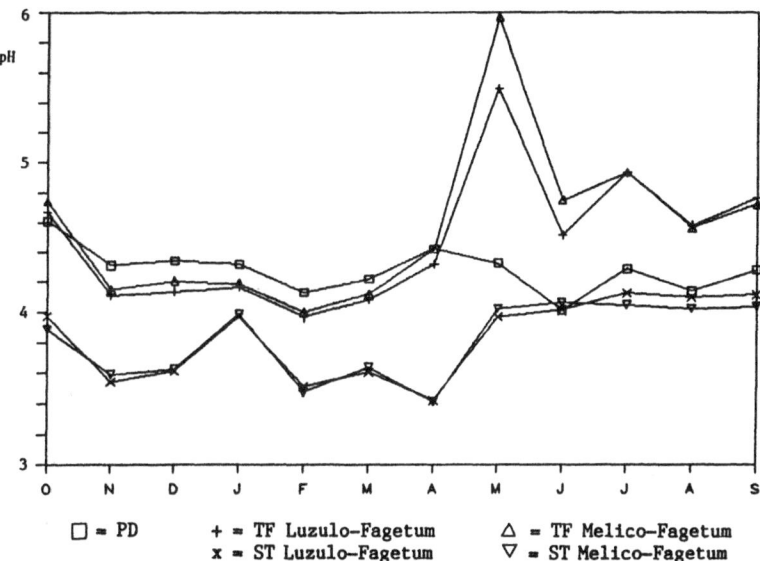

Figure 1. Average pH-values in the precipitation deposition (PD), the stemflow (ST) and the throughfall (TF) of the two compared stands.

TABLE 4. Calculated annual rates of cation exchange and leaching in the canopy (CE) and cation content of the leaves

		CE $kg.ha^{-1}.a^{-1}$	CE kmol eq. $ha^{-1}.a^{-1}$	leaf content $mg.g^{-1}$ dry weight
K	LF	24.2	0.619	8.99
	MF	22.5	0.576	7.69
Ca	LF	6.5	0.326	6.51
	MF	7.4	0.368	11.95
Mg	LF	1.7	0.137	0.79
	MF	1.7	0.138	1.44
Mn	LF	2.20	0.080	1.37
	MF	0.38	0.014	0.36
H^+	LF	−0.83	−0.821	
	MF	−0.79	−0.779	

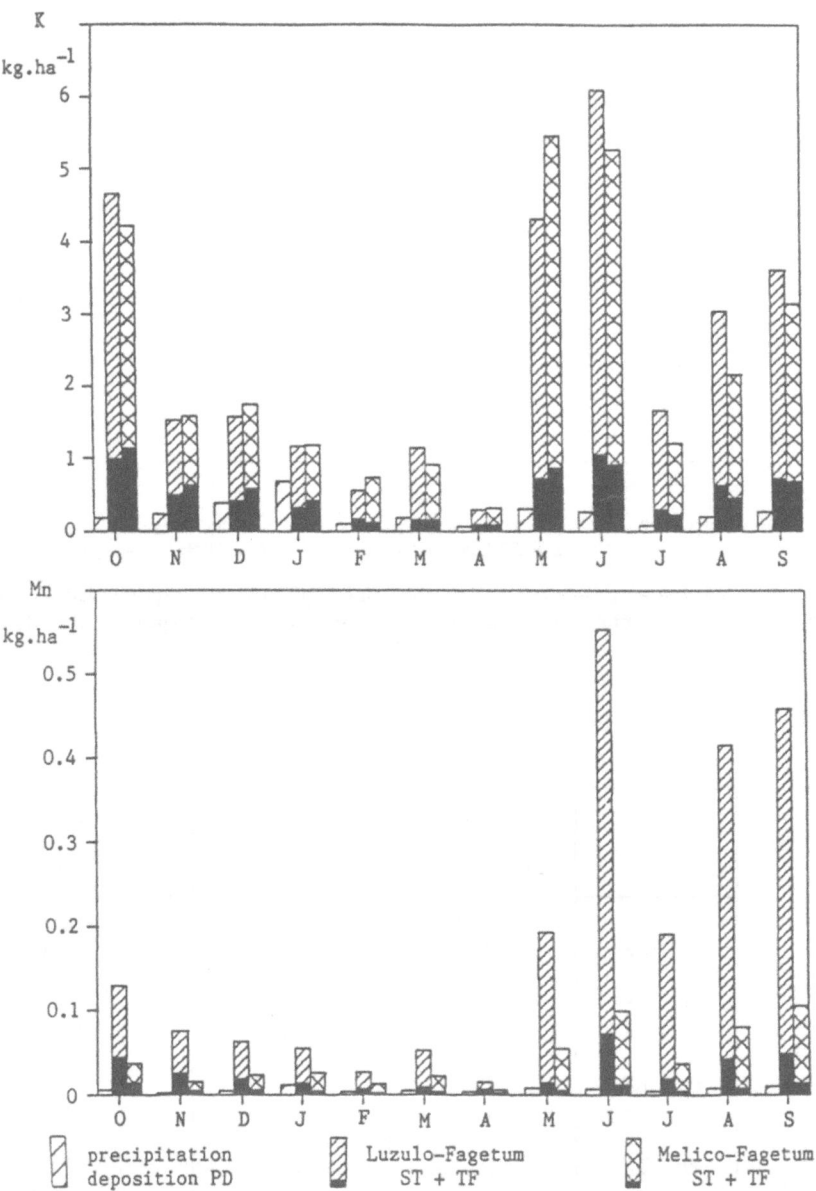

Figure 2. Fluxes of K and Mn in the open (PD) and below the canopy of both stands. The lower black part of the columns represents the stemflow (ST), the upper part the throughfall (TF).

In order to estimate to which extent the different leaching rates were influenced by tree nutrition, samples of the foliage from each stand were analyzed for total cation nutrient content. The leaching rates of cations are reflected by the leaf contents, except for Mg (tab. 4). The K and Mn contents are higher in the foliage of the Luzulo-Fagetum, the opposite was found for Ca and Mg.

TABLE 5. Average pH-values and mean ion concentrations in the soil solution during the sampling period (relative distribution of ions in percentage of total ion equivalent in parentheses)

	pH H+	NO_3	SO_4	HCO_3 mg.1^{-1}	Cl	Na
Luzulo-Fagetum	4.43 (4.7)	15.9 (32.3)	17.3 (45.3)	– –	6.35 (22.4)	2.58 (14.5)
Melico-Fagetum	8.35 –	22.3 (14.4)	13.7 (11.1)	106 (69.5)	4.44 (5.0)	2.22 (3.6)

	Ca	Mg mg.1^{-1}	K	Mn	Fe ug.1^{-1}	Al
Luzulo-Fagetum	8.0 (50.9)	0.49 (5.2)	0.46 (1.5)	340.6 (1.6)	12 (0.1)	1496 (21.5)
Melico-Fagetum	50.4 (91.9)	1.13 (3.4)	1.16 (1.1)	1.3 (0.0)	3 (0.0)	7 (0.0)

The relative distribution of cationic nutrients in the soil solution (tab. 5) and also the NH_4Cl-exchangeable cations from the soil (tab. 2) are able to explain the cationic leaf contents. The soil solution over limestone was dominated by Ca (90 %). Only very low amounts of Mn, Al and Fe were found here. The main anion was HCO_3, presumably caused by the dissolution of $CaCO_3$. Thus, the total ion content in the soil solution of the Melico-Fagetum (2.6 mmol eq.1^{-1}) was about three times as high as in the Luzulo-Fagetum above sandstone (0.8 mmol eq.1^{-1}). In the latter, Ca contributed only about 50 % to the total cation amount. In addition, Mn, Al and H^+ accounted for nearly 30 %. Although the absolute concentrations of K and Mg are lower in the soil solution of the Luzulo-Fagetum, their relative contribution to the total cation amount is higher as compared with the Melico-Fagetum (tab. 5).

4. DISCUSSION

The buffer rates in the canopy of the two compared beech stands are not significantly different. Both stands buffer about 0.8 kmol eq.ha^{-1}.a^{-1}, which represent about 50 % of the total H$^+$ deposition. The quantity of 50 % was assumed for forests on neutral and moderately acid soils with good mineral nutrition (Ulrich 1983b). In both stands, the Melico-Fagetum on limestone and also the Luzulo-Fagetum on sandstone, the cation supply of the trees is obviously sufficient to achieve relative high buffering rates in the canopy. The good nutrition and buffering in the Luzulo-Fagetum is possibly a consequence of the thin loess layer above the bed-rock. Nevertheless, the element concentrations in the seepage water indicate a continuing nutrient loss from the upper soil on this site.

The process of cation exchange during H$^+$-buffering could be accounted for 70 % of the total cation leaching from the canopy. Nitrogen turnover in the canopy was not assessed in our investigations, however, its uptake may also influence the magnitude of exchange processes. Other investigations in deciduous forests attributed between 40 and 60 % of K, Ca and Mg leaching to the exchange precesses (Lovett et al. 1985).

Though there are no quantitative differences in the total cation leaching between the two stands, the particular cations affected could be related to the cation content in the foliage. In the Melico-Fagetum on limestone, the contribution of Ca to the canopy exchange is higher. Simultaneously, K and especially Mn become less frequent. In accordance, nutrient loss from the leaves of beech forests during senescence was higher with increasing leaf contents of the considered cation (Staaf 1982).

The different nutrition of the stands reflects the availability of cations in the soil. Regarding the concentrations in the soil water, this is not true for K. The higher relative contribution of K to the total cation content in the soil solution of the Luzulo-Fagetum may explain the higher contents of the leaves in this community as compared with the Melico-Fagetum. On the other hand, a possible Ca antagonism during the K uptake caused by high Ca-concentrations is to be considered in the calcareous soil of the Melico-Fagetum (Kinzel 1982).

In spite of the annual variation of deposition rates and element turnover in forest ecosystems, the differences in canopy processes between these two stands are obvious. Although the data are not an average of several years, the estimated deposition rates agree well with results from other investigations in north-west Germany (Meiwes 1985, Block and Bartels 1985, Ulrich et al. 1986). To confirm our results, further years of investigations would be needed.

AKNOWLEDGEMENTS: These investigations were supported by the Gesamtverband des deutschen Steinkohlenbergbaus, Essen, FRG.

5. REFERENCES

Block, J. and Bartels, U. (1985) 'Ergebnisse der Schadstoffdepositions-
messungen in Waldökosystemen in den Meßjahren 1981/82 und 1982/83',
Forschung und Beratung Reihe C 39, Landwirt-schaftsverlag, Münster-
Hiltrup, 296 p.

Brechtel, H. M., Balazs, A. and Lehnardt, F. (1986) 'Precipitation input
of inorganic chemicals in the open field and in forest stands' in
H.-W. Georgii (ed.), Atmospheric Pollutants in Forest Areas, D. Reidel
Publishing Company, Dordrecht, pp. 47–67.

Cronan, C.S. and Reiners, W.A. (1983) 'Canopy processing of acidic precipitation by coniferous and hardwood forests in New England',
Oecologia 59, 216–223.

Gehrmann, J. (1987) 'Derzeitiger Stand der Belastung von Waldökosystemen
in Nordrhein-Westfalen durch Deposition von Luftverunreinigungen',
Forst- und Holzwirt 42, 141–145.

Johnson, D. W., Richter, D. D., Lovett, G. M. and Lindberg, S. E. (1985)
'The effects of atmospheric deposition on potassium, calcium and
magnesium cycling in two deciduous forests', Can. J. For. Res. 15,
773–782.

Kelly, J. M. and Strickland, R. C. (1986) 'Throughfall and plant nutrient concentration response to simulated acid rain treatment', Water,
Air and Soil Pollution 29, 219–231.

Kinzel, H. (1982) Pflanzenökologie und Mineralstoffwechsel, Ulmer Verlag, Stuttgart, 534 p.

Lovett, G. M., Lindberg, S. E., Richter, D. D. and Johnson, D. W. (1985)
'The effects of acid deposition on cation leaching from three deciduous forest canopies', Can. J. For. Res. 15, 1055–1060.

Matzner, E. (1988) 'Der Stoffumsatz zweier Waldökosysteme im Solling',
Ber. Forschungsz. Waldökosysteme/Waldsterben 40, 217 p.

Matzner, E. and Ulrich, B. (1984) 'Raten der Deposition, der internen
Produktion und des Umsatzes von Protonen in zwei Waldökosystemen',
Z. Pflanzenernaehr. Bodenkd. 147, 290–308.

Meiwes, K.J. (1985) 'Bioelementbilanz eines Buchenwaldes auf Kalkge-
stein', Mitt. Deutsch. Bodenkdl. Ges. 43, 981–985.

Meiwes, K.J., Hauhs, M., Gerke, H., Asche, N., Matzner, E. und Lammers-
dorf, E. (1984) 'Die Erfassung des Stoffkreislaufs in Waldökosystemen', Ber. Forschungsz. Waldökosysteme/Waldsterben 7, 68–142.

Mengel, K., Lutz, H.-J. and Breininger, M. Th. (1987) 'Auswaschung von
Nährstoffen durch sauren Nebel aus jungen intakten Fichten (Picea
abies)' Z. Pflanzenernaehr. Bodenkd. 150, 61–68.

Staaf, H. (1982) 'Plant nutrient changes in beech leaves during senescence as influenced by site characteristics', Acta Ecol. Plant 3, 161–
170.

Ulrich, B. (1983a) 'Interaction of forest canopies with atmospheric constituents', in B. Ulrich and J. Pankrath (eds.), Effects of Accumulation of Air Pollutants in Forest Ecosystems, D. Reidel Publishing Company, Dordrecht, pp. 33–45.

Ulrich, B. (1983b) 'Stabilität von Waldökosystemen unter dem Einfluß des
Sauren Regens', Allg. Forstzeitschr. 26/27, 668–677.

Ulrich, B., Mayer, R. and Matzner, E. (1986) 'Vorräte und Flüsse der chemischen Elemente', in H. Ellenberg, R. Mayer and J. Schauermann (eds.), Ökosystemforschung, Ergebnisse des Solling-Projekts 1966–1986, Ulmer Verlag, Stuttgart, pp. 375–417.

Wagner-Hucke, A. (1986) 'Vergleich bodenchemischer und -physikalischer Kenngrößen von Böden verschiedener Buchenwaldgesellschaften', Staatsexamensarbeit, Universität Düsseldorf (Bot. Inst. III, Abt. Geobotanik), unpublished.

Waraghai, A. and Beese, F. (1985) 'Pufferverhalten und Ionenstatus von Waldbäumen auf unterschiedlich belasteten Böden', Mitt. Deutsch. Bodenkdl. Ges. 43, 489–494.

SPATIAL VARIATION OF STRESS FACTORS AND THEIR INFLUENCE ON NORWAY SPRUCE NUTRITION

M. Kazda
Institut of Forest Ecology
Universität für Bodenkultur
Vienna, Austria

Present address:
Institut of Ecological Plant Physiology and Geobotany
Dept. Geobotany
Universität Düsseldorf
Universitätsstr. 1
D–4000 Düsseldorf, FRG

ABSTRACT. Three Norway spruce stands of different elevation and exposition were selected for investigations of connections between the atmospheric deposition, soil solution and tree nutrition. On the investigated sites, the pollutants load and its influence on the soil solution chemistry and tree nutrition depends to a large extent on the topography and exposition towards atmospheric input. The previous management practices of forest sites are an additional factor which affects the nutrition of recent stands. Despite a distance of only few kilometers between the stands, the influence of atmospheric pollutants and the impact of previous management practices upon forest stands may vary substantially. In connection with the forest decline, this meso scale variability should be still considered.

1. INTRODUCTION

Pronounced symptoms of Norway spruce (**Picea abies** Karst.) decline have occurred in the Bohemian Forest of Upper Austria during the last decade. Preliminary screenings gave an evidence of calcium and magnesium deficiency, which was also reported from West Germany (Bosch et al. 1983, Zöttl and Hüttl 1985). Foliar leaching of nutrients due to acid deposition was postulated as an important factor which impairs tree nutrition (Tamm and Cowling 1977, Ulrich 1983). Impact of atmospheric pollutants such as sulphur dioxide and ozone can also lead to nutrient losses from the foliage (Prinz et al. 1982, Seufert and Arndt 1986). In the soil, aluminium toxicity may influence the root vitality and the nutrient uptake (Meiwes et al. 1986). An other aspect in the nutritional disorders in Middle European forests is the effect of high nitrogen input. Nitro-

295

H.-W. Georgii (ed.), Mechanisms and Effects of Pollutant-Transfer into Forests, 295–304.

gen input contributes considerably to the soil acidification and enhances the dissolution of Al^{3+} ions in the soil (van Breemen and Jordans
1983). Since the today input levels in Central Europe exceed the annual
need of forest trees considerably they can cause dilution of several
nutrients and increased susceptibility to various adverse effects
(Nihlgard 1985, Glatzel et al. 1987, Roelofs and van Dijk 1987).

In the Bohemian Forest, three Norway spruce stands of different
elevation and exposition were selected for investigations of interactions between the atmospheric deposition, soil solution and tree nutrition. A detailed approach to assess these effects is given in Kazda
(1989). The presented paper is focused to the differences in spatial
meso scale variability in which following factors may affect the nutrition of Norway spruce stands:

  a) Foliar leaching of nutrients
  b) Soil solution chemistry
  c) Probability of aluminium toxicity in the soil solution
  d) Nutrient deficiency and unbalanced nitrogen nutrition

2. METHODS

Three spruce stands (Tab. 1) situated about 4 km from each other were
chosen in the north-south direction in the Bohemian Forest of Upper
Austria. The stand S1 is situated on a lower slope open to the north-
east. The stand S3 was selected on the south-west slope. The stand S2 on
the plateau lies in about half a distance between the two other plots.
The stands of the plots S1 and S3 exhibit less extended damage. The
stands of the site S2 show pronounced signs of forest decline such as
substantially reduced needle mass and needle discoloration. Logging was
the prevailing exploitation of the sites S1 and S3. The sites of the
plot S2 were used as a mowed pasture for several decades. About 90 years
ago, Norway spruce stands were also established on these sites.

TABLE 1. Characteristics of the investigated Norway spruce
stands

Plot	Elevation above sea level	Exposition	Slope %	Age years	Percent needle loss
S1	850 m	NE	10	90-120	10-20
S2	1020 m	plateau	2	80- 90	30-40
S3	900 m	SW	14	90-100	10-20

All plots have the similar soil type namely a podsolic brown earth
(spodosol) developed on "Eisgarner" granite. The soils of the site S2

exhibit severe nutrition depletion due to previous agricultural practices and built-in of nutrients into the forest biomass.

Throughfall was sampled evenly by 15 open funnels (diam. 18 cm). Soil solution was obtained by three ceramic cup tension lysimeters in each 15, 30 and 60 cm soil depth (Soil Moisture Comp.). Data for soil water potential were obtained by tensiometers in each depth. All these equipments were randomly installed on square plots (20x20 m) in each stand. Close to these plots, four trees of various degrees of damage were chosen for needle sampling. Needle samples were taken five times from the upper crown during the vegetation period 1986 and analysed for major nutrients. For more details on the sites and methods see Kazda (1989).

3. RESULTS AND DISKUSSION

3.1. Foliar leaching of nutrients

The pollutant load on the investigated sites depends highly on the topography. The stand S3 on the south-west slope receives the highest sulfate and nitrogen input. The stand S1 exposed towards north-east often exhibits only half of these inputs (Tab. 2).

TABLE 2. Annual fluxes in the throughfall in the investigated Norway spruce stands in Bohemian Forest (Glatzel et al. 1988).

	Ca	Mg	K	Na	SO_4-S	NO_3-N	N	Cl	H^+
					$kg.ha^{-1}$				
S1 / 1987	10	1	11	2	19	5	14	5	0,5
S2 / 1987	10	1	14	2	28	7	18	7	0,7
S3 / 1987	12	2	16	2	30	11	24	8	0,7

Despite the higher deposition with the precipitation on the western sites, significant amount of nutrients can be leached in the other stands exposed to the high pollutants concentration in the air masses coming from the northern directions (Amt der Oberösterreichischen Landesregierung 1986).

In May and June 1985, very high fluxes of cationic nutrients, especially of magnesium in the throughfall were measured on the site S1 (Fig. 1). This foliar losses however, did not occur in the following periods. Since needle damage due to insect attack were not found, this high foliar leaching can be attributed to a pollutants event. The stand S1 is periodically exposed to high SO_2 concentrations in the air. Thus membrane leakage as a consequence of SO_2 and O_3 effects may be responsible for this losses (Prinz et al. 1982; Seufert and Arndt 1986). Also

298

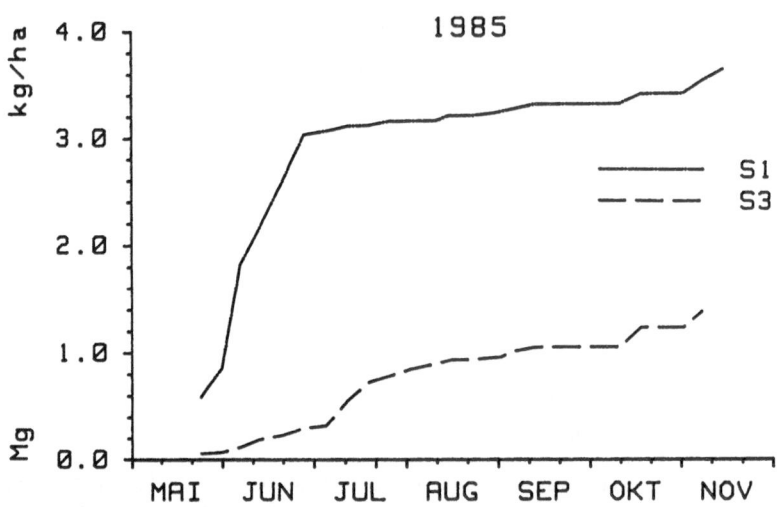

Fig. 1. Cumulative fluxes of magnesium in the throughfall on the investigated plots S1 and S3 during the sampling period 1985

an erosion of cuticular waxes due to high polluted fog events (Kazda and Glatzel 1986) may contribute to the leaching. Other effects which can lead to a damage of cuticular waxes are discussed in Euteneuer et al. (1988). Electron microscopy findings by Kazda and Wolf (unpublished) on needles from a adjoining stand to the stand S1 corroborated this possible link, since extensive erosions of the cuticular waxes were observed on the needles from the vegetation period 1985.

As a result of these magnesium losses from the foliage, retranslocation and in consequence also low nutrition and yellowing of the older needles can be assumed (Mies and Zöttl 1985, Kazda and Weilgony 1988). The results from 1985 showed, that a single events can influence the nutrition of the stands to a great extent.

3.2. Soil solution chemistry

The differences in the element input with the precipitation are also reflected in the composition of the soil solution. On the plot S1, rather low nitrate concentrations were found in the soil solution. However, nitrate concentrations were up to ten times higher in the soil solution of the site S3 on the south-west slope. On this plot, a sharp increase of the nitrate concentrations took place in course of a long dry period in 1986 (Fig. 2). A similar increase as a result of a period with warm and dry weather was described also by Ulrich (1981) and Matzner and Ulrich (1987).

Though the soil water suction potential on the site S3 was at least as high as on the other plots, such an increase did not occurred on these sites. The plots S1 and S2 exhibit substantial nitrogen demand and the nitrate was probably utilized by the roots. No substantial increase of nitrate concentration in the soil solution took place also during the sampling period 1987, since higher amounts of precipitation and lower temperatures discouraged the mineralisation of organic matter. The probability, nitrification exceeds the nitrate uptake from the soil, does not only depend on the weather factors but also on the nitrogen status of the stands.

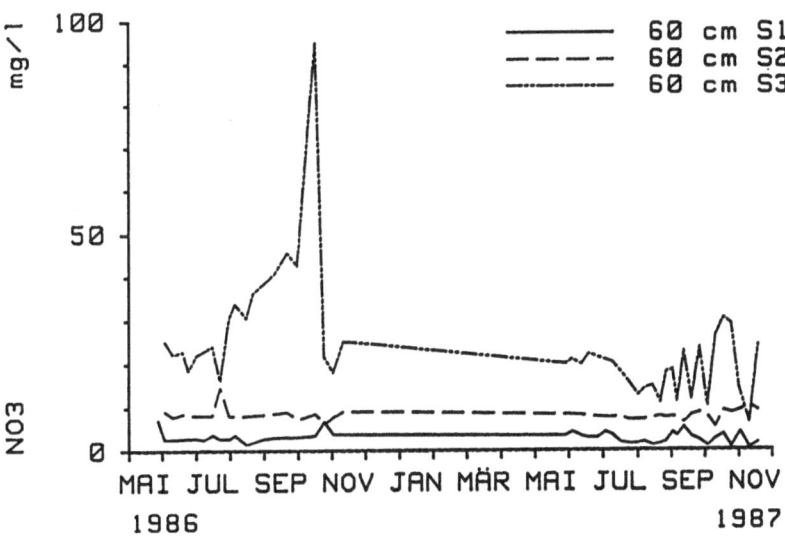

Fig. 2. Means of nitrate concentrations in 60 cm soil depth on the three investigated plots in course of the sampling periods 1986 and 1987

The Table 3 displays the average element concentrations in the soil solution on the three investigated plots during the sampling period 1986. Though the acid input is only moderate on the site S2, the lowest nutrient concentrations were found there. Otherwise, aluminium is the prevailing cation in the soil solution on the plot S2. This is an evidence that the previous management practices of the forest sites largely determine the nutrient status of the recent stands. On sites such as S2 where the nutrient reserves were depleted due to the removal of biomass, the nutrition and growth of the forest stands become very low. In addition, nutrient immobilisation in the biomass enhances the soil acidification (Nilsson et al. 1982). High levels of nitrogen input into the forests during the last decades compensate the nitrogen losses. Due to stimulated growth, built-in of nutrients into the biomass and further soil acidification, they may have led to the dilution of other nutrients, however. This is the appropriate reason for the bad cationic nutrient status and for the high aluminium concentrations in the soil solution of the stand S2.

Table 3. Means of element concentrations in the soil solution on the investigated plots in 15, 30 and 60 cm soil depth during the sampling period 1986 (standard deviation in parenthesis)

a) K, Ca, Mg, Na, Mn

Site Depth	K	Ca	Mg $mg.1^{-1}$	Na	Mn
S1					
15 cm	3.87 (2.49)	1.06 (0.55)	0.69 (0.30)	0.88 (0.37)	0.51 (0.27)
30 cm	2.67 (3.40)	0.88 (0.60)	0.66 (0.33)	0.84 (0.31)	0.44 (0.31)
60 cm	1.05 (0.86)	0.49 (0.49)	0.59 (0.27)	1.33 (0.74)	0.34 (0.29)
S2					
15 cm	1.57 (1.33)	0.41 (0.50)	0.40 (0.25)	1.40 (0.88)	0.11 (0.07)
30 cm	3.03 (2.20)	0.41 (0.41)	0.42 (0.20)	0.63 (0.25)	0.17 (0.09)
60 cm	1.16 (1.79)	0.28 (0.50)	0.28 (0.22)	0.99 (0.39)	0.02 (0.14)
S3					
15 cm	7.45 (4.56)	1.47 (0.97)	0.93 (0.64)	2.06 (1.45)	0.40 (0.39)
30 cm	4.21 (2.08)	2.23 (2.19)	1.29 (0.54)	1.93 (1.20)	0.62 (0.22)
60 cm	5.31 (4.68)	1.51 (1.09)	1.36 (0.75)	1.53 (0.83)	0.59 (0.36)

b) Al, Cl, NO_3, SO_4 and pH

Site Depth	Al	Cl	NO_3 $mg.1^{-1}$	SO_4	pH
S1					
15 cm	1.64 (0.65)	3.05 (2.34)	4.3 (3.5)	14.7 (6.4)	4.50 (0.32)
30 cm	1.27 (0.55)	2.64 (3.14)	4.5 (4.5)	11.7 (5.4)	4.77 (0.31)
60 cm	1.95 (1.24)	1.33 (0.71)	3.4 (2.5)	17.3 (9.7)	4.82 (0.33)
S2					
15 cm	2.50 (0.92)	0.95 (0.82)	15.5 (7.6)	11.3 (9.7)	4.58 (0.29)
30 cm	2.66 (1.57)	1.74 (0.92)	9.6 (6.2)	14.6 (6.3)	4.67 (0.23)
60 cm	2.19 (1.00)	1.41 (0.73)	10.5 (9.0)	10.8 (4.7)	4.68 (0.32)
S3					
15 cm	1.33 (0.81)	3.05 (2.39)	15.9 (12.6)	13.7 (9.4)	4.98 (0.61)
30 cm	1.72 (1.06)	3.18 (3.14)	18.0 (11.2)	17.4 (5.7)	4.56 (0.53)
60 cm	3.33 (1.83)	3.44 (4.03)	30.5 (18.3)	16.9 (8.5)	4.44 (0.23)

3.3. Probability of aluminium toxicity in the soil solution

Due to the high aluminium and low cationic nutrient concentrations in the soil solution on the plot S2, the calcium/aluminium ratio (mol/mol) in the soil solution was very low on this plot and dropped periodically below 0.1 (Fig. 3). This produces high probability of root damage due to aluminium toxicity (Meiwes et al. 1986). Despite different dynamics, aluminium reaches almost identical values in the needles of all stands at the end of the vegetation period. The average values vary between 78 and 83 µg.g^{-1} in one year old and between 116 and 121 µg.g^{-1} in two year old needles respectively and do not show any significant differences between the plots. Therefore needle analysis is not useful tool to estimate possible aluminium effects in the soil. Since the calcium and magnesium supply and nutrition is very poor on the S2 site, it cannot be decided by now, to which extent the aluminium ions contribute to the high degree of tree damage. More details on this investigations in the Bohemian Forest are given in Kazda and Zvacek (1989) and Kazda (1989).

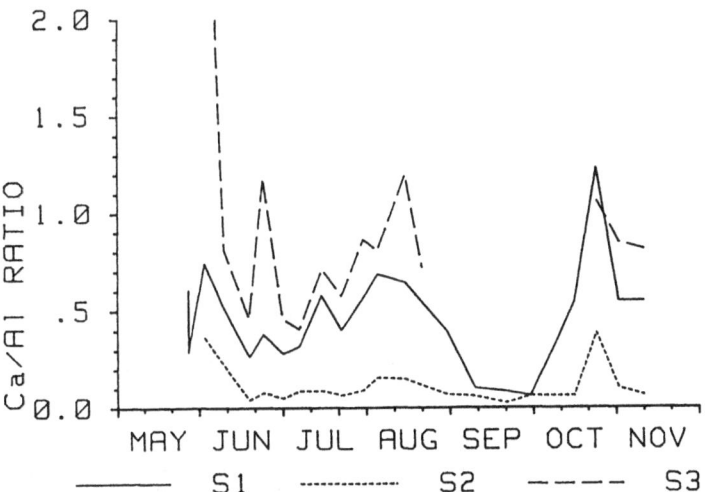

Fig. 3. Calcium/aluminium ratio (mol/mol) in the soil solution from 30 cm soil depth on the investigated plots during the sampling period 1986 (Due to a long dry period in September/October 1986 no samples were obtained on the plot S3.)

3.4. Nutrient deficiency and unbalanced nitrogen nutrition

Needle analysis showed low calcium and magnesium nutrition for all stands, especially in stand S2 (Tab. 3). Comparisons of the distribution of calcium, magnesium and potassium between the soil solution, the xylem sap and the needles showed, that the trees respond to the low nutrient supply from the soil solution in mobilisation and retranslocation of nutrients (Kazda and Weilgony 1988).

Nitrogen nutrition shows an other differentiation of the stands. It is very good in the stands S2 and S3 and low in S1. In the stands S2 and S3, the ratio of calcium and magnesium to nitrogen indicates dilution of these nutrients. The trees of the stand S1 absorb significant amounts of nitrogen in the canopy in order to cover their demand as was indicated by a negative canopy difference for nitrogen compounds during several vegetation periods (Katzensteiner 1987, Kazda 1989), i.e. the nitrogen flux with the throughfall was lower then nitrogen input in the open. Despite this uptake in the canopy, needle analysis showed only low nitrogen nutrition.

TABLE 3. Nutrient contents in one year old needles of the investigated stands and reference values according to Hüttl (1986)

	N	P	K	Ca	Mg	S
			% DM			
S1	1.3	0.15	0.54	0.18	0.08	0.11
S2	1.5	0.15	0.56	0.06	0.05	0.12
S3	1.6	0.16	0.58	0.13	0.07	0.10
Optimum (Hüttl 1986)	1.5	0.15	0.60	0.30	0.10	
Deficiency (Hüttl 1986)	1.2-1.3	0.11-0.12	0.40-0.45	0.10-0.20	0.07-0.08	

The sulfur needle contents show the influence of the topography upon the distribution of SO_2 polluted air masses coming mostly from the northern directions. Therefore, sulfur needle contents are higher on the north-east plot S1 and on the plateau plot S2.

4. CONCLUSIONS

The deposition of atmospheric pollutants affect to a great extent the mineral nutrition of forest stands in Central Europe. The output of cationic nutrients from the system on the one hand and the nitrogen input on the other hand produces pronounced divergences between the demand and availability of cationic nutrients. This processes have an effect especially on sites which are nutrient poor from the outset or due to previous anthropogenic influence.

Since there is only little hope in a substantial reduction of atmospheric pollutants in the next decades, compensative melioration and liming of forest sites become necessary. This can be done only after a comprehensive analysis of the ecosystems and its exposition to the pollutants. It has to be still considered that several factors influence simultaneously and to a locally varying extent the mineral nutrition of the forest sites. The results of these investigations in the Bohemian Forest showed that even within a distance of only few kilometers the influence of atmospheric pollutants and the impact of previous management practices may vary considerably. In terms of interpreting possible causes of the forest decline in the Central Europe, this meso scale variability should always be taken into account.

AKNOWLEDGEMENT: I am grateful to Dr. G. Glatzel for his generous support during these investigations. This research was financed by the Austrian Ministry of Science (Grant No. BMWuF 36.036/2-23/85 and GZ 36.041/2-23/86) and by the Austrian Research Foundation (Grant No. 5169).

REFERENCES

Amt der Oberösterreichischen Landesregierung (1986) 'Meßbericht 4. Über die Ergebnisse des automatischen Luftmeßnetzes des Landes Oberösterreich', Amt der Oberösterreichischen Landesregierung. 161 p.

Bosch Ch., E. Pfankuch, U. Baum and K.E. Rehfuess (1983) 'Über die Erkrankung der Fichte (Picea abies L. Karst) in den Hochlagen des Bayerischen Waldes', Forstwiss. Cbl., 102, 167-181.

Breemen van N. and E.R. Jordans (1983) 'Effects of atmospheric ammonium sulfate on calcareous and non-calcareous soils of woodlands in the Netherlands', in: B. Ulrich and J. Pankrath (eds.), Effects of Accumulation of Air Pollutants in Forest Ecosystems, D. Reidel Publ. Comp., pp. 171-182.

Euteneuer T., L. Steubing and R. Debus (1988) 'Quantitative Beurteilung der Morphologie epikutikularer Wachse von Picea abies (L.) Karst', Angew. Botanik 62, 63-72.

Glatzel G., M. Kazda, D. Grill, G. Halbwachs and K. Katzensteiner (1987) 'Ernährungsstörungen bei Fichte als Komplexwirkung von Nadelschäden und erhöhter Stickstoffdeposition - ein Wirkungsmechanismus des Waldsterbens?', Allg. Forst- u. J.-Ztg. 158, 92-97.

Glatzel G., K. Katzensteiner, M. Kazda, D. Stöhr, G. Markart, and M. Kühnert (1988) 'Eintrag atmosphärischer Spurenstoffe in Wälder, Ergebnisse aus vier Jahren Depositionsmessungen', in: E. Führer and F. Neuhuber (eds.), Symposium 1988. Waldsterben in Österreich; Theorien, Tendenzen, Therapien. Bundesministerium für Wissenschaft und Forschung, pp 60-72.

Hüttl R. (1986) 'Forest fertilisation: Results from Germany, France and the nordic countries', The Fertiliser Society. Proceedings No. 250, 40 p.

Katzensteiner K. (1987) 'Deposition und Umsatz atmosphärischer Spurenstoffe in einem Fichtenwaldökosystem im nordwestlichen Mühlviertel', Forschungsbericht. Institut für Forstökologie, Universität für Bodenkultur, Wien. 101 p.

Kazda M. and G. Glatzel (1986) 'Schadstoffbelasteter Nebel fördert die Infektion durch pathogene Pilze', Allg. Forstzeitschr. 18, 436-438.

304

Kazda M. and P. Weilgony (1988) 'Seasonal dynamics of major cations in xylem sap and needles of Norway spruce (Picea abies L. Karst) in stands with different soil solution chemistry', Plant and Soil 110, 91–100.

Kazda M. and L. Zvacek (1989) 'Aluminium and manganese and their relation to calcium in soil solution and needles of Norway spruce stands', Plant and Soil 114, 257–267.

Kazda M. (1989) 'Zusammanhang zwischen Stoffeintrag, Bodenwasserchemismus und Baumernährung in drei Waldökosystemen im Böhmerwald, Oberösterreich', Forstliche Schriftenreihe der Universität für Bodenkultur, Wien (submitted).

Matzner E. and B. Ulrich (1987) 'Results of studies on forest decline in northwest Germany', in: Hutchinson T.C. and K.M. Meema (eds.), Effects of Atmospheric Pollutants on Forests, Wetlands and Agricultural Ecosystems, NATO ASI Ser. Vol. G16. Springer-Verlag, Berlin, pp. 25–42.

Meiwes K.J., S.K. Khanna and B. Ulrich (1986) 'Parameters describing soil acidification and their relevance to the stability of forest ecosystems', For. Ecol. Man. 15, 161–179.

Mies E. and H.W. Zöttl (1985) 'Zeitliche Veränderung der Chlorophyl- und Elementgehalte in den Nadeln eines gelb-chlorotischen Fichtenbestandes', Forstw. Cbl. 104, 1–8.

Nihlgard B. (1985) 'The ammonium hypothesis - An additional explanation to the forest dieback in Europe', Ambio 14, 2–8.

Nilsson S.I., H.G. Miller and J.D. Miller (1982) 'Forest growth as a possible cause of soil and water acidification: an examination of the concepts', Oikos 39, 40–49.

Prinz B., G.H.M. Krause and H. Stratmann (1982) 'Waldschäden in der Bundesrepublik Deutschland', LIS-Berichte, Essen. Vol. 28.

Roelofs J.M.G. and H.F.G. van Dijk (1987) 'The effects of airborne ammonium deposition on canopy ion-exchange in coniferous trees', in: Commission of the European Communities. Direct Effects of Dry and Wet Deposition on Forest Ecosystems - Particular Canopy Interactions. Air Pollution Research Report, Vol. 4, pp 34–39.

Seufert G. and U. Arndt (1986) 'Untersuchungen zum Einfluß von Luftverunreinigungen auf den Stoffhaushalt von Waldbäumen', in: F. Horsch et al. (eds.), Projekt Europäisches Forschungszentrum für Maßnahmen zur Luftreinhaltung. KfK-PEF, Vol. 4, pp. 239–258.

Tamm C.O. and E.B. Cowling (1977) 'Acid precipitation and forest vegetation', Water, Air and Soil Pollution 7, 503–511.

Ulrich B. (1981) 'Theoretische Betrachtung des Ionenkreislaufes in Waldökosystemen', Z. Pflanzenernaehr. Bodenkd. 144, 647–659.

Zöttl H.W. and R. Hüttl (1985) 'Schadsymptome und Ernährungszustand von Fichtenbeständen im südwestdeutschen Alpenvorland', Allg. Forstzeitschr. 40, 197–199.

EXPERIMENTS ON TREE CANOPY DEPOSITION OF SO_2
AND RESULTING LEACHING EFFECTS

Dr. G. SEUFERT
Plant Ecology at Univ. of Hohenheim
POB 700562
D-7000 Stuttgart 70
FRGermany

ABSTRACT. From 1983-1988 an experiment was running at the University of
Hohenheim to investigate the long term effects of low level exposure
with O_3, SO_2 and simulated acid rain, single and in combination, on
spruce-, fir- and beech seedlings in modelecosystems. Systems consisting of open-top chambers built upon lysimeters were protected against
the intrusion of ambient rain and dust. During the 5 year duration of
the experiment definite effects on mineral cycling were observed. Most
noticeable are throughfall enrichment with sulphate through dry deposition of SO_2 as influenced by duration of needle-wetting and factors promoting SO_2-oxidation. Depending on sulphur deposition leaching of calcium, magnesium, manganese, zink and ammonia from needles was elevated,
in totol leading to an enhanced acid input to the soil.

1. Introduction

Observations in diseased forests show widespread deficiencies of base
cations in trees growing in areas previously thought to be nutritio-
nally satisfactory (e.g. Evers 1981). Furthermore, the natural acidification of forest soils has increased in the last few decades (e.g. Evers
1983). In general terms, the main ecological process affected by the
changed chemical environment of forest ecosystems seems to be mineral
cycling, with close interacting connections between effects in the crown
and in the soil. Looking at possible dose-response relationships between
pollutant deposition and forest disease there are enormous scientific
limits. For example, sensitivity of forest ecosystems to acidification
is dependent on many unknown site characteristics; the total deposition
to forests, both actual and historical, is unknown.
    Acid deposition to soils is enhanced under canopies compared to
wet deposition above, especially in conifer stands (Adam et al. 1987,
rev. in Seufert 1988), but this represents not the total acid load. Protons in the canopy are partly replaced by base cations from inside the
plant, which must be replaced by root uptake, with proton exchange acidifying the rhizosphere. This canopy buffering is difficult to quantify

H.-W. Georgii (ed.), Mechanisms and Effects of Pollutant-Transfer into Forests, 305–313.

under ambient conditions, because looking at throughfall enrichment it is not possible to make a distinction between leached elements from inside, washed off elements from dust deposition and uptake of elements by canopy surface.

Among others, canopy buffering was an important objective of the fumigation/irrigation experiment with young forest trees in modelecosystems described in this paper, which was running from 1983-1988 at the University of Hohenheim. To proof results under ambient conditions a similar experiment is running since 1986, where 19 years old spruce trees in a stressed forest are kept in charcoal-filtered air and ambient air with the help of open-top chambers (Seufert and Arndt 1985, Evers and Seufert 1989).

2. Material and Methods

The method used to give trees a defined atmospheric environment under semi-natural conditions in view of climatic and edaphic parameters is the OTC-technique, as developed by groups in the USA to look at air pollutant effects on crops. However this technique has not been used up to now with forest trees and over a number of years with the same plants. Therefore, various modifications of the common chamber type became necessary to work successfully with groups of 1-2 m tall tree seedlings. A detailed description of chamber design, fumigation-, irrigation- and monitoring-equipment as well as climatic/edaphic site conditions in the chambers is given in Seufert (1988).

The chambers with a height of 4 m and a diameter of 3 m, built upon lysimeters 80-90 cm deep, are protected from ambient rain and dust input by shelters 1 m above the open top. The air is filtered through particle filters and activated carbon, where NO_2, SO_2, O_3 and particles are removed to 80-95 %. Air is supplied with the help of 60 m^3/min-blowers to give 1.5 air changes per minute and enters the chambers by aeration tubes situated on the ground, dividing the ground into three segments. SO_2 is applied from a pressure bomb, O_3 is produced by an ozone generator using pure oxigen, flow is regulated by mass flow controllers. Both gases are introduced through teflon pipes into the stream of purified air. Gas concentrations are monitored using SO_2 analyzers (Beckman 953 and LfE COSO2), O_3- and NO_x analyzers (Monitor Labs model 8810 for O_3, model 8840 for NO_x).

The fumigation history of the experiment is shown in Fig 1. Weekly defined SO_2- and O_3-concentrations were adjusted, depending on the time of the year and the weather conditions. Altogether, the fumigation history can be characterized as a gradual approach to the pollutant situation in remote forest areas of SW-Germany. The problem was that there were no SO_2- or O_3-measurements from forests available at the beginning of the experiment; in the first year the SO_2 concentration was too high, O_3 was too low compared to the situation at forest sites in SW-Germany. In contrast, the fumigation in the last two years represents this situation with a weekly SO_2-average of 15-25 $\mu g/m^3$ in summer, 30-120 μg in winter and an O_3-average of 60-180 μg in summer, 20-60 μg in winter.

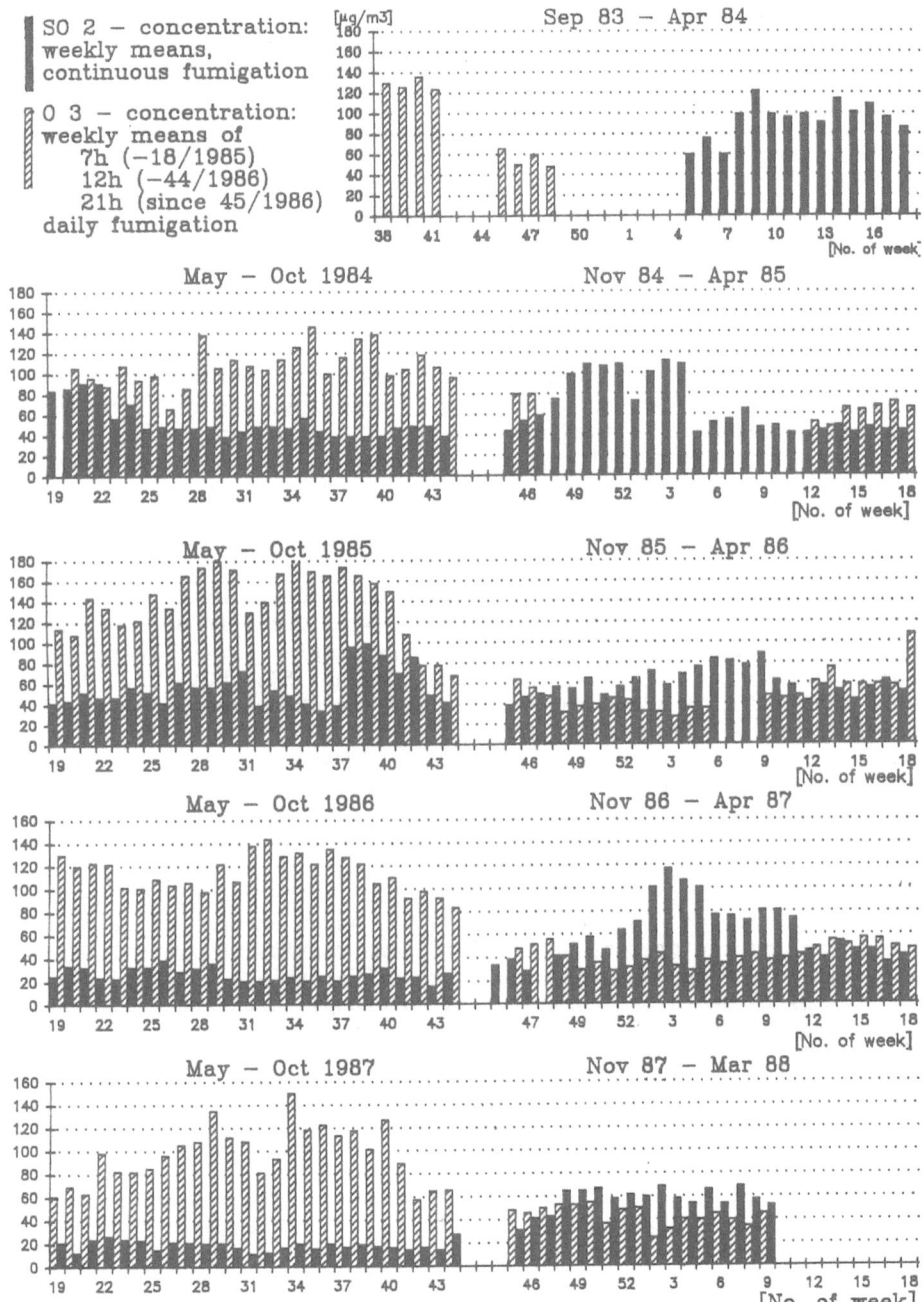

Figure 1.   Annual pattern of the O_3 and SO_2 concentrations (µg/m³) from Sep. 83 - Mar. 88 in the open-top chambers for the O_3, SO_2 and $SO_2 + O_3$ treatments.

Two different rain treatments (pH 4 and pH 5) were both dispensed with separate pumps, the chemical composition of the pH 4 rain follows the situation in S.W.-german forest areas. The pH 5 treatment received the same ions at a dilution of 1:10. 15 l/m² of artificial rain or fog per week were supplied through different nozzle types which have been installed in the chamber top. To investigate the effects of different wet deposition properties, precipitation was given as general rain with a droplet size of 600 μm and an intensity of 20 mm/h until summer 1986; in summer 1986 as drizzle rain (400 μm, 2,5 mm/h); in winter 1986/87 and summer 1987 as fog (70 μm, 0,5 mm/h). With fog application the duration of needle-wetting was about 10times compared to general rain. In winters 1983/84 and 1987/88 the rain shelters have been removed to get ambient rain into the chambers.

Modifications of climatic conditions in the chambers have been moderate enough to allow spruce- and fir- seedlings a normal growth over 5 years, whereas beech showed excessive longitudinal growth and through the years increasing aphide infection. Spruce- (Picea abies L. Karst., cloned material), fir- (Abies alba Mill.) and beech- (Fagus silvatica L.) seedlings were planted in Feb. 1983 in groups to form a closed canopy. The soil in the lysimeters was a restored acid brown earth from the northern part of the Black Forest with a base saturation of 20-30 %.

The samples of rain solutions and throughfall water were collected immediately after the weekly irrigations with the help of rins; after measuring volume, pH and conductivity the samples were filtered and the ion concentrations (Ca, Mg, K, Na, Mn, Zn, AL, NH4, NO3, Cl, PO4, SO4) were analyzed with adequate methods like ion chromatography, atomic absorption and spectroscopy within one week.

The material balance of the modelecosystems as a whole was observed by measuring the input/output-flowrates of the individual elements and was divided into the canopy and soil compartments by measuring throughfall-flowrates, representing canopy output as well as soil input. These investigations on mineral cycling follow the concepts and methods of the compartment model developed in the course of the German Solling-MAB-project (Ulrich et al. 1979). As an excerpt of data on mineral cycling the present paper gives some results on mineral transport with water flow through the canopy, to differentiate uptake-, washoff- and leaching- processes contributing to canopy throughfall alteration under the influence of air pollutants.

3.   Results

With regard to effects of precipitation forms, figure 2 shows the canopy differences for sulphate, calcium, magnesium and manganese in the subsystems spruce in summer and winter seasons in the different treatments. The pattern of throughfall alteration is almost the same for fir and beech, with absolute values being lower in fir- and higher in beech canopies (data in Seufert 1988). As mentioned above, there was ambient rain in the chambers in winter 84, general rain in the next four seasons, drizzle rain in summer 1986, fog application in winter 86/87 and summer 1987, and ambient rain again in winter 1987/88. CTD-values as calculated from weekly throughfall- minus rain input- flowrates have been summed up to canopy throughfall differences for summer- (May-Oct.) and winter-seasons (Nov.-Apr.). With (+) or (-) flowrates an ion is enriched in throughfall or retained in the canopy. Because there is no extra input of elements in the modelecosystems, with the exception of gaseous sulfur in the SO_2-treatments, leaching is predominating plant uptake in the case of enrichment and vice versa in the case of canopy retainment.

As shown in figure 2, there is a parallelity between sulphate as the dominating anion and the cations considered here. The behaviour of manganese is very similar to sulphate, but manganese is never retained in the canopy. It is obvious that the combined exposure with SO_2+O_3 caused the highest sulphate enrichment and cation leaching, while O_3 only increased zinc flow in throughfall, especially in summer, when O_3 was given in higher concentrations than in winter (data in Seufert 1988).

Furthermore, the effects of the different precipitation forms on canopy difference become evident when comparing summer and winter data for the different years. Compared to the seasons with general rain, leaching effects as a result of SO_2-fumigation were higher by drizzle rain (summer 1986), and were most pronounced in winter 1986/87 and summer 1987 when fog was dispensed. At precipitation forms with needle-wetting several times, foliar uptake of sulphate and calcium in SO_2-free atmosphere was observed, but difinite enrichment in SO_2 atmosphere despite of very low concentrations (see Fig 2). There was definite enrichment in all treatments in winter 1984 with ambient rain , where a wash off of material deposited before the experiment started should be involved. In contrast, ambient rain in winter 1987/88 showed throughfall alterations similar to artificial fog. There is no regular difference between summer and winter seasons despite of definite difference in SO_2-concentration.

A summary of all canopy-throughfall observations is given in Tab. 1, where only effects of enrichment (+) or retainement (-) of the different ions are marked, which have been observed at all three tree species. The different treatments (b-f) are related to the pH 4-control (a). The effects of the time needles are wetted have been determined from the seasons with general rain in comparison to seasons with drizzle rain or fog and are described for the variants without (g) and with (h) SO_2. However, the effects of physiological activity (i) and of phloem sucking insects (k), which can be observed over the course of the year, are independent of treatments.

With pH 4-irrigation in filtered air (a) there can be observed moderate leaching of SO_4, Ca, Mg, Zn, Na, and Cl, whereas K and Mn show definite enrichment of rain input by 4-7 ordes of magnitude. Leaching of Ca, Mg, Mn and especially Zn is increased to some extent by proton concentration, in comparison of pH 4- to pH 5 (b) rain. Ozone fumigation (c) did not show any leaching effects, with the exception of Zn. After SO_2-treatment (d), a definite enrichment of sulphate is connected with leaching of Ca, Mg, Mn, Zn and NH_4, altogether leading to an enhanced acid input to the soil. In combination with Ozone (e) SO_4, Ca and Mn are further enriched, in the conifer canopies Mg too. Ambient air (f) shows the same pattern of ion enrichment as SO_2-fumigation, but not as defi-

310

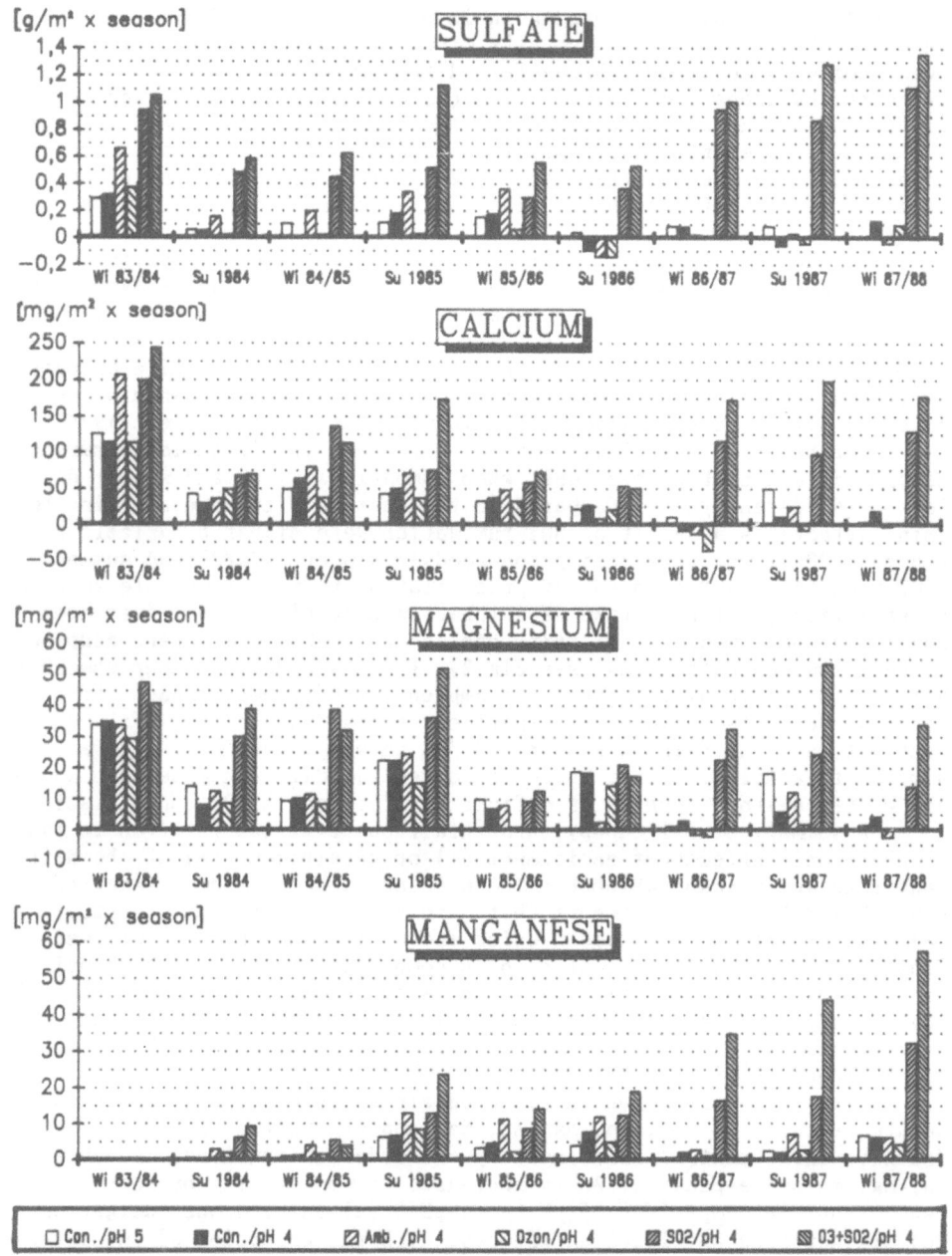

Figure 1. Annual pattern of the O_3 and SO_2 concentrations ($\mu g/m^3$) from Sep. 83 - Mar. 88 in the open-top chambers for the O_3, SO_2 and SO_2+O_3 treatments.

nite. After prolonged needle-wetting because of precipitation form SO_4, Ca and Mg are retained in the canopy of trees growing in SO_2-free air (g); with SO_2 (h), these ions are enriched together with Mn and in particular with protons. In summer- compared to winter seasons (i) K is enriched, whereas NH_4, NO_3 and PO_4 are retained in the canopy. Infestation of phloem-sucking insects (k) shows definite enrichment of K, to some extent of Cl and Mg.

Table 1. General effects of different treatments and influences on mineral transport through spruce-, fir- and beech-canopies

affected ions / affecting parameters	SO₄	Ca	Mg	K	Mn	Zn	H	NH₄	NO₃	PO₄	Na	Cl
a) rain pH 4	+	+	+	+++	+++	+	--	0	0	-	+	+
b) rain pH 5	0	-	-	0	-	--	+	0	0	0	-	0
c) ozone	0	0	-	0	0	++	0	0	-	0	+	0
d) SO₂	++	++	++	0	++	++	++	++	+	0	0	0
e) O₃ + SO₂	+++	+++	++	0	+++	++	++	++	+	0	0	0
f) ambient air	+	+	+	0	++	n	+	++	+	0	+	n
g) wetting without SO₂	-	--	-	0	0	n	0	+	n	n	n	0
h) wetting with SO₂	++	++	+	0	++	n	+++	+	n	n	n	0
i) physiol. activity	0	0	0	++	0	0	0	--	-	-	0	0
k) phloem sucking	0	0	+	+++	0	0	--	0	0	n	0	++

meaning of symbols:
through the influence the ion-flowrate in the canopy will be

	somewhat	definite	strongly	
enriched	+	++	+++	0 = no trend visible
reduced	-	--	---	n = no data

4. Discussion

The results confirm in several cases the observations made in European forests. For example, if one looks at sulphate enrichment in different conifer stands compared to rain input (rev. in Seufert 1988), the question arises where all this sulfur is coming from. Our data suggest that leaching is not important, whereas the dry deposition of SO_2 is followed by a clear washing off of sulphate and is accompanied by canopy leaching of Ca, Mg, Mn, NH_4 and Zn, all this leading to an enhanced acid input.

The definite differences between the SO_2 and SO_2+O_3 treatments are remarkable, since the exposure regime of SO_2 in both cases is identical, but they are difficult to discuss, because ozone alone did not show any clear effects. Possibly the SO_2- driven effects are promoted during the simultaneous presence of ozone, through disturbed stomatal regulation or through increased permeability of the cuticle. SEM investigations on stomatal wax plugs showed melted wax structures in current year needles after pH 4-, but not pH 5-irrigation, with an increasing extent of melting in the order O_3-, SO_2-, O_3+SO_2-fumigation (Schmitt et al. 1986). Increased wettability of fumigated spruce needles, as indicated by the measurement of the contact angles with a goniometer, of needles dipped into water, shows the same order (Biermann 1987).

Apart from observations on enhanced wettability and permeability of the cuticle, there are some theoretical indications to explain the enhanced uptake and turnover of sulfur in the O_3+SO_2-treatment compared to SO_2 alone. After overcoming the atmosperic resistance, the SO_2 deposition is limited by the surface resistance of the receptor. If this is minor when the surface is wet, then further SO_2 uptake is mainly limited by the removal of the physically dissolved SO_2 through oxidation to sulphate (Fowler 1980). It is known in atmospheric chemistry that this process is promoted by ozone as well as by manganese as a catalyst (Dlugi et al. 1987). The SO_2-oxidation by ozone is further accelerated in the presence of terpens as emitted from spruce twigs (Stangl et al. 1988).

The main mechanism underlying leaching and uptake of dissolved substances must be cuticular transport, because penetration of water solutions with natural surface tension through stomata is prevented by the architecture of the stomatal apparatus (Schönherr and Ziegler 1975), with wax plugs in the stomatal antechamber as an additional barrier in conifer species. Leaching could be explained as passive transport of ions through cuticular pores formed by polar functional groups which are hydrated in the presence of water (Schönherr 1982). Cations at these sites are exchanged by hydrogen from solutions wetting the surface and are replaced from exchangable cation pools in free space areas within the plant (Tukey 1970). In contrast to the situation in clean air with carbonic acid as coreactant, cation leaching is promoted after SO_2-deposition because sulphuric acid is always dissociated.

Besides leaching from the exchangable pool in the free space, to some extent cations from within the cells could have been leached because biochemical investigations made by other groups have indicated membrane damage as a direct metabolic effect of SO_2 and O_3, with effects of SO_2+O_3 being more than additive (Bender et al. 1986, Mehlhorn et al. 1986). Especially K and Mn are lost in great quantities after membrane damage, as has been shown in leaching experiments with cut twigs after artificial frost stress (Seufert 1988).

As a conclusion with regard to the extent of dry deposition of SO_2 and subsequent leaching of cations, the data suggest that SO_2 concentration per se is not as important as weather conditions promoting dry deposition and conditions promoting SO_2 oxidation to sulphate such as the presence of ozone, manganese, biogenous emitted terpens, as a prerequisite for further SO_2 deposition.

ACKNOWLEDGMENTS. Funding of the experiment by the Federal Minister of Agriculture and Forestry (BML, Bonn) and helpful discussions with Prof. U. Arndt are gratefully aknowledged.

5. References

Adam, A., Evers, F.H. and Littek, T. (1987) "Ergebnisse niederschlags-analytischer Untersuchungen in südwestdeutschen Wald-Ökosystemen 1981-1986." KfK-PEF Nr. 24, 119 pp.

Bender, J., Jäger, H.J., Seufert, G. and Arndt, U. (1986) "Untersuchungen zur Einzel- und Kombinationswirkung von SO_2 und O_3 auf den Stoffwechsel von Waldbäumen in Open-Top-Kammern." Angew. Bot. 60, 459-477.

Biermann, J. (1987) "Mikromorphologische und physikochemische Veränderungen pflanzlicher Oberflächen unter Immissionseinfluß." Dr. Thesis Univ. Gießen, 122 pp.

Dlugi, R., Jordan, S., Mangold, S. and Mätzing, H. (1987) "Die Entstehung sulfat- und nitrathaltiger Partikel und Tropfen in der Atmosphäre." KfK-PEF Nr. 29, 78 pp.

Evers, F.H. (1981) "Ergebnisse ernährungskundlicher Erhebungen zur Tannenerkrankung in Baden-Württemberg." Forstw. Cbl. 100, 253-265.

Evers, F.H. (1983) "Orientierende Untersuchungen langfristiger Bodenreaktionsänderungen in südwestdeutschen Düngungs-Versuchsflächen." Forst-und Holzwirt 38, 317-320.

Evers, F.H. and Seufert, G. (1989) "Wasser-, nadel- und bodenalalytische Begleituntersuchungen zu den Versuchen mit oben offenen Kammern und dem Luftmeßprogramm beim Edelmannshof." KfK-PEF, in print.

Fowler, D. (1980) "Removal of sulphur and nitrogen compounds from the atmosphere in rain and by dry deposition." in D. Drablos and A. Tollan (eds.), Ecological impact of acid precipitation. SNSF-project, Oslo, ISBN 82-90376-07-3, pp. 22-32.

Mehlhorn, H., Seufert, G., Schmidt, A. and Kunert, K.J. (1986) "Effect of SO_2 and O_3 on production of antioxidants in conifers." Plant Physiol. 82, 336-338.

Schmitt, U., Rütze, M. and Liese, W. (1986) "Rasterelektronische Untersuchungen an Stomata von Fichten- und Tannennadeln nach Begasung und saurer Beregnung." Eur. J. For. Path. 17, 118-124.

Schönherr, J., Ziegler, H. (1975) "Hydrophobic cuticular ledges prevent water entering the air pores of liverwort thalli." Planta 124, 51-60.

Schönherr, J. (1982) "Resistance of plant surfaces to water loss: transport properties of cutin, suberin and associated lipids." Encyclop. Plant Physiol., XII B, Springer Verlag, Berlin, 153-179.

Seufert, G. and Arndt, U. (1985) "Open-top-Kammern als Teil eines Konzepts zur ökosystemaren Untersuchung der neuartigen Waldschäden. Allg. Forstzeitschr. 40, 13-18.

Seufert, G. (1988) "Untersuchungen zum Einfluß von Luftverunreinigungen auf den wassergebundenen Stofftransport in Modellökosystemen mit jungen Waldbäumen." Ber. Forschz. Waldökosys., Göttingen, No.44, 258 pp.

Stangl, H., Kotzias, D. and Geiss, F. (1988) "How forest trees aktively promote acid deposition." Naturw. 75, 42-43.

Tukey, H.B. (1970) "The leaching of substances from plants." Ann. Rev. Plant Physiol. 21, 305-324.

Ulrich, B., Mayer, R., and Khanna, P.K. (1979) "Deposition von Luftverunreinigungen und ihre Auswirkungen in Waldökosystemen im Solling." Sauerländer Verlag, Frankfurt, 291 pp.

EFFECTS OF EXHAUST GAS ON VARIOUS PLANT SPECIES IN FUMIGATION
CHAMBERS UNDER CONTROLLED CONDITIONS

R. JURAT, U. KNORRE, J. DÄHNE, M. KUPRIAN, J. HENRICH,
H. SCHAUB
Botanisches Institut der Johann Wolfgang Goethe -
Universität Frankfurt, Siesmayerstraße 70
D-6000 Frankfurt / Main 11

ABSTRACT. Four species of annual plants were exposed to exhaust gas in
fumigation chambers under controlled conditions. Biomass yield, organic
N-content and activities of N-assimilating enzymes were affected
differently in various plant organs in dependence on species and day-
or night-time fumigation, respectively. Nitrate uptake rates from the
medium were altered when only the shoots were fumigated.

1. INTRODUCTION

Plant stress due to automobile emissions is discussed for many years.
As reported by Nobel and Michenfelder (1987) vegetation near highly
frequented motor roads is affected whereas plants from clean air areas
are not. Different effects of engine emissions on plants in dependence
on the distance from a motorway may be due to concentration gradients
as well as to changes in exhaust composition, climatic conditions and
the additional presence of other air pollutants. NO and NO_2 are components of exhaust gas and known as stressors to vegetation for many
years (Koziol, 1984). However, the mixture of various pollutants may
affect plants synergistically or antagonistically. Thus, only investigations with exhaust gas in controlled fumigation experiments provide a
solid basis for our understanding of the effects observed outdoor. In
our study effects of low levels of exhaust gas on growth and nitrogen
metabolism of various annual plant species were examined.

2. MATERIALS AND METHODS

All plant species except Phaseolus vulgaris were cultivated from seed in
a slightly modified Hoagland nutrient solution. Phaseolus vulgaris was
grown in sand culture and watered daily with 1:10 diluted (v/v) nutrient
solution. 14 days old bean plants, 20 days old plants of Helianthus
annuus, Cucumis sativus and Zea mays and 26 days old plants of Brassica
napus were transferred into the combined growth and fumigation chambers

315

H.-W. Georgii (ed.), Mechanisms and Effects of Pollutant-Transfer into Forests, 315–322.
© 1989 by Kluwer Academic Publishers.

(Jurat and Schaub, 1988). Nutrient solution was separated from the atmosphere by an elastic, gastight plastic material. The plants were exposed to exhaust gas for 10 days each under the following conditions: 16 h photoperiod at 600 $\mu E\ m^{-2}\ s^{-1}$ (PAR), day/night temperature 22°C/ 17°C, rel. humidity 60-65%, CO_2 concentration 300 ppm. Fumigation was either continuous at an NO_x concentration of 30 ppb, or 8 h/d at 90 ppb NO_x during the light and dark period, respectively. Control plants were grown in filtered air.

Constant levels of exhaust gas in the chamber were achieved by continuous monitoring NO_x (NO and NO_2). When the actual NO_x concentration diverged from the set point, diluted exhaust gas was pumped into the chamber until the set point was reached again. Exhaust gas was produced under controlled conditions, diluted 1:12, filled in special bags and analysed by Prof. Dr. Hohenberg, TH Darmstadt. Since the composition of the exhaust gas changed during the first 48 hours, it was used for fumigation earliest from the 3rd day on when the composition was constant (NO_x/HC = 1.87).

During the fumigation period rates of nitrate uptake from the nutrient solution by Helianthus annuus and Cucumis sativus were estimated three times over a period of 24 hours as described by Jurat and Schaub (1988). Fresh weight and water uptake rates of the intact plants were determined by weighing, water culture pots without plants serving as blanks.

At the end of the fumigation period roots, stem parts and leaves were separated, weighed subsequently, then frozen in liquid nitrogen and freeze dried. Enzyme activities were estimated in fresh plant material. N-contents of the plants were determined according to Kjeldahl (Co. Büchi). Enzyme activities were estimated with reference to Agüera et al. (1987) and Guerrero (1982) for nitrate and nitrite reductase and to Manderscheid and Wild (1986) for glutamine synthetase.

3. RESULTS AND DISCUSSION

3.1. Biomass production

At the end of the fumigation period plants did not show any visible injury. However, fumigation during the light period (8 h/d) decreased total dry weight of Brassica napus and Phaseolus vulgaris by 5 and 4%, whereas the other species showed slightly increased values compared to the control. These results are in agreement with Bialobok (1984) who found different sensitivities of various species to NO_x pollution. According to Mansfield (1987) the dry deposition of NO_2 is likely to produce growth stimulation in some circumstances. Continuous exposure (24 h/d) to exhaust gas also slightly increased dry weights of Helianthus annuus, Cucumis sativus and Zea mays, however growth inhibition was found after fumigation during the dark period (8 h/d) (data not shown). Different reactions of Raphanus sativus to NO_2 and SO_2 fumigation during day- and night-time were also reported by Hogsett et al. (1984).

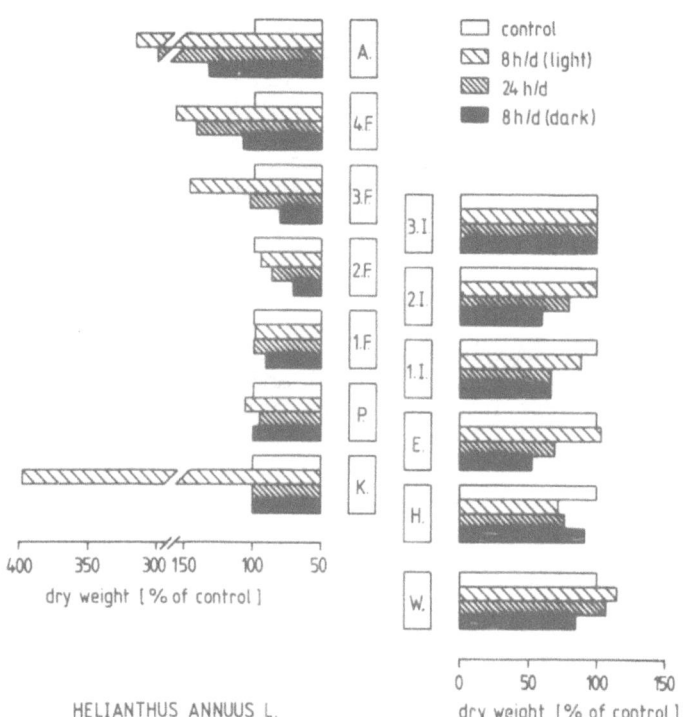

Figure 1. Dry weights of Helianthus annuus (per cent of control) fumigated with exhaust gas for 10 days either continuously (24 h/d) at a
NO_x concentration of 30 ppb or 8 h/d at 90 ppb NO_x during the light and
dark period, respectively. The data represent means of 10 plants for
each treatment. A.: apical leaves, E.: epicotyl, 1.F.-4.F.: 3rd-6th
pair of leaves, H.: hypocotyl, I.: stem parts, K.: cotyledons, P.:
primary leaves, W.: root.

As shown in Figure 1 various plant parts of Helianthus annuus were
affected differently by fumigation. Dry weights of the apical leaves
were drastically increased by 200-216% when plants were exposed to the
gas 24 h/d and 8 h/d during the light period, respectively. Biomass
production of the cotyledons of plants fumigated during the light
period was also increased. However, in the 4th pair of leaves (2.F.)
decreased values were found compared to the control.

Fumigation during the dark period induced only slightly increased dry weights of the apical leaves, but reduced biomass production in the 4th and 5th pair of leaves (2.F. and 3.F.) by 19–29%. Stem and root growth was not affected so drastically as the leaf dry weight. According to Eastham and Ormrod (1986) leaf growth of Populus canadensis was stimulated by NO_2 whereas stem growth was inhibited. Gould and Mansfield (1988) found an increase in shoot-to-root ratios when plants of Triticum aestivum were fumigated with a mixture of NO_2 and SO_2. From the strong increase in apical biomass of Helianthus annuus shown here one might conclude that plant stability is affected.

3.2. Nitrogen metabolism

NO and NO_2 from the atmosphere may be assimilated by plants (Koziol, 1984) and according to Ito et al.(1986) and Rowland (1986) organic N-contents of plants were increased after fumigation with NO_2.

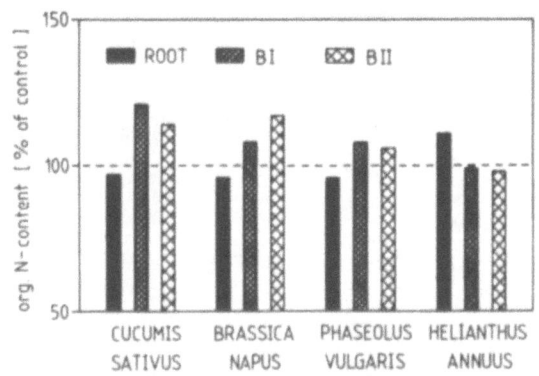

Figure 2. N-contents of four plant species per g dry weight (per cent of control) fumigated with exhaust gas for 10 days 8 h/d during the light period at a NO_x concentration of 90 ppb. The data represent means of 10 plants of each species. BI: basal leaves (cotyledons – 4th pair of leaves), BII: apical leaves (5th pair of leaves – apical leaves).

As shown in Figure 2 exposure to exhaust gas also increased organic N-contents in the leaves of Cucumis sativus, Brassica napus and Phaseolus vulgaris whereas root concentrations were slightly decreased compared to the control. However, N-levels of Helianthus annuus were affected differently: Increased root values and slightly decreased concentrations in the leaves. From these data one might assume that nitrogen transport may be affected differently in various species. As reported by Gould and Mansfield (1988) the distribution of 14C-labelled assimilates in Triticum aestivum was affected by NO_2 and SO_2.

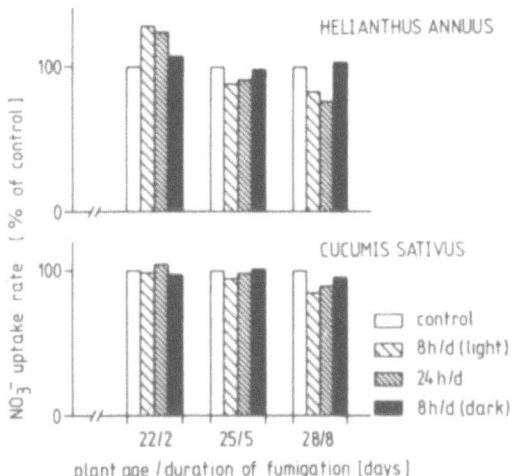

Figure 3. Nitrate uptake rates per g fresh weight of intact plants of Helianthus annuus and Cucumis sativus (per cent of control) fumigated with exhaust gas for 10 days either continuously (24 h/d) at a NO$_x$ concentration of 30 ppb or 8 h/d at 90 ppb NO$_x$ during the light- and dark-period, respectively. The data represent means of 10 plants for each treatment.

Additionally, nitrate uptake rates from the medium were changed by fumigation in dependence on species, plant age and duration of fumigation, respectively (Fig. 3). After a fumigation period of 2 days nitrate uptake of Helianthus annuus was increased compared to the control. Later on uptake rates of both Helianthus annuus and Cucumis sativus decreased under the control values when plants were exposed to the exhaust gas continuously or 8 h/d during the light period. The strongest inhibition was found in nitrate uptake of Helianthus annuus when grown under continuous stress for 8 days.

After a fumigation period of 10 days (continuous treatment) nitrite reductase (NiR) activity of the apical leaves of Helianthus annuus was strongly increased whereas reduced values were found in the root and the basal leaves (Fig. 4). Nitrate reductase (NR) activity was also inhibited in the root and the basal leaves. In contrast, glutamine synthetase (GS) activity was slightly increased in both roots and basal leaves and decreased in the apical part of the shoot.

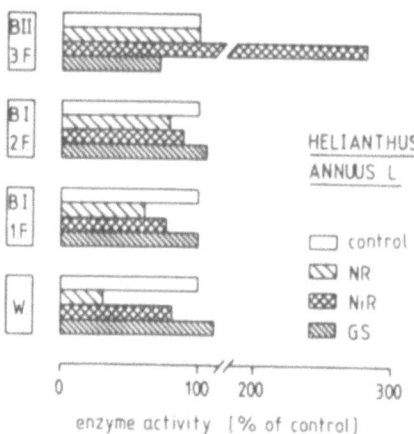

Figure 4. Activities of N-assimilating enzymes of Helianthus annuus (per cent of control) fumigated with exhaust gas for 10 days continuously (24 h/d) at a NO$_x$ concentration of 30 ppb. GS: glutamine synthetase, NR: nitrate reductase, NiR: nitrite reductase.
BI: basal leaves (cotyledons - 4th pair of leaves), BII: apical leaves (5th pair of leaves - apical leaves).

The strong increase in NiR activity may be due to an enhanced nitrogen metabolism in the younger, still expanding leaves exposed to a NO$_x$ containing atmosphere. These data are in agreement with results reported by Takeuchi et al.(1985) and Hisamatsu et al.(1988) who found decreased NR activities after fumigation with NO$_2$. According to Tischner et al. (1988) activities of various N-assimilating enzymes in seedlings of Picea abies were increased in the shoot, however decreased or only slightly altered in the root. Fumigation with NO also enhances nitrogen metabolism as indicated by increased enzyme activities (Wellburn, 1980). However, it has to be considered that exhaust gas is a mixture of various components containing not only NO$_x$ but also hydrocarbons. The results presented in this paper demonstrate that growth and nitrogen metabolism of various plant species is affected when the shoots are exposed to exhaust gas. As the roots were not in contact with the atmosphere, direct effects of the pollutant on the rhizosphere can be excluded.

4. ACKNOWLEDGEMENTS

The authors thank the BMFT and FAT for financial support and R. Pathe for her technical assistance.

5. REFERENCES

Agüera, E., De la Haba, P. and Maldonado, J. M. (1987) 'In vitro stabilization and tissue distribution of nitrogen assimilating enzymes in sunflower', J. Plant Physiol. 128, 443-449.

Bialobok, S. (1984) 'Controlling atmospheric pollution', in M. Treshow (ed.), Air pollution and plant life, John Wiley and Sons Ltd, pp. 451-478.

Eastham, A. M. and Ormrod, D. P. (1986) 'Visible injury and growth responses of young cuttings of Populus canadensis and Populus nigra to nitrogen dioxide and sulfur dioxide', Can. J. For. Res. 16(6), 1289-1292.

Gould, R. P. and Mansfield, T. A. (1988) 'Effects of sulfur dioxide and nitrogen dioxide on growth and translocation in winter wheat', J. Exp. Bot. 39(201), 389-399.

Guerrero, M. G. (1982) 'Assimilating nitrate reduction' in I. Coombs and D. O. Hall (eds.), Techniques in bioproductivity and photosynthesis, Pergamon Press, Oxford, pp. 124-130.

Hisamatsu, S., Nihira, J., Takeuchi, Y., Satoh, S. and Kondo, N. (1988) 'NO_2 suppression of light induced nitrite reductase in squash melon', Plant Cell Physiol. 29(3), 395-401.

Hogsett, W. E., Holman, S. R., Gumpertz, M. L. and Tingey, D. T. (1984) 'Growth responses in radish to sequential and simulta-neous exposures of NO_2 and SO_2', Environ. Pollut. Ser. A Ecol. Biol. 33(4), 303-306.

Ito, O., Okano, K. and Tosuka, T. (1986) 'Effects of nitrogen dioxide and ozone exposure alone or in combination on kidney bean plants: Amino acid content and composition', Soil Sci. Plant Nutr. 32, 351-364.

Jurat, R. and Schaub, H. (1988) 'Effects of sulfur dioxide and ozone on ion uptake of spruce (Picea abies (L.) Karst.) seed-lings.', Z. Pflanzenernährung Bodenkunde 151, 379-384.

Koziol, M. J. and Whatley, F. R. (1984) Gaseous air pollutants and plant metabolism, Butterworths, London.

Manderscheid, R. and Wild, A. (1986) 'Characterization of gluta-mine synthetase of roots, etiolated cotyledons and green leaves from Sinapis alba L.', Z. Naturforsch. 41c, 712-716.

Mansfield, T. A., Whitmore, M. E., Pande, P. C. and Freer-Smith, P. H. (1987) 'Responses of herbaceous and woody plants to the dry deposition of SO_2 and NO_2', in T.C. Hutchinson and K. M. Meema (eds.), Effects of atmospheric pollutants on forests, wetlands and agricultural ecosystems, Nato Asi Series G (16), Springer Verlag, Berlin.

322

Nobel, W. and Michenfelder, K. (1987) 'Untersuchungen über Wirkungen von Automobilabgas auf pflanzliche Bioindikatoren im Umfeld einer verkehrsreichen Straße in einem Waldschadensgebiet', FAT-Schriftenreihe Nr. 63, pp. 276.

Rowland, A. J. (1986) 'Nitrogen uptake, assimilation and transport in barley in the presence of atmospheric nitrogen dioxide', Plant Soil 91, 353–356.

Takeuchi, Y., Nihira, J., Kondo, N. and Tezuka, T. (1985) 'Change in nitrate-reducing activity in squash seedlings with NO_2 fumigation. Plant Cell Physiol. 26, 1027–1035.

Tischner, R., Peuke, A., Godbold, D. L., Feig, R., Merg, G. and Hüttermann, A. (1988) 'The effect of NO_2 fumigation on aseptically grown spruce seedlings', J. Plant Physiol. 133, 243–246.

Wellburn, A. R., Wilson, J. and Aldride, P. H. (1980) 'Biochemical responses of plants to nitric oxide polluted atmospheres', Environ. Pollut. 22, 219–228.

STUDIES ON POPLAR CLONES AFTER EXPOSURE IN OPEN-TOP CHAMBERS TO FLUE GAS FROM AN ATMOSPHERIC FLUIDISED BED-BOILER (AFBB)

JAN MOOI
Research Institute For Plant Protection
P.O. Box 9060
6700 GW Wageningen
The Netherlands

HANS-JOACHIM BALLÁCH
Institute For Applied Botany
University Of Essen
Henri-Dunant-Str. 65
D-4300 Essen
West Germany

ABSTRACT. The first results and an overview of a research project to determine the phytotoxicity of complex mixtures of air pollutants on the basis of biochemical, chemical and morphological investigations are given. Rooted stem cuttings of two poplar clones with differing tolerances to air pollutants (Populus nigra 'Loenen' and Populus maximowiczii 'Rochester') were fumigated in open-top chambers at six-week periods with flue-gas, emitted by a hard-coal-fired atmospheric fluidised bed boiler (AFBB). The computer-controlled fumigation system used permitted the study of flue-gas phytotoxicity in comparison with its main constituents SO_2, NO_2 and NO.

Depending on the pollutant mixtures, the poplars exhibited more or fewer premature leaf drop, morphological changes of the leaf surfaces, increased sulphur contents of the leaves, as well as changes in the contents of carbohydrates, starch and chlorophylls.

A preliminary conclusion is that the phytotoxicity of the flue-gases from an AFBB is, under the experimental conditions chosen, decisively determined by their main constituents SO_2, NO_2 and NO. Admittedly, some individual symptoms are gradually intensified by the flue-gases in proportion to the main constituents, but a different course of damage has not so far become evident.

1. INTRODUCTION

Up to now, the effects of major pollutants (e.g. O_3, SO_2 and NO_x) have been the most frequently studied kind of mixture response, but there has been little research on more complex mixtures involving minor

323

H.-W. Georgii (ed.), Mechanisms and Effects of Pollutant-Transfer into Forests, 323–331.
© *1989 by Kluwer Academic Publishers.*

gaseous pollutants. The flue-gas emitted by the AFBB represents an example of a complex pollutant mixture consisting of SO_2, NO_2 and NO as the main constituents and substances such as HCl, HF and different hydrocarbons as minor pollutants.

The first fumigation experiment was carried out in 1987 using higher pollutant concentrations. In open-top chambers with flue-gases or their main constituents, the average SO_2 concentrations were 125 µg/m³, the NO_2-concentrations 112 µg/m³ and the NO-concentrations 36 µg/m³. In a second experiment, carried out in 1988, the SO_2-, NO_2- and NO-concentrations were reduced by approximately a half. Thus, at least the 1988 experiment was performed with realistic concentrations of air pollutants as measured, for example, in the outer regions of the Ruhr Area (LIS-Monatsberichte, 1983 - 1989).

Poplars have long been known to be suitable for use in research into the effects of air pollution, because, among other reasons, cloned material is available. Various clones show differing degrees of resistance to air pollution. For example, the cuttings of Populus nigra 'Loenen' used in our experiments react more sensitively to pollutant load than the Populus maximowiczii 'Rochester' also used in this study. Rapid growth is also a further criterion for selection in exposure experiments. This characteristic led in the post-war years to an increased cultivation of poplars in the Federal Republic of Germany. Today, the tree plays a subordinate economic role, with a few regional exceptions such as the Rhenish lignite mining areas, where it is used as a pioneer species. This is different to the Netherlands, where the poplar accounts for 30 % of the total timber production.

2. MATERIAL AND METHODS

The exposure system, the atmospheric fluidised bed boiler (AFBB), the methods of pollutant dispensing and monitoring as well as methods of cultivation of the poplars are described in Meijer and Mooi, 1987 and Ballach et al., 1988. The morphological analyses were performed by Prof. Dr. Greven (University of Düsseldorf). Preparation of the leaf samples was accomplished using 4 % glutaraldehyde fixation for several hours in 0.05 M cacodylate buffer. (pH 7.2 - 7.4). Samples were dehydrated, critical-point-dried (CPD) and sputter-coated with Au. All specimens were examined with a Leitz AMR 1000 SEM. The methods of biochemical and chemical analysis are described in Ballach et al. (1988) or will be published (Ballach in preparation).

3. RESULTS AND DISCUSSION

The AFBB and open-top chambers (OTCs) are depicted in Fig. 1 and a schematic diagram is given in Fig. 2.

Figure 1. Open-top chambers and hard-coal-fired plant.

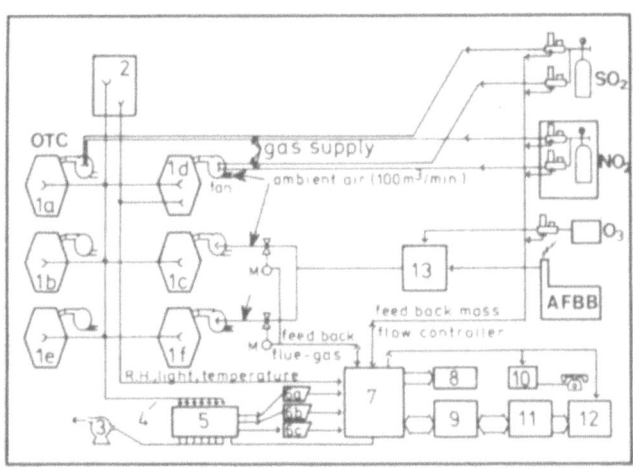

Figure 2. Schematic diagram of the open-top chambers and the AFBB installation.

A detailed survey of the average air pollution concentrations can be seen in Fig. 3 and Tab. 1.

326

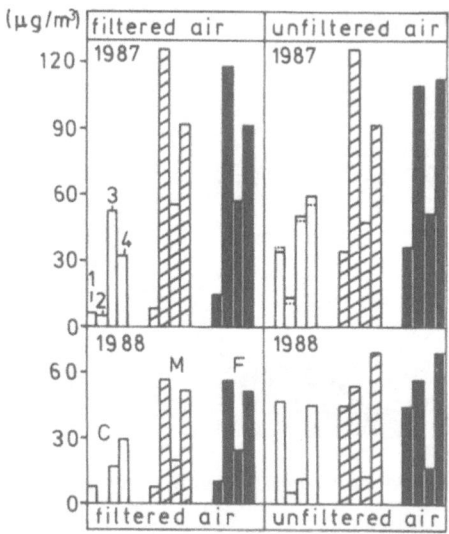

Figure 3. Average air pollution concentrations in 1987 and 1988
in the fumigation treatments with filtered and unfiltered air.
White columns (C): control chambers, hatched columns (M): open-
top chambers with main constituents of flue-gas, black columns (F):
OTCs with flue gas; 1: O_3, 2: SO_2, 3: NO, 4: NO_2.

TABLE 1. Average air pollution concentrations calculated from
876 hourly values in 1987 and from 922 hourly values in 1988.
The exposure periods of poplars were from May 11 to June 22,
1987 and from May 25 to July 6, 1988.

OTC	filtered air						not filtered air					
	control		main components flue-gas		flue-gas		control		main components flue-gas		flue-gas	
year	1987	1988	1987	1988	1987	1988	1987	1988	1987	1988	1987	1988
component ($\mu g/m^3$) SO_2	5	0	125	56	117	56	13	5	125	53	109	56
O_3	6	8	8	8	14	10	34	46	34	44	36	44
NO	52	17	55	19	57	24	50	11	47	12	51	16
NO_2	32	29	91	51	91	51	59	44	91	68	112	68

The concentrations of total hydrocarbons in the open-top chambers with flue-gas were between 5 and 10 ppb.

3.1 Morphological changes and premature leaf drop

The foliar surface of younger and older leaves from both clones exposed to air pollutants in 1988 was studied by scanning electron microscopy (SEM). First results from the 'Loenen' clone show that the flue-gas and its main constituents added to filtered ambient air caused surface changes such as shrunken cell walls of the stomata and damage to the cuticular. Symptoms such as shrunken guard and accessory cell walls of poplar leaves induced by mixtures of air pollutants have been reported by Krause and Jensen (1979) and Krause (1980). It seems worth mentioning that surface changes, even of younger leaves from Populus nigra 'Loenen' which were macroscopically symptomless, were observable by SEM (Fig. 4).

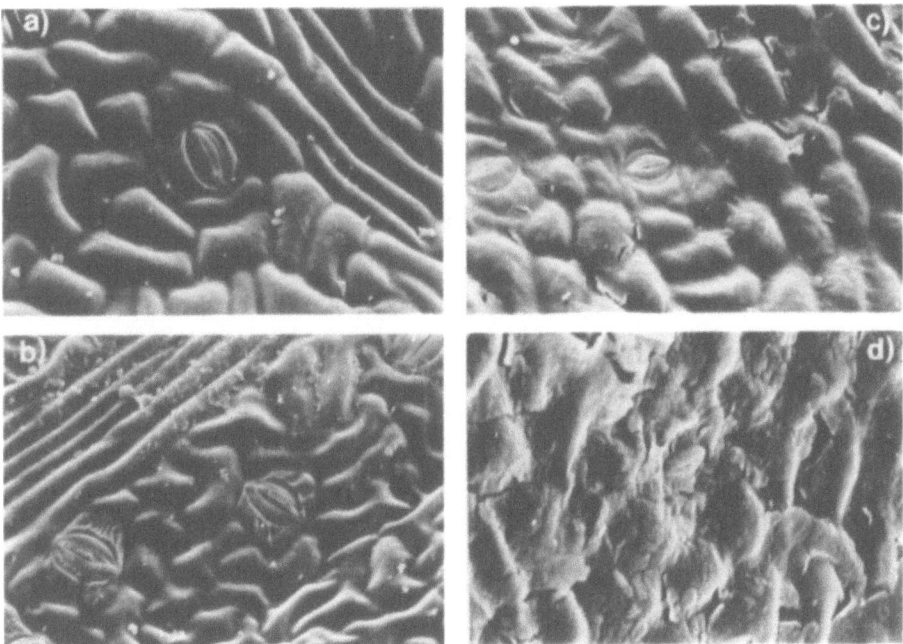

Figure 4. Surfaces of poplar leaves ('Loenen' clone) from control chambers (a,b) and for comparison after exposure to main constituents of flue-gas (c) and flue-gas (d). All OTCs with filtered air, exposure time: May 25 to July 6, 1988. Specimens a and b were from older leaves and c and d from younger leaves.

Severe injury symptoms on poplars, i.e. premature defoliation (Fig. 5) and a reduction of dry matter production, are known to be caused by

ozone alone or in combination with SO_2 and/or NO_2 (Noble and Jensen, 1980; Mooi, 1983, 1984).
Due to the higher pollutant concentrations in 1987, the 'Loenen' clone showed greater leaf drop than in 1988 (Fig. 6).

Figure 5. Extent of premature leaf drop of Populus nigra 'Loenen' from the 1987 studies. F: filtered air, NF: unfiltered air.

Figure 5 shows that an increase in ozone concentration was accompanied by an increase in premature leaf drop. This was true for both clones, but particularly so for 'Loenen' and here especially for the control plants with unfiltered air. The more resistant 'Rochester' clone showed, without exception, significantly lower leaf abscission than the 'Loenen' clone. In the experimental year 1988, but not at the higher pollutant concentrations of the previous year, the flue-gas from the AFBB led, in contrast to its main constituents, to a slight increase in leaf drop of the 'Loenen' clone (Fig. 6).

3.2 Sulphur contents of the leaves

In the poplars of the 'Loenen' clone, fumigated in 1987, there were elevated sulphur contents, especially in the older leaves, which lay above the values of the middle-aged and young leaves. The leaf contents clearly reflect the degree of SO_2-exposure, as a comparison of Fig. 6 and Tab. 1 reveals. Further, it is clear from Fig. 6 that the plants with the highest abscission additionally have the highest sulphur contents in their lower leaves. Only the control plants which were exposed to unfiltered air showed a high degree of abscission caused by ozone and had low sulphur contents in their older leaves.

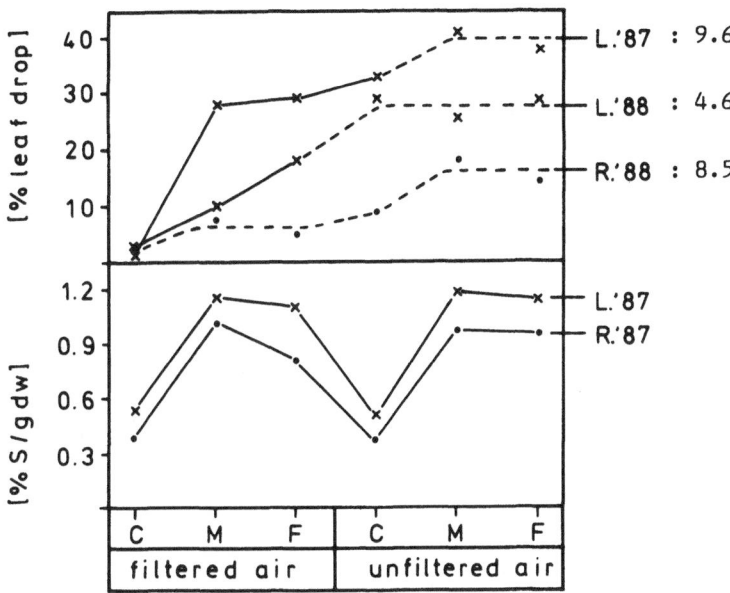

Figure 6. Comparison of premature leaf drop with sulphur contents
(averages of 2 measurements) of the older poplar leaves. LSD: Least
significant difference after analysis of variance (ANOVA) for each
year and each species. L. '87: 'Loenen'clone fumigated in 1987 (n = 8),
L. '88: 'Loenen' fumigated in 1988 (n = 10), R '88: 'Rochester' clone
exposed in 1988 (n = 10); C: control chambers, M: main constituents
of flue-gas, F: flue-gas.

3.3 Biochemical analyses

Those poplars ('Loenen' clone) exposed in 1987 generally revealed in
the older leaves of the fumigated plants higher amounts of soluble
carbohydrates and starch with simultaneously decreased chlorophyll
contents compared with the control plants (Ballach et al., 1988). To
what extent the symptoms of so-called premature senescence occur at
the lower pollutant concentrations in the 1988 fumigation experiments
is being investigated at present. The younger and older leaves and, to
some extent, the roots are being investigated for the following groups
of substances: proteins, amino acids, starch, carbohydrates, caro-
tenoids, chlorophylls and nucleotides. Use of the HPLC-method enables
very detailed investigations to be made as the examples in Figs. 7
and 8 show.

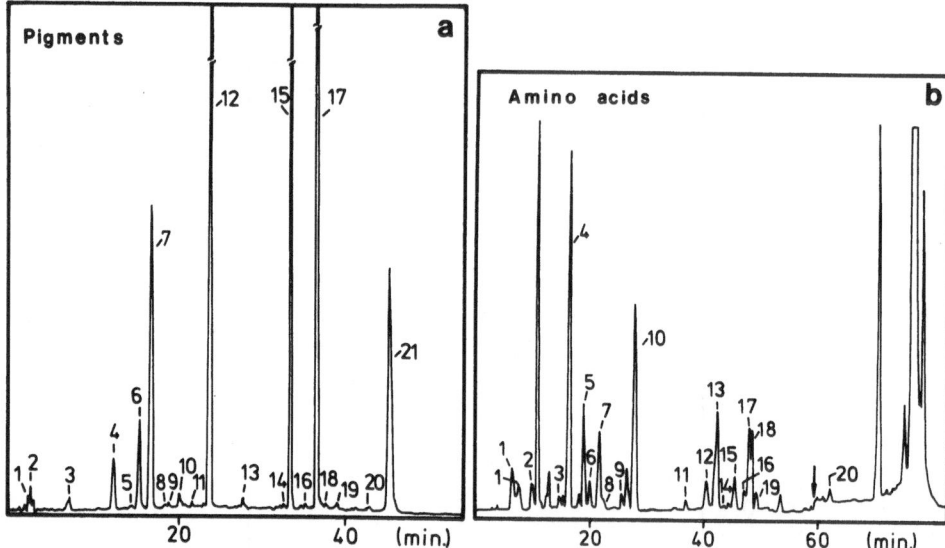

Figure 7. HPLC-profiles of pigments (a) and amino acids (b) from poplar leaves. Explanation of peak-numbers in Fig. 7a: 1= chlorophyllide a, 2= pheophorbide b, 3= ethyl-chlorophyllide b, 4= ethyl-chlorophyllide a, 5= xanthophyll, 6= neoxanthin, 7= violaxanthin, 8, 9= neochromes, 10, 11= xanthophylls, 12= lutein/zeaxanthin, 13= xanthophyll, 14= chlorophyll a, 15= chlorophyll b, 16= unknown chlorophyll, 17= chlorophyll a$_p$, 18= chlorophyll a'$_p$, 19= xanthophyll, 20= pheophytin a, 21= ß-carotene; p= phytol. Peak-numbers in Fig. 7b: 1= Asp, 2= Glu, 3= Asn, 4= Ser, 5= Gln, 6= His, 7= Gly, 8= Thr, 9= Ala, 10= Arg, 11= Tyr, 12= Val, 13= Met, 14= Gaba, 15= Ile, 16= Trp, 17= Phe, 18= Leu, 19= Orn, 20= Lys, 21= Pro.

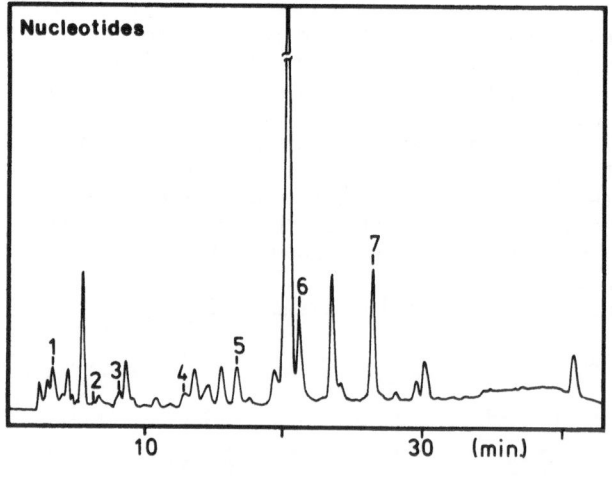

Figure 8a. HPLC-chromatograms of nucleotides from poplar leaves. 1= CMP, 2= UMP, 3= GMP, 4= AMP, 5= GDP (I.S.), 6= ADP, 7= ATP; I.S. = internal standard.

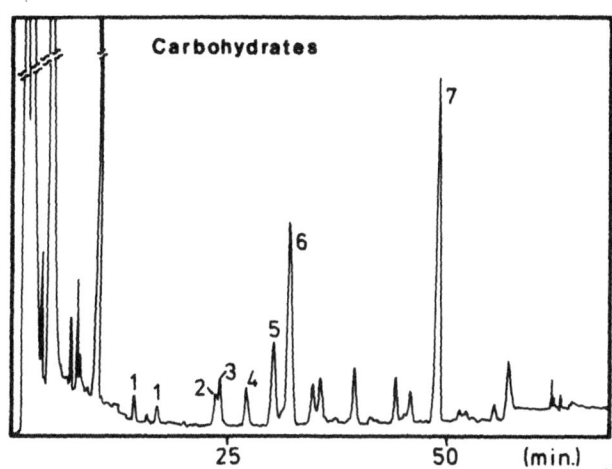

Figure 8b. HPLC-chromatograms of carbohydrates from poplar leaves.
1= fructose, 2= glucose, 3= arabitol ?, 4= myo-inositol, 5= sorbitol,
6= salicin, 7= sucrose.

References

Ballach, H.-J., J. Mooi and J. Bücker (1988) 'Injury of Populus nigra
    'Loenen' caused by flue-gas from an atmospheric fluidised bed boiler
    (AFBB)', Acta Biol. Benrodis 1, 69 - 79.
Krause,C.R. and K.F. Jensen (1979) 'Surface changes on hybrid poplar
    leaves exposed to ozone and sulfur dioxide', SEM-79, III, SEM Inc.,
    AMF O'Hare, IL 60666, USA, 77 - 80.
Krause, C.R. (1980) 'Scanning electron microscopic detection of injury
    to hybrid poplar leaves induced by air pollution', SEM-80, III, SEM
    Inc., AMF O'Hare, IL 60666, USA, 591 - 594.
Landesanstalt für Immissionsschutz des Landes Nordrhein-Westfalen (LIS),
    (1983 - 1989) 'Berichte über die Luftqualität in Nordrhein-Westfalen'.
Meijer, E. and J. Mooi (1987) 'Technische beschrijving / handleiding
    OTC-begassings installatie', IPO-Report R 353, 75 pp.
Mooi, J. (1983) ' Responses of some poplar species to mixtures of SO_2,
    NO_2 and O_3', Aquilo Ser. Bot. 19, 189 - 196.
Mooi, J. (1984) 'Wirkungen von SO_2, NO_2, O_3 und ihrer Mischungen auf
    Pappeln und einige andere Pflanzenarten', Der Forst- und Holzwirt
    39, 438 - 444.
Noble, R.D. and K.F. Jensen (1980) 'Effects of sulphur dioxide and
    ozone on growth of hybrid poplar leaves', Am. J. Bot. 67, 1005 - 1009.

INPUT OF SULPHUR AND ACIDITY IN THREE SPRUCE STANDS IN THE
VICINITY OF A STRONG SO_2-SOURCE

Godt, J. and M. Weyer
Gesamthochschule/Universität Kassel
Landscapeecology/Soil Science
Henschelstr. 2
D - 3500 Kassel

Abstract. Deposition measurements (flux rates) in three
spruce stands in the vicinity of a strong SO_2 source are
discussed concerning the role of long range transport/local
sources, input of air pollutants and reactions of trees on
input of H^+. High total deposition rates of H^+, deriving
from the SO_2 emissions of the paper mill to a great extent,
are resulting in high leaching rates of Ca and K (Mg,Mn) out
of the canopy. Although $CEC_{eff.}$ of soils indicate strong
acidification of all sites, the exchangeable pool of cations
(Ca, Mg, K, Mn) seems to be sufficient for buffering 20-35 %
of total H^+ input in the canopy. Total deposition rates of S
and H^+ exceed critical load levels by far. Official air
pollution measurements according to TA Luft are insufficient
for a complete description of the stress situation of forest
ecoysystems due to air pollutant intake.

1. Introduction

In the vicinity of a strong SO_2 source, a paper mill close to
the town of Hann. Münden FRG, air pollution control has been
carried out according to the TECHNISCHE ANLEITUNG ZUR REIN-
HALTUNG DER LUFT (TA Luft, Bundesminister des Inneren 1983).
As the TA Luft refers to measurements of gases and particles
in the atmosphere and to open field deposition only, additional
measurements, using an ecosystemare approach (analysis of
flux rates) (Ulrich 1983), have been carried out in three
spruce stands, differently exposed with respect to the SO_2
source. The measurements carried out in our investigation (Godt
1988) aimed at the assessment of total deposition of substances
(air pollutants including nutrient elements) to the forest
canopy, the reaction of forest vegetation onto the impact of
pollutants, and at an estimate of the relative contribution
of local sources versus long range transport to the pollution
load.

333

H.-W. Georgii (ed.), Mechanisms and Effects of Pollutant-Transfer into Forests, 333–343.
© *1989 by Kluwer Academic Publishers.*

2. Site description

In tab.1 site charateristics are listed. The three spruce stands are located in a forested mountain area (from 280 to 320 m above sea level) in Southern Niedersachsen, FRG (Fig.1).

Tab. 1 : Site characteristics

District	exposition	soil parent material	type of soil	age (yr) of spruce trees	$l/(m^2.a)$ open field precipitation	pH (soil) in water
1	plateau SSW windward	loess over sand stone	pods. Braunerde	76	844	0-5 cm: 3.3 30-40 cm: 4.3
53	mountain ridge close to SO_2-source	loess over sand stone	pods. (Para-braunerde) Pseudogley	74	927	0-5 cm: 4.3 30-40 cm: 4.4
101	mountain ridge (leeward)	loess over sand stone	pods. Para-braunerde	110	901	0-5 cm: 3.3 55-65 cm: 4.2

The paper mill is located in a narrow valley on the border of the river Fulda. Meteorology is characterised by a high frequency of temperature inversion, south-westernly winds are predominating. One spruce stand is located in the windward (distr. 1) 3 km away from the SO_2 source. The spruce stand in distr. 53 is located in the direct neighbourhood, one spruce stand in the leeward (distr. 101) 3 km away from the paper mill (fig. 1).

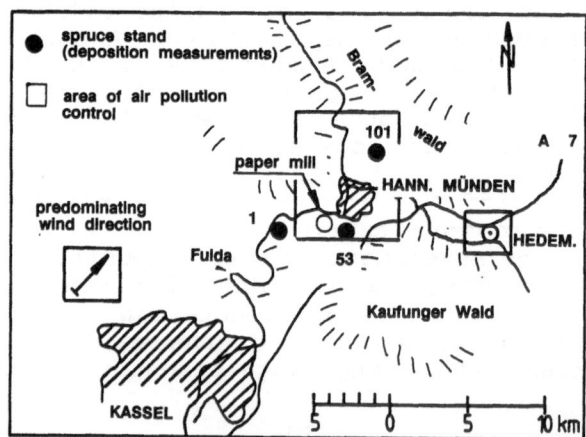

Fig. 1: Location of the three deposition measurement sites (spruce)

In the surrounding of district 53 and 101, SO_2 concentrations
from 66 to 105 μg SO_2/m^3 (long time value, I1) and from 136
to 727 μg SO_2/m^3 (short time value, I2) have been measured in
1985 (INHAK 1986) with highest concentrations in the vicinity
of the paper mill (annual emission 6.800 t SO_2, TÜV Hannover
1987). While exposition towards the source is different, all
other parameters, relevant for deposition processes, can be
considered to be about the same. Information on cation exchange
capacity ($CEC_{eff.}$) of the soils is given in fig. 2. All soils
are in the Al/Fe- or Al-buffer range, characterised by relatively high quantities of exchangeable H^+ (particularly in
the upper soil layers) and Al (Fe) and low quantities of exchangeable Ca and Mg (< 15 %). In Dec. 1985, the spruce stand
in district 1 has been limed from the air with 2 t/hectare.

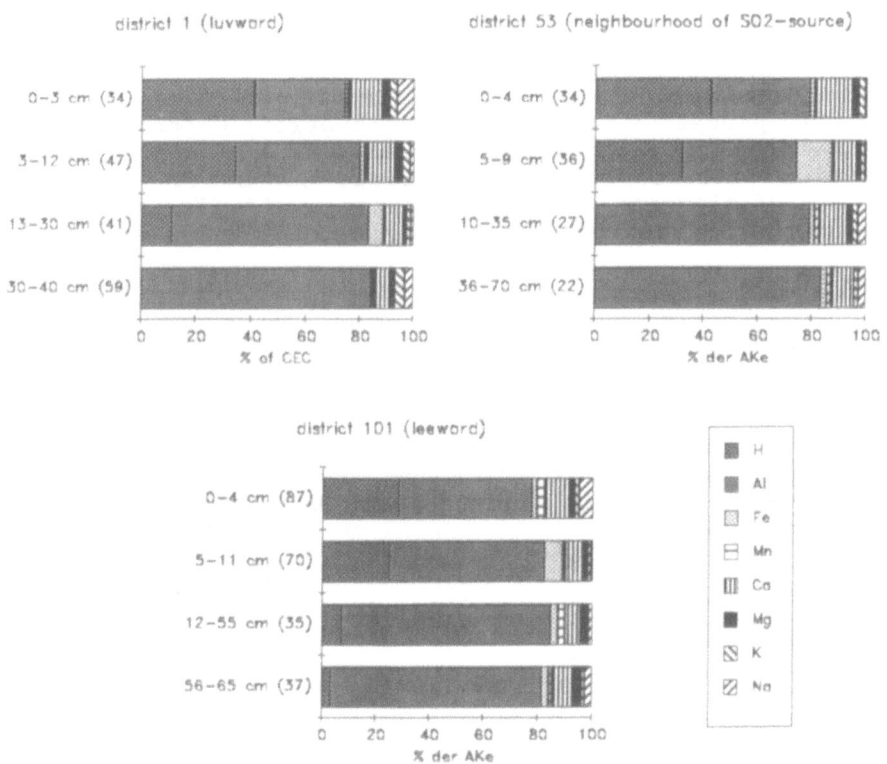

Fig. 2: Relative quantities of exchangeable cations and cation
exchange capacity ($CEC_{eff.}$, μmol IÄ/g, in brackets)
in soil profiles of the spruce sites (Ließ 1986)

336

3. Methods

Bulk samplers (4 x open field deposition, 6 x canopy drip/site) have been used for collection of open field deposition (bulk) and deposition in canopy drip. Samples have been taken on a weekly (May 1985 -April 1986) and two weeks basis (May 1986- April 1987) and analysed for pH and major cations (AAS, IC).

Flux rates have been calculated according to Mayer (1987), Ulrich (1983) and Meiwes et al. (1984), disregarding leaching of S in autumn. In this approach **Na** is used as a 'tracer element' for which as well as for S and Cl no leaching or ab/adsorption effects in the canopy have been assumed (apart from low leaching rates of S in autumn). So for these elements, interception deposition (dry deposition of gases and fine particles in the canopy) can be calculated by substracting flux rates in canopy drip from those in deposition to open field (bulk). For calculation of particle bound interception deposition of other macroelements (Mn, Fe, Ca, Mg, K, H^+) interception deposition rate : deposition rate in open field for Na is used. By balancing flux rates in the canopy of trees, different processes such as interception deposition (dry deposition of fine particles and gases in the canopy), total deposition (interception + open field deposition), leaching (wash-out of elements out of the canopy) and buffering processes (reduction of H^+ activity) can be quantified.

4. Results

In fig. 3 mean deposition of **sulphur**/(ha.a) over the period of 1985-1987 is given. Comparing open field deposition of S at the three sites, no distinct differences can be observed.

Fig. 3: Sulphur deposition (kg S/(ha.a), total deposition, interception deposition and open field deposition) in spruce stands in Stadtforstamt Hann. Münden, May 1985- April 1987

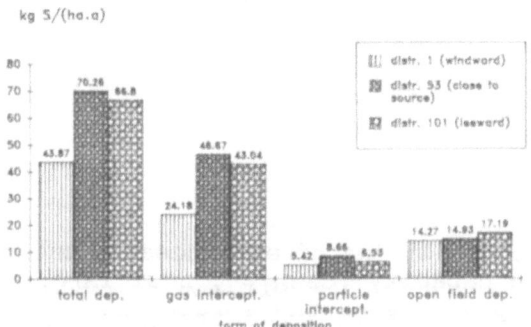

In contrast to this, bigger differences can be seen by comparing interception deposition rates. Whereas interception deposition of S is about the same in the neighbourhood and in the leeward of the source, interception deposition of S in the windward stand is about half of that in the exposed stands. High interception deposition rates of S result in higher total deposition rates (significantly different at p=0.05) in the district 53 (neighbourhood of source) and 101 (leeward). Interception of gaseous S (SO_2) is five times higher than interception of particle bound S (SO_4^{2-}).

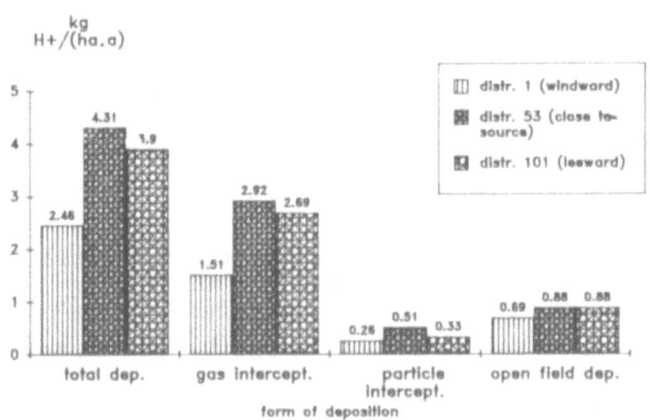

Fig. 4: H^+ (kg/(ha.a) deposition (total deposition, interception deposition and open field deposition) in spruce stands in Stadtforstamt Hann. Münden, May 1985 - April 1987

For H^+ (fig. 4) again, open field deposition does not show great differences, but differences can be seen when comparing interception deposition rates and total deposition rates in the windward stand with those in the exposed stands. Interception deposition of H^+ bound in particles is of minor importance compared to the H^+ interception, deriving from SO_2 interception.

In tab. 2 total deposition rates of macroelements for Northwest Germany are compared to the data from the spruce stands near Hann. Münden. In Hann. Münden S deposition in the vicinity of the source shows similar loads compared to other stands with high intake. On the other hand, total deposition of H^+ reaches highest values in Hann. Münden (district 53). In district 1, total deposition rates of H^+ and S range at the mean (H^+) or lowest levels (S).

338

Tab. 2: Total deposition rates of macroelements in different
spruce stands of Northwest Germany (kg/ha.a)

site	year	H⁺	Ca	Mg	K	Mn	S	authors
Hann. Münden, 1	1985-1987	2.46	12.64	2.77	5.19	0.36	43.87	this work
Hann. Münden, 53	"	4.31	11.12	1.67	5.02	0.41	70.26	"
Hann. Münden,101	"	3.90	10.43	1.61	4.08	0.40	66.80	"
Witzenhausen	V/83-X/85						60.9	HLFU (1986)
Grebenau	"						49.9	"
Königsstein	"						48.8	"
Xanten	"	2.6					65	Gehrmann (1987)
Haard	"	4.1					80	"
Paderborn	"	2.5					67	"
Glindfeld	"	2.1					41	"
Monschau	"	1.7					42	"
Olpe	"	3.5					68	"
Elberndorf	"	3.4					67	"
Teutoburger Wald	1982-1984		18.5	4.72	4.4	0.91	81.4	Godt (1986)
Reinhardswald	1981-1983						89	Brechtel et al. (1986)
Wingst distr. 221	1983-1984	1.6	10.15	7.4	9.9	0.65	56.3	Büttner et al. (1986)
Wingst distr. 28	1983-1984	1.4	12.0	8.6	11.6	0.35	62.9	"
Spanbeck	1982-1985	2.66	16.24	2.13	6.05	0.79	57.74	Bredemeier 1986
Hils	1983-1985	2.1	11.0	1.8	4.7	0.3	53.1	Widey and Gerriets (1986)
Westharz (5 sites)	1981-1982	2.0	19.1	4.0	9.7	0.4	48.2	Matzner in Hauhs (1986)
Solling F1	1969-1983	3.81	21.1	4.2	8.1	0.9	85.3	Matzner et al. (1984)

Fig. 5 and tab. 3 show the total deposition rates of H⁺, and
the buffering and leaching rates in the canopy of the three
spruce stands of Hann. Münden compared to the Solling area
(tab.3). In relation to the input, the highest buffering rate

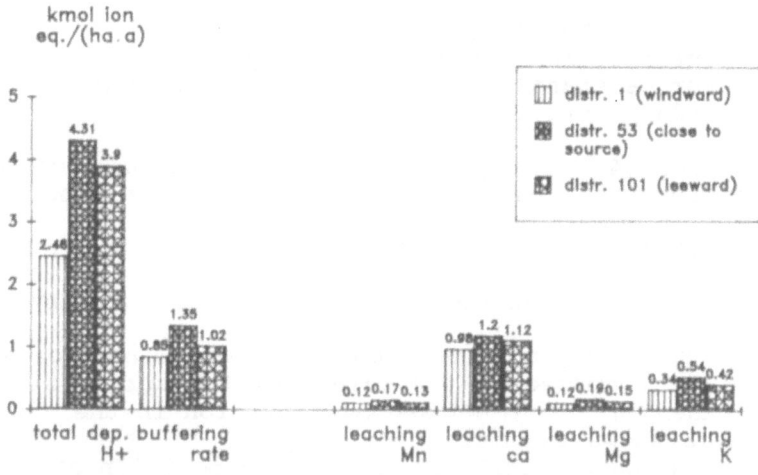

Fig. 5: Total deposition rates of H⁺, buffering rate and
leaching of major cations out of the canopy of spruce
stands in Stadtforstamt Hann. Münden, May 1985 - April
1987

Tab. 3: Comparison of total H^+ deposition rate, buffering rate, leaching of cations and dissolved organic anions (calculation procedure see Matzner 1986)in three spruce stands in Hann. Münden and in a spruce stand in the Solling area (Matzner 1986), keq/(ha.a)

	distr. 1 1985-1987	Hann. Münden distr. 53 1985-1987	distr. 101 1985-1987	Solling F 1 1963-1982
(1) total H^+-deposition	2.46	4.31	3.90	3.81
(2) buffering rate	0.85	1.35	1.02	0.71
(3) leaching of Ca, Mg, K, Mn	1.56	2.10	1.82	1.29
(4) leaching of dissolved org. anions	0.59	0.85	0.71	0.40
(5) 2 - (4 - 5)	-0.12	0.10	-0.09	-0.18

(35 % of total H^+ deposition) is reached in district 1 (windward), lowest buffering rate has been calculated for the spruce stand in district 101 (leeward, 26 % of total H^+ deposition). Total buffering rate and leaching of cations (keq/(ha.a) is highest in the spruce stand in the neighbourhood of the SO_2 source. The element that has been leached most is Ca, followed up by K.

4. Discussion

As the CEC_{eff}. data show (fig. 2), soil acidification, characterised by relatively high exchangeable quantities of H^+, Al and Fe, is present even in deeper soil layers at all three stands to about the same extent. So for all stands, apart from the exposition towards the SO_2 source, soil conditions, which are expected to influence flux rates, can be considered as ceteris paribus conditions.

Statistically significant differences of S input in the exposed spruce stands, compared to the stand in the windward, indicate a clear influence of the paper mill (1/3 of total S input) on stress situation of forest ecosystems in the surrounding. When comparing total deposition of H^+, the influence of the SO_2-source is even more evident (up to 43 % of total deposition) due to high interception deposition rates of SO_2. From official measurements according to TA Luft (discont. measurements of SO_2) and additional continuous measurements of SO_2, conclusions have been drawn that 20 % of the high SO_2 concentrations in the atmosphere (10 % paper mill, 10 % diffuse sources) are related to local sources only (INHAK 1986). For a complete characterisation of the impact of air pollutants on forest ecosystems, informations on input rates are indispensable. SO_2 concentrations in the atmosphere can not be compared to S deposition rates directly; nevertheless from input rates presented here, contradictory conclusions to those that have been drawn from official measurements (TA Luft) concerning the role of local sources/long range transport must be pointed out.

In spite of the close distance to the SO_2 source, total S-deposition rates do not exceed those measured in areas far away from local sources. The reason for this might be that absorption of SO_2 in water films on vegetational cover will be limited by pH, which is in turn determined by H^+ input and buffering processes. F.i. in wash-off liquids from spruce trees, very low pH values down to pH 2.9 have been found (Godt 1986, Godt et al.1986, see also Frevert and Klemm 1984).

Conclusions on the buffering reactions in the spruce canopy can be drawn from fig. 5 and tab. 3. Cole and Johnson (1977) and Cronan and Reiners (1983) describe buffering reactions as cation exchange processes on leaf surfaces, resulting from leaching of Ca, Mg, K and Mn. Kaupenjohann et al. (1988) found higher leaching rates of Ca, Mg and K out of three year old spruce trees, treated with acidic rain solutions of different pH. Similar effects have been described by Mengel et al. (1986). For beech, Glavac (1986) found high leaching rates of cations out of leafs in relation to age and pH of washing solution. Cation loss as resulting from leaching processes has been quantified by Ulrich (1983) and Matzner (1986). The authors point out that buffering processes in the canopy of trees are resulting in acidification of the rhizosphere. Similar quantities of cations have been leached in district 1 Hann. Münden, compared to the spruce stand in the Solling. For the exposed stands in Hann. Münden, higher leaching and buffering rates have been calculated in relation to the higher H^+ intake. In spite of the acidic soil conditions in Hann. Münden, the pool of available cations (Ca and K most of all) seems to be sufficient for buffering 26 % (district 101) to 35 % (district 1) of total H^+ input in the canopy. In district 1 the highest buffering rate of 35 % is probably influenced by liming.

When trying to judge the consequences of deposition of air pollutants in forest ecosystems, Nilsson and Grennfelt (1988) defined critical loads for several soil classes. For forest soils, the critical load is 'the highest deposition of acidifying compounds that will not cause chemical changes in soil, leading to longterm harmfull effects in ecosystems'. For soil conditions such as in Hann. Münden, the critical load (total deposition) for H^+ has been set at 0.2-0.5 kg/(ha.a) and at 3-8 kg/(ha.a) for S respectively. In Hann. Münden mean input

rates in spruce and beech (Godt et al. 1989) stands are several
times higher. Even higher input rates than those inside the
stand can be expected at the edge or at exposed parts of a
stand (Godt and Mayer 1988).
For SO_2, acting on its own, **critical levels** ('concentrations
of pollutants in the atmosphere above which direct adverse
effects on receptors, such as plants, ecosystems or materials,
may occur to present knowledge') have been set at 20 $\mu g/m^3$
(annual mean value) for sensitive plants and at 70 $\mu g/m^3$ for
24h mean value (United Nations 1988). In Hann. Münden, reported
SO_2 concentrations (INHAK 1986) again are far above these
critical levels.

6. Acknowledgement

For excellent technical assistance we would like to thank B.
Köhne and P. Heede.

7. References

Brechtel, H.M., A. Ballazs, F. Lehnhardt (1986) 'Precipitation
    input of inorganic chemicals in the open field and in forest
    stands - results of investigations in the state of Hesse',
    in H. W. Georgii (ed.), Atmospheric Pollutants in Forest
    Areas, D. Reidel Publishing Comp., Dordrecht, pp. 47-68
Bredemeier, M. (1986) 'Ergebnisse der Messungen auf den Stand-
    orten Lüneburger Heide, Spanbeck und Harste' in B. Ulrich
    (ed.), Raten der Deposition, Akkumulation und des Austrages
    toxischer Luftverunreinigungen als Maß der Belastung und
    Belastbarkeit von Waldökosystemen, Berichte des Forschungszen-
    trums Waldökosysteme/Waldsterben Reihe B, Band 2, pp. 11-
    25
Büttner, G., N. Lamersdorf, R. Schultz and B. Ulrich (1986)
    'Deposition und Verteilung chemischer Elemente in küstennahen
    Waldstandorten - Fallstudie Wingst, Abschlußbericht', Berichte
    des Forschungszentrums Waldökosysteme/Waldsterben, Reihe B,
    Band 2, pp.11-25
Bundesminister des Inneren (1983) 'Technische Anleitung zur
    Reinhaltung der Luft, Verwaltungsvorschrift zum Bundes-
    immissionsschutzgesetz', Gemeinsames Ministerialblatt Nr. 6,
    28.02.1983, pp. 94-111
Cole, D. W., D. W. Johnson (1977) 'Atmospheric additions and
    cation leaching in a Douglas Fir ecosystem', Water Resources
    Research 13, pp. 313-317
Cronan, C.S. and W.A. Reiners (1983) 'Canopy processing of
    acidic precipitation by coniferous and hardwood forests in
    New England', Oecologia (Berlin) 59, pp. 216-223
Frevert, T. und O. Klemm (1984) 'Wie ändern sich pH-Werte in
    Regen- und Nebelwasser beim Abtrocknen auf Pflanzenober-
    flächen?', Arch. Met. Geoph. Bioel. Ser. B 34, pp. 75-81

Gehrmann, J. (1987) 'Derzeitiger Stand der Belastung von Waldökosystemen in Nordrheinwestfalen durch Deposition von Luftverunreinigungen', Der Forst- und Holzwirt No.6, pp. 141-145

Glavac, V. (1986) 'Ist die Abnahme der Ca-, Mg-, K- und Zn-
Gehalte in Blättern immissionsgeschädigter Altbuchen die
Folge vergrößerter Blattauswaschung oder verminderter Mineral-
stoffversorgung', Verhandlungen der Gesellschaft für Ökologie,
Gießen, B. XVI, pp. 253- 266

Godt, J. (1986) 'Untersuchungen von Prozessen im Kronenraum
von Waldökosystemen und deren Berücksichtigung bei der
Erfassung von Schadstoffeinträgen - unter besonderer Beachtung
der Schwermetalle', Berichte des Forschungszentrums Waldöko-
systeme/Waldsterben 19, 265 pages.

Godt, J., M. Schmidt and R. Mayer (1986) 'Processes in the
canopy of trees: Internal and external turnover of elements',
in H.-W. Georgii (ed.), Atmospheric Pollutants in Forest
Areas, D. Reidel Publishing Company Dordrecht, pp.263-274

Godt, J. (1988) 'Immissionsbelastungen und deren Auswirkungen
im Stadtwald von Hann. Münden - eine landschaftsökologische
Studie im Auftrag der Stadt Hann. Münden', Gesamthoch-
schule/Universität Kassel, FB Stadt - und Landschaftsplanung,
Feb. 1988, 116 pages

Godt, J. and R. Mayer (1988) 'Deposition rates of airborne
substances to forest canopies in relation to surface structure', in M.H. Unsworth and D. Fowler (eds.), Acid deposition
at high elevation sites, Kluwer Academic Publishers Dordrecht,
pp. 593-606

Godt, J., M. Weyer und R. Mayer (1989) 'Schadstoffbelastungen
von zwei Buchenaltbeständen in Nordhessen: Ein Vergleich der
Säurebelastung und des Puffervermögens an einem Standort über
Kalk und über Buntsandstein', Mitt. der Gesellschaft f. Ökologie, 17. Jahrestagung Göttingen 1987 (in print)

HLFU (Hessische Landesanstalt für Umwelt (ed.)(1986) 'Waldbelastungen durch Immissionen - 3. Zwischenbericht', Umwelt-
planung und Umweltschutz, H. No. 38, 135 pages

INHAK (Institut für Umweltschutz GmbH) (1986) 'Bericht über
die im Auftrag des Niedersächsischen Ministers für Bundesan-
gelegenheiten im Raum Hann. Münden in der Zeit vom Januar
bis Dezember 1985 durchgeführten Immissionsmessungen',
Bd 1+2, Selbstverlag Bückeburg

Kaupenjohann, M., B.U. Schneider, R. Hantschel, W. Zech and
R. Horn (1988) 'Sulfuric acid rain treatment of Picea abies
(Karst L.): Effects on nutrient solution, throughfall chemistry, and tree nutrition', Z. Pflanzenernähr. Bodenk., 151,
pp. 123-126

Ließ, S. (1986) 'Immissionsbelastung von Böden im Hann. Mündener
Stadtwald', Projektarbeit FB Stadt- und Landschaftsplanung,
GhK/Universität Kassel, 72 pages

Matzner, E., P.K. Khanna, E. Cassens-Sasse, M. Bredemeier und
B. Ulrich (1984) 'Ergebnisse der Flüssemessungen in Waldökosystemen', Berichte des Forschungszentrums Waldökosysteme/

Waldsterben 2, pp. 29-49

Matzner, E. (1986) 'Deposition/canopy interactions in two forest ecosystems of Northwest Germany', in H.-W. Georgii (ed.), Atmospheric Pollutants in Forest Areas, D. Reidel Publishing Company Dordrecht, pp. 247-262

Mayer, R. (1985) 'Verfahren zur Erfassung der Schadstoffzufuhr in Waldökosystemen', Staub Reinhaltung der Luft 45, Nr. 6, June 1985, pp. 267-292

Mayer, R. (1987) 'Die Zufuhr von Stickstoff zum Boden durch trockene Ablagerung und mit den Niederschlägen', Arbeitsergebnisse der AG für Ländliche Entwicklung, 3, Gesamthochschule/Universität Kassel FB Stadt- und Landschaftsplanung, pp. 13-19

Meiwes, K.-J., M. Hauhs, H. Gerke, N. Asche, E. Matzner und N. Lamersdorf (1984) 'Die Erfassung des Stoffkreislaufes in Waldökosystemen', Berichte des Forschungszentrums Waldökosysteme/Waldsterben, Bd. 7, pp. 68-141

Mengel, K., M. Th. Breininger and H. J. Lutz (1986) 'Effect of acid fog on nutrient leaching and needle condition of picea abies', in Commission of the European Communities (ed.), Direct effects of dry and wet deposition on forest ecosystems - in particular canopy interactions, Proceedings of a workshop 19-23 October 1986, Lökeberg, pp. 25-33

Nilsson, J. and P. Grennfelt (eds.)(1988) 'Critical loads for Sulphur and Nitrogen', Report from a workshop held at Skokloster, Sweden 19-24 March 1988, Nordic Council of Ministers

TÜV Hannover (1987) 'Emissionskataster Hann. Münden, im Auftrag der Stadt Hann. Münden', Selbstverlag 62 p.

Ulrich, B. (1983) 'Interaction of forest canopies with atmospheric constituents: SO_2, alkali and earth alkali cations and chloride', in B. Ulrich and J. Pankrath (eds.), Effects of Accumulation of Air Pollutants in Forest Ecosystems, D. Reidel Publishing Company Dordrecht, pp. 33-45

United Nations (ed.) (1988) 'ECE Critical levels workshop', Economic Commission for Europe, Bad Harzburg FRG 14 - 18 March 1988, Final Draft Report

Wiedey, G. and M. Gerriets (1986) 'Ergebnisse der Messungen im Hils', in B. Ulrich (ed.), Raten der Deposition, Akkumulation und des Austrags toxischer Luftverunreinigungen als Maß der Belastung und Belastbarkeit von Waldökosystemen', Berichte des Forschungszentrums Waldökosysteme/Waldsterben, Reihe B, Band 2, pp. 26-54

LIST OF PARTICIPANTS

A. Arends
Netherlands. Energy Research
Foundation
Postbus 1
NL-1755 ZG Petten
Netherlands

H. Bachmann
Inst. f. Bodenkunde und
Bodenerhaltung
Universität Gießen
Wiesenstr. 3-5
6300 Gießen

Prof. Dr. K. Bächmann
Technische Hochschule Darmstadt
FB Anorg. Chemie und Kernchemie
Hochschulstr. 4
6100 Darmstadt

F. W. Badeck
Universität Frankfurt
Inst. f. Physik. u. Theor. Chemie
Niederurseler Hang
6000 Frankfurt/M 50

Dr. H. Ballach
Universität Essen
Angewandte Biologie
Universitätsstr. 5
4300 Essen

S. Barth
INHAK
Inst. f. Umweltschutz GmbH
Postfach 1328
3012 Langenhagen

M. Bauer
Bayer. Forstl. Versuchs- und
Forschunsanstalt München
Schellingstr. 14
8000 München 40

Dr. G. Baumbach
Universität Stuttgart
Inst. für Verfahrenstechnik
und Dampfkesselwesen
Pfaffenwaldring 23
7000 Stuttgart

K. Baumann
Universität Stuttgart
Inst. für Verfahrenstechnik
und Dampfkesselwesen
Pfaffenwaldring 23
7000 Stuttgart 80

Dr. S. Beilke
Umweltbundesamt
Pilotstation Frankfurt
Frankfurter Str. 135
6050 Offenbach

Dr. E. Bieber
Umweltbundesamt
Pilotstation Frankfurt
Frankfurter Str. 135
6050 Offenbach

U. Bieder
Inst. für Kernenergetik u.
Energiesysteme
Pfaffenwaldring 31
7000 Stuttgart-80

B. Biggemann
Universität Gießen
Inst. f. Bodenkunde u. Bodenerhaltung
Wiesenstr. 3-5
6300 Gießen

Dr. B. Bockholt
Landesamt für Umweltschutz
und Gewerbeaufsicht Rheinland-Pfalz
Postfach 119
6504 Oppenheim

346

Dr. H. Borchert
Landesamt für Umweltschutz
und Gewerbeaufsicht Rheinland-Pfalz
Postfach 119
6504 Oppenheim

H. Braun
Universität Frankfurt
Inst. für Biologiedidaktik
Sophienstr. 1-3
6000 Frankfurt/M.

U. Bressen
Hess. Forstl. Versuchsanstalt
Prof. Oelkers Str. 6
3510 Hann. Münden

S. Bürgermeister
Universität Frankfurt
Inst. für Meteorologie und Geophysik
Feldbergstr. 47
6000 Frankfurt/M.

G. Butterweck
Isotopenlabor für biolog. u. med.
Forschung
Burckhardtweg 2
3400 Göttingen

J. Constantin
Universität Göttingen
Inst. für Bioklimatologie
Büsgenweg 1
3400 Göttingen

G. Dortmann
Fraunhofer Institut
für Atmosphärische Umweltforschung
Kreuzeckbahnstr. 19
8100 Garmisch-Partenkirchen

F. Dröscher
Universität Stuttgart
Inst.f. Verfahrenstechnik
und Dampfkesselwesen
Pfaffenwaldring 23
7000 Stuttgart 80

J. Eichhorn
Hessische Forstliche
Versuchsanstalt
Prof. Oelkersstr. 6
3510 Hann. Münden

K.-H. Eickel
VDI-Kommission Reinhaltung der Luft
Postfach 1139
4000 Düsseldorf

W. Elbert
Max Planck Inst. für Chemie
Saarstr. 23
6500 Mainz

K.-H. Enderle
Universität Frankfurt
Zentrum für Umweltforschung
Robert-Mayer Str. 5-7
6000 Frankfurt/M.

Dr. G. Enders
Universität München
Inst. für Bioklimatologie
und angewandte Meteorologie
Amalienstr. 52
8000 München 40

H. G. Ernst
Universität Frankfurt
Zentrum für Umweltforschung
Robert-Mayer Str. 5-7
6000 Frankfurt/M

A. Esch
Universität Gießen
Inst. für Pflanzenernährung
Südanlage 6
6300 Gießen

B. Fähnrich
Universität Frankfurt
Inst. für Meteorologie und Geophysik
Feldbergstr. 47
6000 Frankfurt/M.

W. Forschner
Universität Mainz
Inst. für Botanik
Saarstr. 21
6500 Mainz

M. Fricke
Universität Frankfurt
Inst. für Meteorologie und Geophysik
Feldbergstr. 47
6000 Frankfurt/M.

Dr. W. Fricke
Umweltbundesamt
Pilotstation Frankfurt
Frankfurter Str. 135
6050 Offenbach

Prof. Dr. H.-W. Georgii
Universität Frankfurt
Inst. für Meteorologie und Geophysik
Feldbergstr. 47
6000 Frankfurt/M.

Prof. Dr. T. Gies
Universität Frankfurt
Inst. für Biologiedidaktik
Sophienstr. 1-3
6000 Frankfurt/M.

Dr. J. Godt
Gesamthochschule Kassel
FB Stadt- und Landschaftsplanung
Landschaftsökologie/Bodenkunde
Henschelstr. 2
3500 Kassel

Prof. Dr. G. Gravenhorst
Universität Göttingen
Inst. für Bioklimatologie
Büsgenweg 1
3400 Göttingen

Dr. H. D. Gregor
Umweltbundesamt Referat I 3.3
Allg. Wirkungsfragen Waldschäden
Bismarckplatz 1
1000 Berlin 33

Prof. Dr. K.-O. Groeneveld
Universität Frankfurt
Inst. für Kernphysik
August Euler Str. 6
6000 Frankfurt/M.

S. Grosch
Universität Frankfurt
Inst. für Meteorologie und Geophysik
Feldbergstr. 47
6000 Frankfurt/M.

C.-E. Grüneklee
Universität Gießen
Inst. für Bodenkunde u. Bodenerhaltung
Wiesenstr. 3-5
6300 Gießen

G. Halbig
Universität Frankfurt
Inst. für Meteorologie und Geophysik
Feldbergstr. 47
6000 Frankfurt/M.

Dr. K. Hanewald
Hessische Landesanstalt für Umwelt
Unter den Eichen 7
6200 Wiesbaden

W. Hartmann
Max Planck Inst. für Chemie
Saarstr. 23
6500 Mainz

W. Haunold
Universität Frankfurt
Inst. für Meteorologie und Geophysik
Feldbergstr. 47
6000 Frankfurt/M.

J. Hauptmann
TH Darmstadt FB 8
Hochschulstr. 4
6100 Darmstadt

Dr. E. Hegewald
KfA Jülich
Inst. für Biotechnologie 3
Postfach 1913
5170 Jülich

Dr. G. Helas
Max Planck Inst. für Chemie
Saarstr. 23
6500 Mainz

J. Henrich
Universität Frankfurt
Inst. für Botanik
Siesmayerstr. 70
6000 Frankfurt/M.

G. Henrici
Universität Frankfurt
Inst. für Meteorologie und Geophysik
Feldbergstr. 47
6000 Frankfurt/M.

U. Herrmann
Universität Frankfurt
Zentrum für Umweltforschung
Robert Mayer Str. 7-9
6000 Frankfurt/M.

Dr. D. Hofmann
Universität Frankfurt
Inst. für Kernphysik
August Euler Str. 6
6000 Frankfurt/M.

P. Hofschreuder
Landwirtschaftl. Hochschule
Abt. Reinhaltung der Luft
Postfach 8129
NL-6700 EV Wageningen
Netherlands

A. Hogrebe
Universität Gießen
Inst. für Pflanzenernährung
Südanlage 6
6300 Gießen

A. Ibrom
Universität Göttingen
Inst. für Bioklimatologie
Büsgenweg 1
3400 Göttingen

Prof. Dr. R. Jaenicke
Universität Mainz
Inst. für Meteorologie
Postfach 3980
6500 Mainz

Dr. W. Jaeschke
Universität Frankfurt
Zentrum für Umweltforschung
Robert-Mayer Str. 7-9
6000 Frankfurt/M.

R. Jurat
Universität Frankfurt
Inst. für Botanik
Siesmayerstr. 70
6000 Frankfurt/M.

B. Kalte
Universität Frankfurt
Inst. für Meteorologie und Geophysik
Feldbergstr. 47
6000 Frankfurt/M.

Dr. M. Kazda
Universität Düsseldorf
Abteilung Geobotanik
Universitätsstr. 1
4000 Düsseldorf

Prof. Dr. D. Klockow
Universität Dortmund
Inst. für Chemie
Postfach 500 500
4600 Dortmund 50

Dr. A. Knorr
Bayer. Forstl. Versuchs- und
Forschungsanstalt München
Schellingstr. 14
8000 München 40

Prof. Dr. G. Kohlmaier
Universität Frankfurt
Inst. für Phys. und Theor. Chemie
Niederurseler Hang
6000 Frankfurt/M.-50

M. Krämer
Universität Mainz
Inst. für Meteorologie
Postfach 3980
6500 Mainz

G. Kramm
Fraunhofer Institut für
Atmosphärische Umweltforschung
Kreuzeckbahnstr. 19
8100 Garmisch-Partenkirchen

G. Kroll
Deutscher Wetterdienst
Meteorologisches Obs. Hamburg
Frahmredder 95
2000 Hamburg 65

G. Krückemeier
Universität Essen
Inst. für Landschaftsökologie
Postfach 103764
4300 Essen

Dr. G. Lammel
KfA-Karsruhe
Laboratorium für Aerosolphysik
und Filtertechnik
Postfach 3640
7500 Karlsruhe

H. J. Lehnert
Universität Frankfurt
Inst. für Biologiedidaktik
Sophienstr. 1-3
6000 Frankfurt/M.

S. Ließ
Gesamthochschule Kassel
FB 13 Landschaftsökologie
Henschelstr. 2
3500 Kassel

Prof. Dr. G. Masuch
Universität Paderborn
FB 13 Angewandte Chemie
Postfach 1621
4790 Paderborn

Prof. Dr. W. Michaelis
GKSS Forschungszentrum Geestacht
Inst. für Physik
Postfach 1160
2050 Geestacht-Tesperhude

Dr. J. Müller
Umweltbundesamt
Pilotstation Frankfurt
Frankfurter Str. 135
6050 Offenbach

H. Müller
Fraunhofer Inst. für
Atmosphärische Umweltforschung
Kreuzeckbahnstr. 19
8100 Garmisch-Partenkirchen

N. Neikes
Universität Düsseldorf
Abt. Geobotanik Botanik III
Universitätsstr. 1
4000 Düsseldorf

Dr. M. Nestlen
Universität Göttingen
Inst. für Bioklimatologie
Büsgenweg 1
3400 Göttingen

Dr. V. D. Nguyen
KFA Jülich
Inst. für Chemie ICH 4
Postfach 1913
5170 Jülich

J. Nitzsche
Universität Frankfurt
Zentrum für Umweltforschung
Robert-Mayer Str. 7-9
6000 Frankfurt/M.

R. Obermaier
Battelle Inst. e.V
Abtl. Technische Risiken
Am Römerhof 35
6000 Frankfurt/M.

Dr. G. Ockelmann
Universität Frankfurt
Inst. für Meteorologie und Geophysik
Feldbergstr. 47
6000 Frankfurt/M.

P. Otto
Universität Frankfurt
Inst. für Meteorologie und Geophysik
Feldbergstr. 47
6000 Frankfurt/M.

M. Plöchl
Universität Frankfurt
Inst. für Physik u. Theor. Chemie
Niederurseler Hang
6000 Frankfurt/M.-50

J. Polzer
TH Darmstadt FB 8
Hochschulstr. 4
6100 Darmstadt

Prof. Dr. H. Puxbaum
Technische Universität Wien
Getreidemarkt 9/151
A - 1060 Wien
Austria

Dr. F. Queirolo
KFA Jülich
Inst. für Chemie ICH 4
Postfach 1913
5170 Jülich

Dr. P. Rademacher
GKSS Forschungszentrum Geestacht
Institut für Physik
Postfach 1160
2054 Geestacht

Dr. J. Reindl
Bayer. Forstl. Versuchs- u.
Forschungsanstalt
Schellingstr. 12-14
8000 München 40

Prof. Dr. H. Rennenberg
Fraunhofer Institut für
Atmosphärische Umweltforschung
Kreuzeckbahnstr. 19
8100 Garmisch-Partenkirchen

Dr. J. Schabronath
Gesamtverband des
Deutschen Steinkohlenbergbaus
Friedrichstr. 1
4300 Essen 1

B. Schäfer
Universität Frankfurt
Inst. für Meteorologie und Geophysik
Feldbergstr. 47
6000 Frankfurt/M.

Prof. Dr. H. Schaub
Universität Frankfurt
Inst. für Botanik
Siesmayerstr. 70
6000 Frankfurt/M.

D. Schell
Universität Frankfurt
Inst. für Meteorologie und Geophysik
Feldbergstr. 47
6000 Frankfurt/M.

Dr. G. Schmitt
DEKRA - Inst. für Sicherheit
Umweltschutz und Energie
Abt. Meßstelle für Umweltschutz S 42
Schulze-Delitzsch Str. 49
7000 Stuttgart 80

Dr. R Schmitt
Meteorologie Consult
Auf der Platt 47
6246 Glashütten

M. Schönburg
GKSS Forschungszentrum Geestacht
Inst. für Physik
Postfach 1160
2050 Geestacht-Tesperhude

Dr. L. Schütz
Universität Mainz
Inst. für Meteorologie
Postfach 3980
6500 Mainz

G. Seufert
Universität Stuttgart-Hohenheim
Inst. für Landeskultur u.
Pflanzenökologie
Postfach 700562
7000 Stuttgart 70

T. Spranger
Projektzentrum Ökosystemforschung
Schauenburger Str. 112
2300 Kiel

352

R. Staubes
Universität Frankfurt
Inst. für Meteorologie und Geophysik
Feldbergstr. 47
6000 Frankfurt/M.

D. Vogler
Universität Frankfurt
Inst. für Meteorologie und Geophysik
Feldbergstr. 47
6000 Frankfurt/M.

G. Stamm
Universität Göttingen
Inst. für Bioklimatologie
Büsgenweg 1
3400 Göttingen

A. Waraghai
Universität Göttingen
Inst. für Bioklimatologie
Büsgenweg 1
3400 Göttingen

A. Steiger
GSF - BIOP
Ingoldstädter Str. 1
8042 Neuherberg

Dr. H. R. Wegener
Universität Gießen
Institut für Bodenkunde und
Bodenerhaltung
Wiesenstr. 3-5
6300 Gießen

S. Stegen
KFA Jülich
Inst. für Chemie ICH 4
Postfach 1913
5170 Jülich

A. Wessler
Universität Mainz
Inst. für Allg. Botanik
Zur Laubenheimer Höhe 18
6500 Mainz 42

Dr. R. P. Stössel
GKSS Forschungszentrum Geestacht
Inst. für Physik
Postfach 1160
2050 Geestacht

M. Werscheck
Meteorologie Consult
Auf der Platt 47
6246 Glashütten

Dr. U. Teichmann
Universität München
Inst. für Bioklimatologie
und angewandte Meteorologie
Amalienstr. 52
8000 München 40

I. Wettlaufer
Hess. Landesanstalt für Umwelt
Unter den Eichen 7
6200 Wiesbaden

A. Vermetten
Landwirtschaftl. Hochschule
Abt. Reinhaltung der Luft
Postfach 8129
NL - 6700 EV Wageningen
Netherlands

Prof. Dr. A. Wild
Universität Mainz
Inst. für Allg. Botanik
6500 Mainz

Dr. P. Winkler
Deutscher Wetterdienst
Meteorologisches Obs. Hamburg
Frahmredder 95
2000 Hamburg 65

D. Wintermeyer
Universität Dortmund
Inst. für Anorganische Chemie
Postfach 500 500
4600 Dortmund 50

Dr. W. Wobrock
Universität Frankfurt
Zentrum für Umweltforschung
Robert-Mayer Str. 7-9
6000 Frankfurt/M.

R. Zimmermann
Enquetekommission
Vorsorge zum Schutz der
Erdatmosphäre
Görresstr. 15 - Bundeshaus
5300 Bonn

SUBJECT INDEX

A

Abies alba Mill. 305, 308
absorption 63
accumulation 184
accumulation units 23
acetate 114
acid
  acetic~ 111, 113
  amino~s 329
  carboxylic~s 113
  ~deposition 285
  dicarboxylic~s 113
  formic~ 111, 113
  hydrochloric~ (HCL) 64, 66, 113
  inorganic~s 111, 116
  nitric ~ (HNO_3) 64, 113, 260
  organic~s 111
  quinic~ 267
  shikimic~ 267
  sulfuric ~ (H_2SO_4) 113, 260
acidic fog 259, 262, 267
acidification 267
ACIFORN 62
active string collector 222
actual absorption by the needle 175
adaption capacity 251
air pollutants 169, 308, 333
air pollutants, gaseous and particulate~ 99
air pollutant uptake
  of an enclosed branch 169
air pollution monitoring 62, 64
aluminium 286, 308
aluminium toxicity 295, 301
amino acids 329
ammonia (NH_3) 61, 63, 97
ammonium (NH_4) 45, 53, 64, 207, 218,
  225, 308
ammonium nitrate (NH_4NO_3) 97
amounts of rainfall 112, 116

amounts of organic anions 116
anions and cations in rainwater 8
Annual Diffusion Denuder 97
annual
  ~growth rings 193
  ~plants 315
  ~variation 45
antioxidants 253
apoplastic 251
aromatic hydrocarbons (AHC) 133
artificial surface 69
ascorbate 253
atmospheric
  ~deposition 295, 296
  ~HNO_3/NH_3 system 97
  ~pollution 200
  ~residence time 137, 143
Atomic Absorption Spectroscopy
  (AAS) 53, 286, 308
Automated-Thermal-Desorber
  Chromatograph 133
automobil exhaust 158
average diurnal variations 39

B

background air masses 99
beech *(Fagus sylvatica* L.) 285, 305, 308
below-cloud-scavenging 57
benzene 134
bulk precipitation 207

C

cadmium (Cd) 53, 193, 196
calcium (Ca) 87, 218, 285, 297, 308, 333
calcium/aluminium ratio 301
calibration chamber 71
cambium 269
canopy
  ~buffering 305
  capacity of~ 214

356